Physical Limitations of Semiconductor Devices

T0137459

V.A. Vashchenko • V.F. Sinkevitch

Physical Limitations of Semiconductor Devices

 Springer

V.A. Vashchenko
National Semiconductor
Santa Clara, CA
USA

V.F. Sinkevitch
Pulsar R&D Manufacturing Company
Moscow
Russia

ISBN 978-1-4419-4505-1   e-ISBN 978-0-387-74514-5

Printed on acid-free paper.

9 8 7 6 5 4 3 2 1

springer.com

To our families and friends

Preface

Since the beginning of semiconductor era in microelectronics the methodology of reliability assessment became a well established area. In most cases the reliability assessment involves statistical methods for safe operating area and long term reliability parameters at the development of semiconductor processes, components and systems. At the same time in case of catastrophic failures at any development phase the major practical method is failure analysis (FA). However FA is mainly dealing with detection of consequences of some irreversible event that already happened.

This book is focused on the most important and the less summarized reliability aspects. Among them: catastrophic failures, impact of local structural inhomogeneities, major principles of physical limitation of safe-operating area (SOA), physical mechanisms of the current instability, filamentation and conductivity modulation in particular device types and architectures.

Specifically, the similar principles and regularities are discussed for electrostatic discharge (ESD) protection devices, treating them as a particular case of pulsed power devices. Thus both the most intriguing applications and reliability problems in case of the discrete and the integrated components are covered in this book.

In general, the failure of a single device in application circuit could be rather expensive and even critical. Consider for example space application at Moon or Mars mission where a single device failure may result in failure of the critical system for the whole rather expensive mission. In this case the cost of individual reliability assessment for every individual component is relatively low. It is hardly possible to guarantee the reliability of every component using only statistical methods. There are several reasons for this for example a finite probability of local structural defects, or a single upset event related to cosmic particles. However using the physical approach to reliability provision even the impact of local inhomogeneities on safe operating area and possible physical mechanisms can be understood (Chapter 3–5) followed by development and application of the experimental methods of diagnostic control (Chapter 8).

In case of already happened failure event the conclusion about a possible cause often does not provide a direct immediate answer about a better final solution even. The reason of this is a lack of adequate information about involved physical processes that resulted in the failure and the basic failure mechanisms.

For an engineer it is rather hard to "navigate" in the "sea" of numerous publications in the field of reliability towards understanding of a specific reliability problem or a failure reason in a particular case study. To overcome this barrier further steps are required. The overwhelming majority of the publications provide either rather general or a very particular case study analysis that can't be applied to a new practical case directly.

At the same time in real business cases one may expect from the involved specialist a qualified and clear presentation of the conclusions regarding the particular reason and scenario of failure, revealing of the physical mechanism followed by motivated straight forward recommendations of the measures towards improvement of the situation with the final goal to overcome and eliminate the problem in the nearest future.

This book, in particular, is composed to help dealing with such a scenario. Instead of making recommendations with an unreasonable declaration of pseudo scientific conclusions, relying mainly on common sense and previous case studies, this book is offered another practical alternative. The alternative is engineered upon an in depth understanding of the major physical principles and regularities of electro-thermal limitation of semiconductor devices, the methodologies for additional learning using the measurement and simulation tools and implementation of the physical approach for reliability and ESD design. These advanced components of the understanding are delivered in the book material both at different levels: the generic level for semiconductor devices, and the component level for particular discrete and integrated device. For major book topics the material is presented at different parallel levels: phenomenological description, analytical description and numerical physical process and device simulation.

In spite of 21st century outdoor up to now the systematic consistent presentation of this book topic can not be found. The most of the relevant studies are scattered in literature mostly in the form of papers across the time scale of almost 40 years. Up to now the majority of cases of semiconductor device behavior near in the operation regime limits are frequently unexpected and hard to deal with. At the same time failure analysis deals with the consequences and thus often rather low informative to identify the cause.

The work on this have been initiated several years ago with the primary focus and inspiration to overcome mentioned problems by means of summarizing their total lifetime experience in the field of reliability and ESD design. Over 30 years ago in the foreword to special topic paper collection on reliability D.S. Peck has noticed that namely a limitation of our insight itself results in formation of a particular set of the practical approaches to reliability problem. Today we believe that there is no alternative to the physical approach and the material of this book provides a further step towards the physical approach implementation in various practical methods.

Within the framework of one edition it is impossible to provide an equivalent consideration of all physical mechanisms related to the semiconductor devices failures. Therefore this book deals with a preferred specific set of problems, related to the semiconductor devices, transistors breakdown, thermoelectrical instabilities and burnout. There are several reasons why this set of problems is selected.

First of all, the long-term experience of semiconductor device application demonstrates that an overwhelming majority of failures has namely a catastrophic nature. To some extend this fact follows automatically from the conditions when the long-term reliability parameters and operation regimes are provided at the development phase. Thus it can be assumed that the dominant physical reason practically of catastrophic failures is an electrical breakdown and various thermo-electrical current instabilities the devices prior their failure.

Similar physical phenomena are limiting the transistor capabilities especially in pulsed condition of electrical overstress. In these pulsed regimes the impact of the long term reliability factors is minor. Usually the physical mechanisms of elec-tromigration and hot carrier degradation remain inactive in the uniform current distribution state along the contacts. However these mechanisms can be activated in case of high current density realized in ESD protection devices.

As for the degradation mechanisms themselves they can develop instability too.

Chapter 1 provides an introduction to the most important technological peculi-arities of semiconductor devices and basic concepts of reliability related to failure physics.

Basic phenomena of thermoelectric instability in semiconductor devices are thoroughly addressed in four following Chapters 2–5. Only Chapter 6 deals with the most exotic scenario when the degradation leads into catastrophic failure.

Specific of ESD devices within the focus of this book and physical principles of robust ESD device design are presented in Chapter 7. The final Chapter 8 summarizes the practical methodology for of the physical approach application in the reliability problems.

The book is written to be a helpful source of deep understanding of major physical principles of physical limitation of safe operating area, physical mechanisms and scenario of failures in different application aspects, including device level ESD design. Upon the authors successful industry experience this understanding is di-rectly useful towards everyday practical work on solutions for reliability prob-lems, ESD protection and application cases, system level individual reliability provision and many others adjacent areas.

The book is primarily intended for engineers and students working in the field of microelectronics on process and device development of both discrete and inte-grated power components, circuit design and application in high variety of products from integrated analog power up to high speed digital design, ESD protection, reliability, physical process and device simulation. At the same due to a very cross-disciplinary subjects the book might provide a significant interest for much wider circle of engineers and researchers in the field of solid-state physics, non-linier physics and could be a very useful material for students in electrical engineering and semiconductor device physics. Finally, due to a universal nature of non-linier process in active distributed media the researchers in the adjacent fields may find this book as a source of rather important knowledge that will help them to address their everyday professional problems on a different level of greater confidence using the physical approach. The regularities outlined in the book are often de-scribed by similar equations that are suitable for different application and physical problems in electronics and even chemistry and biology.

Many of our colleagues contributed to the material of this book through par-ticipation in discussion, technical and management support it would be difficult to thank them all within the limited space. Nevertheless, the authors would like to acknowledge the valuable contributions of our former colleagues from SRI "Pulsar: A.M. Nechaev, N.A. Kozlov, Y.B. Martynov. We'd like to thank our former colleagues from Research Institute of Applied Physics V.V. Osipov (AMES NASA) and B.S. Kerner (Daimler Benz) for the discussion in the field of non-linier physics.

V.A. Vashchenko is sincerely thankful for ESD discussions his many colleagues from National Semiconductor Corp. especially P. Hopper and A. Concannon. He also would like to acknowledge stimulating discussion with his colleagues in ESD field Ph.D student N. Olson and prof. E. Rosenbaum from University of Illinois; prof. G. Menegesso from University of Padova; prof. M. Bafleur from LAAS-CNRS; prof. J. Liou from University of Central Florida; IMEC team P. Jansen, M. Sholz, D. Linten, G. Groesenekin and his colleagues form Texas Instrument Corp. A. Amerasekera, C. Duvury and V. Vassilev.

Prof. Sinkevitch would like to mention specially the valuable support provided by General Director SRI "Pulsar" Y.P. Dokuchaev.

Finally the authors V.F. Sinkevitch and V.A. Vashchenko are deeply acknowledging the patience and support our lovely wife's Irina and Liliya and our bright children's Anna, Maria and Yana, Denis, respectively.

Contents

Chapter 1
Failures of Semiconductor Device

Several typical failure analysis photographs are presented in Fig. 1.1 regarding failed transistor structures. As usual they present somewhat useful information related to the location of the structural damage that identifies the failed devices especially in the case of integrated circuits. Additional information can be derived about the damage location inside the device itself that might reveal or provide an idea regarding some topology defects of the device.

However, perhaps only one of them is directly informative about the physical processes that caused the failure. It is the photo of the GaAs Schottky gate field effect transistor (MESFET) (Fig. 1.1c) that represents a failure due to the tests under electrical load and increased ambient temperature. The formation of emptiness in metallization at the beginning of the source "fingers" and corresponding metal accumulation at the end of the fingers can be observed. The picture is typical of the phenomenon of metal electromigration. In this case it is not essential to be an expert in the field of reliability to diagnose the reason for failure.

In most practical cases the final damage quite rarely reveals a direct physical failure mechanism. Often the original cause or complete scenario of failure is hidden by secondary postdamage processes. Thus, it is extremely difficult to restore the failure scenario and particular physical mechanism in order to understand step by step scenario of events resulted to catastrophe. In many cases this problem is excessive even for the experienced researcher.

A physical approach to reliability is dictated by a necessity for reliability assessment and maintenance at all stages of the device lifetime cycle. The standard statistical methods are based on confirmation of the required middle time before failure and safe operating area parameters using subsequent statistical processing of the temperature accelerated test and electrical measurement results. Thus, a detailed study of not only direct failure causes, but the dominant physical mechanisms and phenomena that result in particular device failure mode is required during the process, device, and even product development phases.

An example was suggested by W. Shockley and D. Scarlet in their famous study on thermal breakdown in power transistors in 1963 [1]. Since then the priority of application of physical research methods to reliability problems has become more critical.

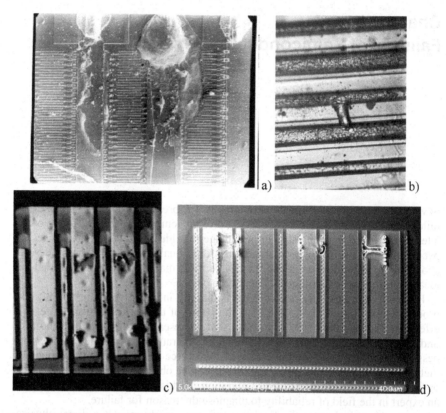

Fig. 1.1 Microphotographs of failed structures: Silicon bipolar transistor (a), SiMOSFET (b), GaAs MESFET (c), snapback NMOS at ESD pulse (d)

Experience demonstrates that the range of physical mechanisms responsible for semiconductor device failure is limited. Moreover, new unique mechanisms are discovered rarely. As a rule it happens in connection with development of a new transistor or process. For this reason the result of unique research of the specific device type can leave the framework of a particular research problem solution.

On the other hand, the particular failure scenario in semiconductor devices is rather specific to the design and operation conditions. A small change in the structure parameters or topology variation might strongly impact the failure mode. Examples include change of the characteristic time, critical conditions, and damage localization in the structure. Therefore, application of general results or previous case studies in each particular case is rather ineffective in general.

Another important factor is a short timeline of practical decision-making process regarding device redesign. In most production cases the decision should be made within several weeks when usually there is no time for an additional learning cycle with test chip experimentation.

Thus, the advanced study of the regularities of failure mechanisms, integration and transformation of the research results into a source of valuable information is a critical way to overcome and address the challenges of modern microelectronics. This approach provides a rapid and effective way to analyze and eliminate the failures by changing manufacturing conditions and transistor operation. An intelligent application of strategic learning enables a reliability-oriented design of new device architectures and processes.

1.1 Catastrophic and Degradation Failure Mechanisms

A traditional approach to define the mode of observed failures in semiconductor devices is to subdivide them into (a) instantaneous and (b) long term. With a better accuracy this classification can be defined as an instantaneous immediate and a gradual progressive change in the transistor parameters above certain critical limits, respectively. The definition is first of all related to an ability of a typical consumer to evaluate and detect the transition of the semiconductor device into a new state when the device parameters go out of the specified parameter range on a timely basis. If the transition time is smaller than the time required for parameter measurement, then the failure is considered sudden or instantaneous. If the failure event can be supported by a set of consequent measurements to demonstrate a gradual walk-out of the parameters from the limits, then this failure mode is considered gradual or long-term. This understanding first of all provides external regularities of the observed failure.

The physical approach, presented in this book, requires a shift of this point of view toward the focus on physical processes inside the transistor structure. In this case the subdivision of failures into "instantaneous" and "long-term" is not adequate at all. For example, one of the typical scenarios for electromigration failure is a long-term transport of metallization material that finally results in a short circuit formation between the contact regions at a certain moment of time. In this case, gradual changes in metallization structure cannot be detected by the conventional direct measurements of the device electrical parameters up until the end result of final irreversible failure. Thus, in this particular case of electromigration failure, it is essentially a degradation mechanism while the failure mode itself according to the formal criterion above can be classified and qualified as instantaneous. At the same time, this failure scenario is essentially different from truly instantaneous irreversible failure mechanisms that may occur as a result of either electrical or thermal breakdown due to conductivity modulation and current instability phenomena.

From the practical point of view, the classification of failure would be much more valuable if it were done with self-representation of the involved physical mechanisms and phenomena responsible for the transition of the device structure into failure mode. In some cases, the failure mechanism can be reduced to a single unique phenomenon. However, most often the failure scenario presents itself as a complex superposition of interactive physical and chemical processes that result in certain transformations of the device structure into a new state. Allocation of the basic failure scenario in general is not a trivial problem. Respectively, a complete classification of the failures based on certain physical mechanisms is extremely difficult too. However, in most practical cases it is possible to allocate at least two independent groups of phenomena in order to identify an elementary classification of the particular failure mode.

The first group of phenomena is related to a qualitative redistribution of the energy state in the transistor structure. A typical result of such energy redistribution is a significant change of the internal thermoelectrical field that may lead to a total and local overheating in the structure, increase of the electric field up to the avalanche breakdown level, redistribution of the current density, hot spot and current filament formation. Due to the relatively small size of microelectronic transistor structures, the energy redistribution processes can be realized in a very fast time domain within the 10^{-10} to 10^{-3} second range.

Another major feature of this group of phenomena is the presence of a critical device state that represents certain critical threshold conditions of the operation regime. Operation above this critical regime inevitability results in significant energy redistribution inside the semiconductor structure. This redistribution often results in device burnout or local accelerated degradation of the structure parameters. These effects are responsible for limitation of the absolute maximum rating of the particular device parameters and physical limitation of the safe operating area.

This set of phenomena corresponds to a typical understanding of "catastrophe." The phenomenon of catastrophe is related to significant abrupt instantaneous changes, swiftness and destructiveness of the object state. The theoretical basis of this phenomenon is studied by the theory of catastrophe.

In semiconductor devices, this group of catastrophic phenomena corresponds to a set of specific failure mechanisms. Usually the term "catastrophic failure" is used in reliability publications as a synonym of "instantaneous failure." In this book, it is not so important that in most cases the catastrophic failure is sudden. The emphasis is given instead to a necessary feature in the form of certain catastrophic phenomenon involved in such failure scenario.

The change in energy state of the structure itself is reversible in general. It may not result in failure. The consequent direct failure cause is the increase of temperature and electric field levels above the material limits that cause the physical change of the device structure. The direct consequences are, for example, metallization melting, cracks and damage of the dielectric isolation and structure of semiconductor material.

Thus, not every catastrophic phenomenon in the transistor results in failure that could be defined as an irreversible change of the device parameters above a certain range. In contrast, in some particular practical cases the catastrophic-type effect of reversible energy redistribution can be reliably observed experimentally. For example, in the short pulse regime it is not only reversible, but even widely used to provide functionality for device-level solutions of electrostatic discharge (ESD) protection.

In opposition to the first group, the second group of failures is called degradation. This corresponds to a large set of diverse phenomena that involve a gradual change in the mechanical or trapped charge state of the transistor structure. In this book, the mechanical state is intentionally treated separately from the energy state. This is done in order to emphasize that the degradation phenomena are first of all related to the transport and redistribution of matter and substance, while the essence of catastrophic phenomena is primarily related to the field and energy redistribution.

The following phenomena are further understood as a change in the mechanical state: a migration or diffusion of atoms and ions, chemical reactions, occurrence and disappearance of diverse charge centers, generation of crystal defects, i.e., the processes that result in a change in the initial mutual arrangement of atoms and molecules or charge state inside the semiconductor device structure, surface and interface regions after removing the electrical regime.

As opposed to catastrophic failures, degradation failures as a rule evolve gradually with a typical time exceeding the characteristic time of catastrophic failures in a few orders of magnitude. Unlike catastrophic failures, the degradation phenomena obey acceleration of the degradation upon given conditions, for example lattice temperature.

In the majority of cases, catastrophic failure is complete and irreversible. It results in a partial or complete loss of the device capability that cannot be restored. In contrast, degradation failures are usually parametric and preserve the basic functionality of the device while bringing some of them above the specified minimum and maximum range including significant deviation. Thus, in this sense, degradation failures are conditional and parametric. They depend on the specified requirements, limits, and figures of merit for the device operation.

Degradation failure can be both reversible and irreversible. Depending on its nature, a complete or partial recovery can be achieved for example in the case of anneals or the device with deep trap centers captured the charge during the reliability test.

One of the "exotic" scenarios of degradation failure is so-called relaxation failure. In this case, the degradation changes in the structure are collected slowly and might not even be possible to detect by standard electrical tests; however, the final irreversible failure occurs quickly, suddenly, instantaneously.

In a general case of semiconductor device failure, the complexity of the design makes it difficult to identify those failure mechanisms that are purely degradation or purely catastrophic based. The redistribution of thermal and electronic flows may provide a significant change in internal field distribution and generate optical phonons that can stimulate and significantly accelerate degradation processes moving atoms, ions, generating and moving of crystal defects. For example, a complex electrothermomechanical stress might be responsible for creation of cracks in gate oxide on CMOS devices followed by creation of a leakage path. There are several mixed mechanisms that cause the degradation and catastrophic phenomena in close interrelation.

1.2 Transistor: Structure, Operation Regimes, Failure Mechanisms

The major principles of transistor design and operation are widely presented in different books in this area, for example [2–7]. The purpose of this section is to summarize those principles of transistor design, process technology, and operation conditions that are critical for further understanding of the reliability and failure physics presented in this book.

1.2.1 Bipolar Transistors

From the reliability complexity point of view, the most challenging types of bipolar transistors are power, switching, high-voltage, and power microwave devices. In these devices, the maximum output power P_{OUT} and the cutoff frequency f_T are limited by the various thermoelectrical phenomena.

Usually the power transistor is considered in devices with the maximum rating of dissipation power over 100 mW. Transistors are considered microwave if f_T exceeds 300 MHz. However, these definitions depend on the application, process, and tradition in different application areas.

Simplified structure of power n-p-n high-frequency transistors is shown in Fig. 1.2. The device is formed on silicon p^+- or n^+-substrate with epitaxial film of n-type. After all process steps in the case of power devices, the initial 400–500 μm substrate thickness is reduced to 100–30 μm to improve heat dissipation.

There are many different architectures of NPN BJT devices starting from topology, subcollector and isolation region design up to implementation of various rather sophisticated self-aligned architectures for polyemitter region in the devices with SiGe base. In the case of power n-p-n with N-epi substrate, the base diffusion or SiGe epitaxial base layer is formed on the epitaxial collector layer that might have an additional collector implant to achieve proper doping profile. The n^+-emitter diffusion in the case of low-cost processes or a polyemitter region are formed on the top of the diffusion or epitaxial P-base using complex spacer techniques to reduce the emitter window size [7]. To bring the collector electrical connection, the subcollector regions are formed by N-buried layer (NBL) and highly diffused N-sinker regions using required planarization, implantation, oxidation and diffusion processes. Finally, the ohmic contact regions and metallization are formed.

In spite of the diversity of alternative semiconductor processes and device designs, the area of active region usually presents itself as an array with the longitudinal dimension of the transistor structure that significantly exceeds the substrate thickness multiple times. This array architecture is implemented to achieve a high total collector current and required high power.

To implement a low on-state resistance, the form of emitter region is significantly changed. A combination of the large emitter region and high density of the injected carriers may result in current shift into the emitter edge and reduced efficiency. To avoid this effect, a backend current balance is widely used in power transistors in the form of backend metallization and distributed poly regions: edge (Fig. 1.2a) or overlay (Fig. 1.2b).

In overlay structures, a ring or rectangular shape single emitters are formed with regular intervals across the whole active region area. Typical size of the

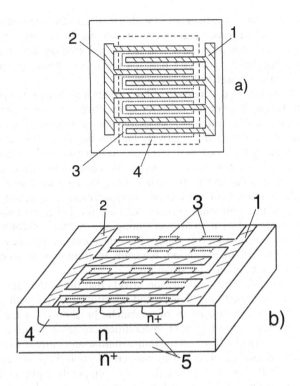

Fig. 1.2 Structure of bipolar transistors: (a) comb-shaped, (b) overlay (1, emitter metallization; 2, base metallization; 3, elementary emitters; 4, base; 5, collector).

emitters and space between them depends on the process and varies from several to fractions of micrometers. Similar distributed structures can provide total operation current up to 10 A with the potential to dissipate DC power above 100 W and operation at maximum chip temperature below 150–200°C.

Catastrophic phenomena of thermal and injection or isothermal instability are observed in the distributed power transistor structures in critical regimes. A number of factors can be noted that are essentially facilitating the development of these dangerous phenomena.

The high-frequency properties of the power transistors are provided mainly by reduction of the emitter–collector spacing. In state-of-the-art microwave structures, the base thickness under the emitter is less than 0.5 μm with the corresponding collector–base p-n junction depth. The basic heat generation in the active operation regime is occurring in the space charge region of the collector junction in the lightly doped n-region of the epitaxial layer. Thickness of this layer depends on the process voltage. For example, for 50-V BiCMOS (Bipolar and Complementary MOS) processes, it is usually less than 10 μm. Thus, the heat source in power transistors is a thin subsurface layer with the linear size a few times higher than the thickness. Such heat source geometry means that the temperature increase on one side of the transistor practically will not be detected on the other side. A weak thermal feedback in the collector junction plane is combined with a strong thermal feedback in the perpendicular direction.

In these conditions due to dependence of the emitter injection on temperature, a slight increase in temperature of the collector space charge region is immediately transmitted into the nearest emitter and respectively amplifies its injection ability. Since the regions of the device are relatively heavily doped, it provides a strong electrical feedback along the whole active region. At constant total current through the device, the current increase through one of the emitters results in an instant corresponding current reduction in the other emitters. Due to a weak thermal feedback along the emitter, the local increase in temperature may result in an immediate sourcing of the additional current into this local region. This positive feedback may be followed by a local overheating and further current redistribution into a localized current filament state. This self-maintained process is called an emitter thermal current filamentation. It results in formation of so-called hot "spots" or hot "dots" in the transistor structure.

Another class of conductivity modulation effects that limits the safe operating area of BJT devices is related to avalanche and injection current instabilities. The instabilities are observed in conditions of avalanche breakdown of the collector junction. This electrical type of current instabilities creates a positive feedback that also may be accompanied by current redistribution and current filament formation. These effects can be observed in isothermal conditions for example in short pulses of ~ 0.2–100 ns. In this case the lattice temperature plays only a secondary role that may impact the initial breakdown voltage level, while thermal heating during the instability is localized in the subsurface region. Avalanche breakdown of the p-n junction is a rather stable process. Some semiconductor devices, for example IMPATT diodes or avalanche photodiodes, are designed for long-term operation in avalanche breakdown conditions.

In some cases, however, depending on the device geometry and doping profiles, the additional thermal current generated in the substrate may be multiplied in the avalanche breakdown region and thus provide the conditions for thermo-electrical instability.

In microwave transistors, the collector and emitter junctions are separated by a spacing of a few micrometers. However, the avalanche processes on the collector junction can provide an excessive hole charge to reduce the emitter potential and cause a strong electron injection from the emitter followed by the avalanche-injection current instability and spatial current filamentation. In spite of the pure electrical nature of the current filamentation, the level of fast local overheating is high and may result in irreversible breakdown and burnout of the device. These mechanisms will be discussed further in Chapter 3.

In most practical cases, both thermal and electrical overloads are not so exotic phenomena in application conditions of power microwave transistors. The devices are intended to operate in generators, amplifiers, voltage regulators, and other power applications. Certainly for reliable operation, the absolute maximum voltage of products is limited and overloads are not expected events. However, the operation in nonlinear regimes with high output power on reactive load may involve a short time entering into the prebreakdown area during tuning or single overload events. In addition, an unpredictable variation of the electromagnetic field on output of transmitting devices (load mismatching) is capable of causing high-intensity avalanche process in the transistor.

High current density in power bipolar transistors also results in degradation phenomena related to electromigration of the metallization material. Electromigration in typical aluminum metallization is a more intensive process in comparison with electromigration in gold or copper. Critical current density for Al migration due to "electron wind" action is $\sim 10^5 \text{A/cm}^2$. This current density level is frequently achieved in thin-film emitter "fingers." Transport of atoms is observed not only along the surface, but also in a perpendicular direction in the structure depth. Moreover, the electron flow in emitters stimulates mutual diffusion of silicon. All of these processes are accelerated in device structure by the lattice temperature.

1.2.2 Silicon Field-Effect Transistors

The first transistor invented by W. Shockley was a field-effect device. However, practical application of the field-effect transistors was delayed due to several technological difficulties. Progress in MOS and diffusion technology finally provided essential components for new design solutions that enabled a competitive level in comparison with bipolar transistors. Then, during rather long period of time it was practically no alternative to low-power metal oxide semiconductor field-effect transistor (MOSFET) combined with both high operating frequencies and high maximum drain voltage. There was even a period when it was expected that the power MOSFETs would push bipolar "competitors" out from the consumer market.

Such forecast has been provided not only by the advantages of power MOSFETs in speed and simplicity of the application circuits, but also by an obvious superiority in reliability.

A first serious step on the path to creation of modern MOSFETs was introduction of the lightly doped drain (LDD) n-region for the n-channel transistors (pldd for P-channel devices respectively). The LDD region essentially increases the drain-source breakdown voltage due to penetration of electric field into the nldd region and appropriate field decrease under the gate electrode (Fig. 1.3a). The LDD region has also dramatically improved the long-term reliability parameters by reduction of the hot carrier degradation rate. However, use of the lightly doped n-region resulted in an undesirable increase of the on state resistance (R_{DSON}) per unit area. R_{DSON} is the major operation parameter of power MOSFETs and a key important parameter for power applications. Reduction of the device on state resistance is further possible by increase in total channel width under corresponding increase in silicon area.

In discrete MOSFETs with V-recess (Fig. 1.3b) or U-recess, the drain electrode is located on the bottom plane of silicon n^+-substrate. In these devices, the n-epilayer thickness corresponds to the desired breakdown voltage value and the consecutive diffusion forms a thin channel p-region and n^+-source. The elementary gate V-recess is possible due to anisotropy in silicon etching. Similarly to overlay bipolar structures, an extremely high density of packing is reached in V-recess MOSFETs too for high current switching.

Other types of switching high-voltage devices—double diffusion MOS (DMOS) and lateral DMOS (LDMOS), drain extended MOS (DeMOS)—will be discussed in Chapter 4. Several other architectures are based on implementation of different junction termination [12], reduced surface field (RESURF) techniques and multi-RESURF or so-called super-junction techniques [13, 14].

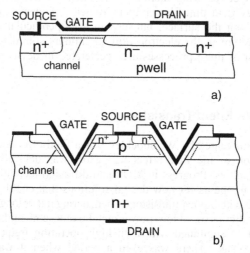

Fig. 1.3 Simplified cross section of the planar (a) and V-recess (b) MOSFET structures

In spite of the large variety of different designs and architectures of power MOS-FETs, a major design feature that is responsible for physical limitation in pulsed safe operating regime, failure mechanisms, catastrophic phenomena, and ESD operation remains the same. This feature presents itself as a "parasitic" bipolar n-p-n transistor formed in the MOSFET structure and connected in parallel to the MOS channel (Fig. 1.3).

The presence of this parasitic structure certainly does not mean directly that properties of MOSFET failure are exactly the same for bipolar transistors. The gain of this parasitic bipolar transistor is significantly reduced due to MOSFET design. It normally provides butted source and bulk regions (p-well or P-body). Thus, in MOSFETs the "emitter" and "base" regions of parasitic NPN (represented by the source and P-well or P-body, respectively) are shorted (Fig. 1.3). Even in the case of nonbutted MOSFET devices the application circuit is usually designed to keep the well tap potential low.

Nevertheless, even under these conditions the source n^+-p junction preserves certain injection capability. This capability increases with channel length reduction and p-well doping level increase. Due to parasitic transistor, the avalanche multiplication in the drain space charge region might be accompanied by an avalanche-injection process. This process is similar to the one observed in power bipolar transistors under corresponding conditions. However, as will be shown in Chapter 4, the isothermal instability in MOSFETs has numerous peculiarities.

The possibility of thermal instability in MOSFETs is much lower than in power bipolar devices. Absence of injection from the source p-n junction excludes the pure "bipolar" type of thermal current filamentation in MOSFETs in spite of a comparable intensity of thermal heating and topology of both transistors. However, power MOSFETs are designed for operation at high current density. Therefore, at some high drain current range a positive feedback between the surface temperature and the channel current may occur depending on the device design. The current increase with temperature is observed up to a level of so-called thermostable current I_D*. At this current level, the derivative of the drain current in lattice temperature $\partial I_D / \partial T$ changes sign and becomes negative.

As shown in several experimental studies, the positive feedback between the drain current and temperature is capable of causing the phenomenon of hot spot formation.

Another catastrophic phenomenon in MOSFETs is related to electrical breakdown of the gate dielectric. Gate oxide is usually formed by a thermal oxidation. Depending on the desired device voltage, the gate oxide thickness can be varied from a few angstroms up to several hundred angstroms. The gate region design in power MOSFETs with small oxide thickness and large total gate width of the power array automatically provides a relative protection against a small level of electrostatic discharge due to high value of the gate input capacitance. However, the same gate features increase the region "sensitivity" to diverse structural and technological defects, gate oxide integrity. As a result, the local electrical breakdown voltage in the oxide might impact the yield of the discrete or integrated power devices.

Presence of MOS structure creates the possibility of various degradation failures. The physical mechanisms of these failures are related to different trapped charge instabilities of the gate and thick field oxide dielectric and hot carrier degradation of the major device parameters (threshold voltage, on-state resistance, breakdown characteristics). This includes such phenomena as migration of various ions under action of the electric field, free electron capture on deep levels, trap recharging and formation of trap centers, various electrochemical processes, connected with neutralization of mobile and built-in charge.

1.2.3 Compound Semiconductor Field-Effect Transistors

The class of GaAs field-effect transistors with Schottky gate (MESFET) and modulation doped FETs (MODFET) or high electron mobility transistors (HEMT) took over the leading position for power microwave application for the frequency range from 5 GHz up to millimeter-wave range.

In spite of the achievements in SiGe HBT [7] especially for integrated components, MESFETs and MODFETs are still critical discrete and monolithic components for amplification of low-power signals due to low level of intrinsic noise and other advantages. The advantage in high-speed performance of MESFETs is first of all provided by the properties of compound semiconductor material (GaAs, InP, and others). GaAs semiconductor material has a much higher value of electron mobility, drift velocity, and nonlocal dependence of the velocity on the electric field (velocity overshoot). Due to bigger band gap (1.4 eV versus 1.1 eV) in comparison with silicon, the thermogeneration level in GaAs is significantly lower. This results in both low internal noise level and higher stability to thermal breakdown. However, at the same time the thermal breakdown effect is enhanced by lower heat conductance of GaAs versus Si.

There are many technological solutions for MESFET fabrication. Typical process steps include a compensated semi-insulating substrate with grown epitaxial semi-insulating buffer layer ($n^- = 10^{14} cm^{-3}$) and a thin (0.1–0.3 μm) active epilayer with the donor concentration $N_D \sim 1$–$3 \times 10^{17} cm^{-3}$. The contact regions are formed either by ion implantation or by an additional epi for contact n^+-region ~ 0.1 μm with $N_D = 1$–$3 \times 10^{18} cm^{-3}$. Electron lithography is used to form the channel and gate recess profiles followed by the source and drain ohmic contact formation using for example Au-Ge-Ni metallization system (Fig. 1.4). The gate region is Schottky contact. It is formed by a metallization system that uses titanium, platinum, and gold (or aluminum) in a self-aligned window. In contrast to MOSFETs, in MESFETs the conducting channel is located at a certain depth from the semiconductor surface. This provides an additional stability of major physical parameters.

The epitaxial buffer layer protects the channel region from penetration of impurities from the substrate during epitaxial growth. It is necessary to eliminate electron mobility reduction. However, according to experimental data the influence of the substrate cannot be excluded completely. Several types of substrate-related instabilities of parameters in GaAs MESFETs have been reported.

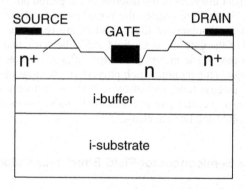

Fig. 1.4 Simplified cross section for GaAs MESFET structure

The catastrophic phenomena in GaAs MESFETs and modulation doped FETs (MODFETs or HEMTs) are related to phenomena of avalanche ionization and injection in strong electric field conditions (Chapter 5). Preconditions for local avalanche breakdown are created by the small spacing (< 0.5 μm) of the Schottky gate metallization edge. A considerable peculiarity occurs by formation of a high electric field domain in GaAs film. The beginning of impact ionization in the domain results in excitation of isothermal instability. The instability, as a rule, results in local burnout. Under certain conditions in GaAs MESFETs, a thermal breakdown can also be observed.

1.2.4 Specifics of Transistor Applications

The peculiarities of a particular failure mechanism depend not only on the specific transistor design, but also on application conditions. For example, in the case of microwave signal amplification the major operation regimes are usually subdivided into three regimes: A, B, and C.

In regime A, the transistor is conducting a current during the whole period of amplified microwave signal. This regime is the most useful for linear amplification application. It is combined with another regime when the collector (or drain) voltage always remains above the saturation voltage. In the case of properly designed class A linear amplifiers, both the collector (drain) operation current and voltage are usually selected below the corresponding maximum rating values in order to avoid nonlinear distortions.

In regime B, the current is conducted only for a half-period of microwave signal with the device remaining in the off-state during the rest of the period. This mode is more energy efficient, but provides a higher level of nonlinear distortions in comparison with regime A.

In regime C, the transistor is in the on-state for less than a half-period of time of the amplified signal. This regime is frequently combined with an overload regime, when the transistor is in the saturation state during part of the operation time. This key operation regime has high efficiency but relatively strong distortions. Constant switching from the off-state to the saturation state at high current levels creates in this key regime a precondition for catastrophic failure. In these regimes, the maximum rating of dissipated power parameter cannot be exceeded. However, exceedingthe breakdown voltage value can cause an isothermal current instability and burnout.

In regime A, the value of dissipated heat power in the structure is high. Therefore, the processes of thermal breakdown and instability must be considered. In comparison with other regimes (B and C), the degradation phenomena in this regime (electromigration, mutual diffusion, and charge instabilities) are more probable due to acceleration by current and temperature.

At the same time, the thermal filamentation phenomenon is more probable in the DC regime or at low operation frequencies. In this case, the oscillation period exceeds a typical thermal time of transistor structure. Microwave regimes are relatively safe with respect to the thermal catastrophic phenomena. However, load mismatching conditions in the microwave regime might produce a significant excessive generation of the dissipated power and voltage in the device junction, exceeding maximum ratings followed by current instability and burnout effects.

1.3 Operation Regime and Reliability

1.3.1 Time Before Failure and Failure Rate

The concept of transistor reliability is based on the regularities derived for a number of reliability parameters or figures of merit. The basic parameter of the semiconductor device is the time before failure (t_{BF}). On a randomly selected set of similar devices, the operation time before failure is distributed according to some casual law. Apparently, the distribution function of failures versus the number of devices depends on the transistor operation regime too. In particular, it is always possible to set an electrical or thermal operation regime that will result in an immediate failure of the device at the very beginning of operation. In this case, the middle time before failure (t_{MTBF}) approaches zero. So it is clear that in the low electric load regime, the majority of transistors will operate practically for an unlimited time.

A problem of operation time without failure has a principal interest in the area of electrothermal regime application. Since an "absolutely reliable" transistor is impossible in principle, certain indirect guarantees are required for consumer electronics design. This explains the occurrence of technical requirements related to the transistor failure rate λ as a quantitative reliability parameter. Failure rate has both a precise mathematical definition and physical sense [2, 11]. The value λ is the probability of the event when the device will not fail during some initial period of time t, but then will fail during the following very small time interval Δt. In general, λ is a time-dependent function $\lambda \equiv \lambda(t)$. Several other important mathematical reliability parameters are related to λ. One of them is the probability of nonfailure operation $P(t)$ that can be defined as

$$P(t) = \exp\left[-\int_0^t \lambda(t)dt \right].$$ (1.1)

An assumption of constant failure rate (λ = const) is often used in semiconductor device reliability studies. This results in an exponential distribution $P = \exp(-\lambda(t))$ (Fig. 1.5a). Constant λ is extremely convenient for various technical estimations and reliability calculations. In addition, the value λ characterizes the time of nonfailure operation directly as an exponential distribution of $\lambda = \left(\bar{t}_{MTBF} \right)^{-1}$, where $\bar{t}_{MTBF}$ is the middle time before failure.

In the first approximation, if λ of the semiconductor device is constant in time, it is possible to provide an estimation of reliability for a more complex system composed of N similar devices. Assuming no dependence of each device failure, the failure rate for the whole system is equal to λN due to multiplication of the appropriate probabilities $P(t)$. For example, at $\lambda = 10^{-6}$ h^{-1} the middle operation time before failure of a separate device is rather high, e ~10 million hours. However, an electronic system or integrated circuit can incorporate thousands and more semiconductor devices. Considering for example the CPU in your computer, at $N = 10^4$ the failure rate of the system will be equal to 10^{-2} h^{-1} and $\bar{t}_{MTBF} = 100$ hours only. Thus, the real operation time of separate devices can be significantly less than $\bar{t}_{MTBF}$.

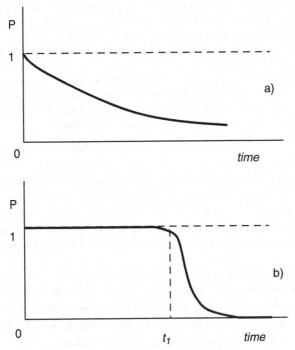

Fig. 1.5 Dependence of the probability of failure-free operation on time: at constant λ (a) and at variable λ due to a degradation process (b)

Thus, the rather logical on first sight assumption results in a paradoxical conclusion, that practically any rather complex system should be functionally unreliable. Experience, however, convinces us that this is not true. This paradox can be clarified as follows.

One of the reasons is that the failure rate in real operation conditions depends on time, i.e., is not constant. Over a rather long period of time, it might not exceed 10^{-8}–10^{-9} h^{-1} until the activation of dominant degradation processes. The degradation processes may considerably increase the failure rate to a value a few orders of magnitude higher, for example 10^{-6} h^{-1}. It can be explained using the example. A hypothetical bipolar junction transistor can be considered with two kinds of failure mechanisms: catastrophic due to irreversible breakdown at collector-emitter $U > 30$ V, and degradation due to electromigration of metallization at I > 2 A. In other words, if the transistor collector-emitter voltage exceeds 30 V, this results in immediate "burnout." If $U < 30$ V and $I > 2$A, the transistor fails eventually too, but after a much longer period of time. For DC dissipated power $P = 30$ W, two operation points can be considered: (1) $U = 30$ V, $I = 1$ A and (2) $U = 10$ V, $I = 3$ A.

In the first case, the transistor will be in a state where the slightest voltage increase results in an instantaneous failure. Any real circuit can experience such positive voltage fluctuations. The frequency of their occurrence p is defined by the circuit and does not depend on the transistor design. Obviously, λ of the transistor in this case will be precisely equal to p, if p = const and λ = const. Increase in the amount of transistors in the circuit results in a proportional failure rate increase and decrease of system reliability as a whole.

In the other operation point at $U = 10$ V, the breakdown is practically excluded. However, high current density in the metallization region ($I = 3$A) causes an intensive mass transport that may result in a breakage of the conductor material and transistor failure. The metallization breakage occurs if the conductor cross-sectional area is decreased down to some critical value S_{CR}. There is a certain guaranteed time t_{INS} of absence of electromigration failure.

Obviously, $t_{INS} \sim \dfrac{S - S_{CR}}{v}$, where S is the initial section area and v is the speed of mass transport. At $t < t_{INS,}$ the failure rate is close to zero and the appropriate probability of operation without failure is not practically different from case 1 (Fig. 1.5b). At $t \geq t_{INS}$, λ sharply increases and P decreases.

Thus, the first of the discussed cases (λ = const) corresponds to influence of random factors that are independent of the transistor design. The second case corresponds to a degradation process that is regularly developed in transistor structures.

There is a practice to guarantee λ in the area of operation regimes. The paradox of this practice is obvious. Indeed, low values of λ suitable for application (10^{-7}–$10^{-9}\,h^{-1}$) cannot be confirmed by tests at the development phase at least during the first years of the released product lifetime. For example, in order to confirm $\lambda = 10^{-8}\,h^{-1}$ a test volume is required of over 10^8 device-hours.

A basic approach is suggested in the standards of the International Electrotechnical Commission. It consists in presentation of abrupt but feasible requirements for manufacture of devices: certification of manufacturing and basic technological processes, fulfillment of the requirements to the device design and materials, serial acceptance of production with reasonable volumes and test results. These tests are conducted in manufacturing conditions using allowable regimes as long as practically possible (usually, during 1000 hours). As a result, the bottom level of test failure rate λ_T is usually confirmed by a group of the devices and it should not exceed 10^{-4}–$10^{-5}\,h^{-1}$ The λ_{OP} value is defined on the basis of information about the device failures under application conditions. It is presented as reference data for circuit designers and system developers. Usually the value λ_{OP} is lower (by at least one to two orders of magnitude) than λ_T for manufacturing conditions.

1.3.2 Safe Operating Area

To guarantee reliable transistor operation to consumers, a number of limitations on the application operation regime are stated by the manufacturer. In the specification on transistor application, maximum ratings of operation currents I_{max}, voltage U_{max}, dissipated power P_{max} and other parameters are specified for given frequency, time domain, ambient temperature, and other conditions. For example, some area can be plotted on the I–U coordinate plane for the device regimes. This area is usually called the Safe Operation Area (SOA). By definition, a reliable operation is provided inside SOA. For example, in simplified cases the SOA of output characteristics of a power transistor in the case of positive bias (Fig. 1.6) is limited by three conditions: $I \leq I_{max}$; $U \leq U_{max}$; $I\,U \leq P_{max}$.

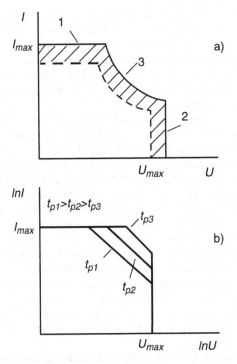

Fig. 1.6 SOA for DC (a) and pulsed (b) regimes: 1, $I = I_{max}$; 2, $U = U_{max}$; 3, $P = P_{max}$ (where t_p is the power pulse duration)

The current limitation is dictated as a rule by capabilities of metallization and contact regions. The voltage limitation is provided by the avalanche breakdown effect; power limitation is caused by Joule heating. It dependson the device ability to provide heat dissipation. These are ordinary typical limitations. There are also other limitations. These limitations are introduced to eliminate various undesirable phenomena, for example various thermoelectrical instabilities (Chapters 2–5) or hot carrier degradation.

The important definitions are a maximum rating and an absolute maximum rating for example for operation voltage. In an operation regime below the first maximum rating, long-term reliability parameters for the device are guaranteed. In the parameter range between maximum and absolute maximum ratings, short-term survivability is guaranteed, but long-term reliability parameters are not confirmed, i.e., the device can experience a short-term overload without harm for functionality. Finally, in the regime above the absolute maximum rating, even a shortterm operation is not guaranteed and the device may be susceptible to an immediate instantaneous burnout or significant degradation of the time before failure. Most of the integrated products and systems have maximum and absolute maximum ratings of their own that are selected below the rating of individual devices and depending on the circuit specification.

There is a methodological difficulty for direct SOA measurements. Apparently, the concept of SOA by itself makes sense if there is an operation regime having a certain parameter of time before failure. Apparently, that SOA for device operation in a continuous regime with for example $t_{mtbf} \sim 10^5$ h can be significantly different from the SOA of the same transistor for application in the pulsed regime for example with pulse duration ~1 µs and low 0.1% duty factor even if the same current–voltage regime is realized.

The method of SOA determination includes the primary operation area. This area is limited by currents and voltage in the short-term regime. For example, several hours of operation in this primary SOA regime should be sufficiently far from the basic instantaneous catastrophic failures. Then, account of different degradation processes imposes the secondary limitations on T_{max} (i.e., on P_{max}) and on I_{max}.

Thus, primary SOA should be reduced down to some area (shaded in Fig. 1.6a). While in the first case the technique of SOA delimitation requires short-term measurements [2, 4], in the second case the long-term tests are used to measure appropriate reliability parameters. It is clear that for every particular device type, it is practically impossible to verify all reliability parameters. In this case, analytical models of various degradation processes are used.

On the basis of the given reliability parameters (for example, λ value during given operation time), the final SOA boundary is established. For final verification of given reliability parameters, reliability tests at one SOA boundary point are carried out for example for the case of maximum substrate current when dealing with NMOS and NLDMOS devices.

As was already discussed, the SOA depends on the given operation regime of the transistor. At transition from DC to pulsed regime, the SOA is usually extended due to reduction of heat dissipation limitations. For example, in accordance with reduction of pulse duration the SOA aspires to the rectangle shape limited by the current and voltage maximum ratings (Fig. 1.6b). Thus, the power limitation criterion is related to some limiting constant temperature of transistor structure chosen beforehand for all pulse durations.

If a single rectangular current pulse regime is used for transistor reliability estimation, the difference in the SOA is determined only by the pulse duration. With pulse duration decrease (below units of microseconds), a new value of limiting regime essentially exceeds the originally defined SOA.

This new SOA will depend on the pulse duration t_P. The real SOA boundary at $t_P < 1$ µs can give valuable information about transistor reliability in the microwave regime and for various switching processes. For example, in the bipolar transistor switching regime at operation on reactive load, the operating point trajectory can essentially differ from load line during the switching (Fig. 1.7). At the transistor turn-on, the inductance interferes with the collector voltage decrease. Therefore, the instant maximum current and dissipated power values can be far beyond the SOA boundary (trajectory 1, Fig. 1.7). With the turn-off process, the load inductance can exceed the voltage maximum rating. However, if the time duration of the operation point outside the SOA is less than 1 µs, the transistor reliability is still provided as illustrated in Fig. 1.7.

With very short pulse regimes, an important consideration is displacement currents and other dynamic coupling effects in the device structure. These effects can significantly reduce the maximum voltage of the device due to engaging of other parasitic components of the device structure. For example, in the case of operation on inductive load or in the case of fast rise time ESD pulse, the parasitic NPN bipolar in MOSFETs may significantly limit pulsed SOA due to the so-called dV/dt effect.

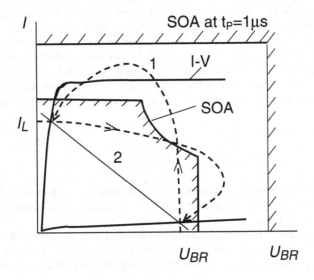

Fig. 1.7 Possible trajectories for operation point at bipolar transistor turn-on ("1") and turn-off ("2")

1.4 Breakdown and Instability

The practice of field application of semiconductor devices and integrated circuits provided the insight that catastrophic phenomena formed the basis of the overwhelming majority of observed failures. For some modern devices, for example compound semiconductor microwave FETs, the share of catastrophic failures at field application exceeds 90%. This fact does not mean a lower "importance" of the degradation failures [9–11].

High speed and irreversible features of the majority of catastrophic phenomena require corresponding high-level complexity of experimental studies. For theoretical analysis a deep knowledge in different areas is required from the physics of hot electrons in semiconductors at high injection up to the nonlinear theory of stability in active systems.

In the previous sections the phenomena of irreversible breakdown, burnout, thermal and isothermal instability, and current filamentation were mentioned. Some understanding of these phenomena is available in numerous referenced publications. These notions are widely used in publications in the field. Now they are used in testing and application of transistors. However, interpretations of these phenomena are frequently rather arbitrary. Therefore, before addressing particular effects in various types of devices, it is necessary to specify the sense and interrelation of these physical notions.

In reliability studies, the terms "thermal breakdown," "secondary breakdown," "injection breakdown," and "irreversible breakdown" are widely used. Sometimes authors assign to each of these concepts a different physical sense.

In this book, the notion of "breakdown" preserves its elementary primary sense. Breakdown is treated similarly to the effect in p-n semiconductor junctions as a process of sharp current increase that is caused by carrier generation current. For example, at avalanche breakdown of the reverse-biased p-n junction the current I is increasing with the voltage U increase according to known empirical formula [2]:

$$I \sim \frac{1}{1 - \left(U/U_{BR}\right)^n},$$

where $n = 4$–6 for Si material and U_{BR} is the avalanche breakdown voltage. The dependence $I(U)$ in this case of "classical" avalanche breakdown is strictly monotonous (Fig. 1.8a).

A different kind of breakdown could be observed in bulk semiconductor structures. If at a constant voltage $U \ll U_{BR}$ the device is heated by some external current source, then from some temperature level the current through the sample will grow sharply according to an exponential dependence $I \sim \exp\left(-E_G/kT\right)$, where E_G, T, and k are the energy of band gap and lattice temperature of semiconductor material and Boltzmann's constant, respectively. This thermogeneration process for carriers is considered a thermal breakdown.

Unfortunately, in a number of publications, thermal instability is defined as thermal breakdown. In this book, thermal instability is defined as a process of uncontrollable current increase with the voltage decrease and in general it may or may not result in device failure.

Thus, the physical sense of the breakdown definition is just a sharp current increase under positive differential conductivity. In other words, the breakdown itself does not include any positive feedback. The major breakdown mechanisms in semiconductors are the avalanche and the thermal, although several additional mechanisms can be found, for example the breakdown related to the change in trap charge state or breakdown in dielectric.

In opposition to breakdown, different thermoelectrical instabilities addressed in this book include a positive feedback. The thermoelectrical instability phenomena in semiconductor devices are rather complex. To provide a "quick start" understanding about the physical sense of these phenomena, the example of thermal instability in semiconductor structure is presented below.

It is assumed that on a uniform sample of bulk semiconductor having length l and area S, the voltage U is supplied. The current density j through the sample is equal to $j = \sigma E = \sigma U/l$, where E is the electric field in the sample and σ is the conductivity of semiconductor material. The conductivity of semiconductor material in this case is expressed by $\sigma \sim \exp(-E_G/kT)$. Then the generated heat per volume unit is equal to σE^2. It is further assumed that heat dissipation is provided by exchange with ambient space of fixed temperature T_S. Then heat dissipation from the surface will be proportional to $(T - T_S)^\beta$, where T is the temperature of semiconductor region and $\beta = 1$–2. The heat balance equation is

$$\exp\left(-\frac{E_G}{kT}\right)\frac{U^2}{l^2} = K(T - T_S)^\beta, \tag{1.1}$$

where the proportionality factor K depends on physical and geometrical parameters of the device structure. In a certain range of values U, this equation can have two solutions: T_1 and T_2 (T_1, $T_2 > T_S$). This corresponds to two different current values I_1 and I_2.

The I–V characteristic of such structures is not simple and has an S-shape (Fig. 1.8b). The physical sense of the I–V characteristic can be explained as follows. n- the initial region "OC" the current increase obeys dependence close to Ohm's law. An appreciable deviation from linear dependence begins when the heating reaches a higher level. This temperature is above 300°C. The conductivity increase is related to intensive thermogeneration of electrons and holes in the sample. Insignificant increase of the voltage in the thermal breakdown region "CA" produces a sharp increase of heat generation.

Up to a certain limit the heat generation is balanced by an increase in temperature since the heat dissipation is proportional to $(T - T_S)^\beta$. However, at some $U > U_{CR}$ the exponential increase of the heat generation cannot be compensated anymore by heat dissipation. In this state, equation (1.1) has no solutions and the semiconductor sample has no stationary states. This means that in the voltage source regime at $U > U_{CR}$, a sample will be uncontrollably self-heated up to the destruction or other physical limitations.

A similar loss of thermal stability or an uncontrollable process of transition into a new state usually means instability. In this case, the instability is of a thermal nature. From Fig. 1.8b the stable states of a sample in the case of thermal instability can be achieved only at corresponding voltage decrease U. If the circuit provides a sufficient load resistance, then the thermal instability may finally evolve into a stable state that corresponds to the I–V characteristic with negative differential conductivity (NDC) (Fig. 1.8b, state R).

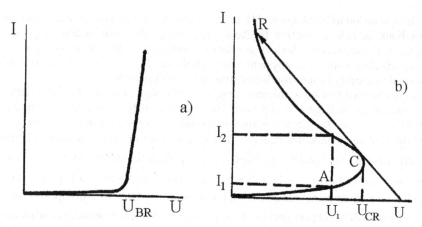

Fig. 1.8 *I–V* characteristic at avalanche breakdown of p-n junction (a) and at the thermal instabilities in semiconductor resistor (b)

One of the major peculiarities of thermal instability is stabilization of the final state due to sharp temperature redistribution inside the semiconductor sample at corresponding voltage drop. Due to this nonlinearity, a hot "spot" or filament formation improves heat dissipation conditions: The higher the temperature in the spot, the more effective temperature gradients. Certainly the temperature redistribution in the semiconductor is impossible without an appropriate current redistribution. Thus, thermal instability is accompanied by current filamentation processes.

If the sample dimension and its physical properties allow filamentation, then the corresponding region must be presented in the *I–V* characteristic (Fig. 1.9). To reflect this statement, the branch "AN" (Fig. 1.9a) is added at lower voltage than the branch "AB" for uniform state.

It is important to emphasize that the scenario of thermal instability can vary depending on the external circuit. In the voltage source regime at $U > U_{CR}$ the device transits into an unstable state. In application this state is usually limited by an irreversible failure mode. In the current source regime, the operation point is single

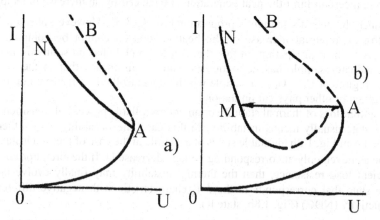

Fig. 1.9 Typical *I–V* characteristics at current filamentation in semiconductor structure

and stable (Fig. 1.9a). In the case of filamentation, the current source regime can provide a stable transition into the filament state (Fig.1.9a). However, in certain samples the filamentation process may result in an uncontrollable transition even in the current source regime (A→M, Fig. 1.9b).

Unfortunately, thermal instability in real device structures is not always simple to interpret in such a way. Serious difficulties are caused by identification of the instability as a failure mechanism. For example, Fig. 1.1a represents typical consequences of a catastrophic failure in power bipolar transistors. Aluminum metallization is melted at 700°C. The presented "landscape" in Fig. 1.1a is evidence that the temperature at least exceeded this critical level. Unfortunately, this sometimes remains the only unique conclusive fact in the further chain of possible options.

If the ambient temperature is stable, an overheating up to 700°C is the result of either aluminum metallization self-heating or heating inside the silicon structure. Metallization self-heating is in this case less probable due to localization and silicon melting (Fig. 1.1a). Presence of original "craters" supports the idea of hot spot formation in the silicon surface. Usually, a hot spot is a narrow region having dimensions of a few micrometers with concentrated current and elevated temperature. In bipolar transistors, similar current localization results in thermal breakdown. In this case, a sharp temperature increase may result in metallization melting.

What could provide a reason for hot spot formation on bipolar transistors? The most probable of course is excessive dissipated power over some critical level. There are several probable causes: electrical overload; presence of structural or die attachment defects; and evolution of some degradation processes in the structure.

A typical set of conclusions provides a set of versions for failure mechanisms. However, in order to confirm or deny any of them, much time and effort are usually required. It is considered good luck if the search circle is immediately narrowed down and the particular reason is reliably identified.

The example of thermal instability was provided above to facilitate a primary understanding of the processes that are extremely widespread in real device structures. The instability mechanisms in real transistor structures are rather diverse and complex and often they are far from a simple mathematical interpretation. For example, there is a group of current filamentation mechanisms that cannot be placed in conformity with an S-shape I–V characteristic. This instability is realized at positive differential resistance. In real semiconductor structures, basic changes in thermoelectrical instability are provided by an inhomogeneity or structural defects. The analysis of these structures sometimes can be carried over using numerical simulation only. The regularities of thermal and isothermal instabilities in semiconductor devices are discussed in the following four chapters.

1.5 Summary

In this introductory chapter, degradation failures are distinguished from catastrophic failures through the corresponding differences in change in matter versus energy state, long-term versus short-term evolution, the partly reversible parametric failure scenario versus sudden, abrupt irreversible scenario.

It was further discussed that since the methodology of long-term reliability predictions using accelerated tests is well established, validated, and can be controlled by the device architecture at the development stage, the major challenges and the majority of failures in the field belong to catastrophic phenomena.

Thus, this book is mainly focused on the involved physical mechanisms and phenomena responsible for transition of the device structure into failure mode and scenario due to certain phenomena of catastrophic, instantaneous nature that provide for a specific set of failure mechanisms.

Since in most practical cases the final damage very rarely reveals a direct physical failure mechanism due to hidden secondary postdamage processes, the original cause or complete scenario of failure requires an advanced understanding and certain methodology of analysis, integration, and transformation of the research results into a source of valuable information.

The common applied principle elaborated in this book is that the origin of these phenomena is based on change in the energy state of the structure followed by nonuniform increase in all or several of these parameters: electric field, current, temperature, and dissipated power levels above the material limits that cause the physical change of the device structure. This primary change provides the reason for further secondary changes in the device structure due to rapid metallization, melting, cracks, and damage of the dielectric isolation of structure materials.

Finally, this chapter established major terms and definitions for some physical parameters and phenomena based on their physical sense and primary understanding through the phenomenological: maximum rating, absolute maximum rating, irreversible breakdown, burnout, thermal and isothermal instability, current filamentation, thermal breakdown, secondary breakdown, injection breakdown, and irreversible breakdown.

Chapter 2
Theoretical Basis of Current Instability in Transistor Structures

From the physical point of view, a semiconductor device during operation presents itself as an open, distributed, and dissipative system. This means permanent exchange of energy and matter between the transistor and the ambient space. Therefore, the transistor's state at any moment of time can be described by a number of the distributed physical parameters: the temperature $T(x,y,z)$, electric field $E(x,y,z)$, electron–hole current density $j(x,y,z)$, and other in general distributed parameters. Distribution of these parameters depends on time, due to applied external conditions, and on the internal processes in the transistor itself. Under electrical load the transistor is a nonequilibrium system in principle. The dynamic equilibrium of the semiconductor device under operation is far from thermodynamic equilibrium. This fact results in the possibility of formation of rather complex multiple thermoelectrical instabilities in the device. In particular, the transition of the semiconductor device from one state to another may become accompanied by a strong current redistribution or filamentation [15–17].

In spite of a wide variety of observed current filamentation scenarios and mechanisms, to some extent all of them are based on the same physical principles. Theoretical analysis of transistors as a distributed nonlinear system was first completed in [17–19]. It has been particularly shown that the phenomenon of S-shape I–V characteristic formation in semiconductor structures presents itself as a particular case of the fundamental behavior of a nonlinear dynamic system in nonequilibrium conditions [19]. This chapter focuses only on general principles of current filamentation phenomena in phenomenological semiconductor structures followed by a discussion of a number of general problems for different types of semiconductor devices, conductivity modulation, and spatial current instability in the following chapters.

2.1 Basic Equations

Unlike several decades ago, engineering analysis with numerical simulation tools has significantly evolved today. In some cases, rather accurate solutions can be obtained. However, even today the complexity of simulation for thermoelectrical instabilities and three-dimensional analysis still involves a rather simplified approach. It requires rather sophisticated calibration and yet is not accurate for big structures due to total grid point limitation.

At the same time, numerical simulation results are not always easy to interpret. The analytical description provides a very important in-depth understanding of the major physical principles of spatial current instabilities that could be further refined using numerical simulation methodology.

In most practically important cases, the current filamentation phenomena can be mathematically described using only two parameters. This description is independent from the exact physical mechanism. One of the parameters is distributed in the structure, while the other is spatially independent and controlled by the external circuit. For the case of thermal filamentation mechanisms, the distributed parameter is the temperature; for electrical mechanisms, a similar adequate parameter is the electrical potential. Spatially independent value in most cases is the voltage or the current that is supplied to the external terminals.

Current filamentation in semiconductor structures can be described by two differential equations of reaction-diffusion type. As an example, the thermal filamentation mechanism is discussed below in order to demonstrate the major regularities. Similar approach can be applied for electrical current filamentation.

Stationary temperature distribution $T(x,y,z)$ in semiconductor structure is determined by the thermal conductivity equation with the heat source Q:

$$\frac{\partial^2 T}{\partial x^2} + \frac{\partial^2 T}{\partial y^2} + \frac{\partial^2 T}{\partial z^2} = Q(T,U), \qquad (2.1)$$

where the nonlinear function Q is determined by generation and dissipation of the heat in the semiconductor structure; U is the externally measured voltage on the structure. Although in general thermal conductivity is some function of lattice temperature, in this introductory chapter for simplification it is assumed constant $\kappa = \text{const}$. More accurate solution is possible using numerical simulation. The voltage U is spatially independent and is not a fixed value. For example, it could correspond to a certain stationary $T(x,y,z)$ after achieving thermal equilibrium in the structure. The value U entirely represents a state of the semiconductor structure in the case of equilibrium and a linear term Q. Equation (2.1) is true inside a semiconductor chip that can be represented by a parallelepiped with a heat sink on the bottom edge with a fixed temperature T_0 (Fig. 2.1). This simplified approximation is rather accurate in the case of a transistor with constant temperature fixed in the heat sink.

$T\big|_{Z=W} = T_0$ is one of the boundary conditions of (2.1). Other boundary conditions reflect an absence of heat dissipation from other sides of the structure:

$$\frac{\partial T}{\partial y}\bigg|_{Y=0,L_Y} = \frac{\partial T}{\partial x}\bigg|_{X=0,L_X} = 0. \qquad (2.2)$$

In most submicrometer transistor structures, excluding vertical high-voltage power devices, the heat is generated mainly in the surface region of the chip near the side $Z = 0$ (Fig. 2.1). Therefore, the heat source Q in (2.1) can be assumed as an energy flow in the upper plane of chip:

$$\left.\frac{\partial T}{\partial z}\right|_{Z=0} = -\frac{q(U,T)}{\kappa}, \tag{2.3}$$

where $q(U,T)$ is some thermal generation function and κ is the thermal conductivity coefficient. For this case, (2.3) simply represents the sixth boundary condition of (2.1) with the right side equal to zero.

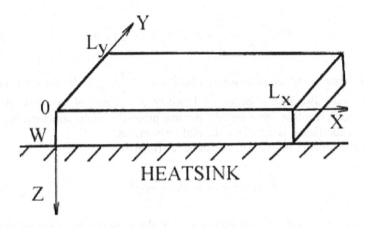

Fig. 2.1 Thermal model for semiconductor chip representation in analysis

The following simplification will include a convergence of the three-dimensional (2.1) onto the one-dimensional case. Another simplification is that in the assumed model the current and temperature redistribution are possible in the whole plane $Z = 0$. However, it is assumed that the redistribution takes place only in a single direction, for example the direction OX. In other directions the distribution is uniform, therefore $\dfrac{\partial^2 T}{\partial y^2} = 0$ is true. In spite of the last true assumption in rare cases, for example $L_Y \ll L_X$ and $L_Y \approx W$, the major regularities of current filamentation are the same.

The second derivative $\dfrac{\partial^2 T}{\partial z^2}$ is given by

$$\frac{\partial^2 T}{\partial z^2} = \left(\left.\frac{\partial T}{\partial z}\right|_{Z=W} - \left.\frac{\partial T}{\partial z}\right|_{Z=0}\right)\Big/ W \tag{2.4}$$

The average value of thermal gradient in the bottom side is $\dfrac{\partial T}{\partial z}\Big|_{z=W} = -\dfrac{T - T_0}{W}$. Using the boundary condition (2.3), (2.4) can be rewritten as

$$\frac{\partial^2 T}{\partial z^2} = \frac{q(U,T)}{\kappa W} - \frac{T - T_0}{W^2}. \tag{2.5}$$

Finally, (2.1) is transformed to

$$W^2 \frac{\partial^2 T}{\partial z^2} + \frac{W}{\kappa} q(U,T) - \left(T - T_0\right) = 0. \tag{2.6}$$

Under nonuniform temperature distribution at $\partial T / \partial x = 0$, the case is similar to the case discussed in Section 1.4: where q is analog of the power density and the coefficient W/κ represents the thermal properties of the structure. In application this coefficient is called the thermal resistance R_T.

For the following analysis, (2.6) can be rewritten as

$$l_T^2 \frac{\partial^2 T}{\partial x^2} + \varphi\ (U,T) = 0, \tag{2.7}$$

where φ is the nonlinear function and l_T is the typical dimension of thermal changing.

There are several ways of simplification of (2.1). However, the final view (2.7) is typical for a wide class of semiconductor systems with S-shape I–V characteristic and current filamentation.

The second equation for complete description of stationary states $T(x)$ and U is the Kirchhoff equation for circuit with the sample

$$\int_0^{L_X} F(U,T)dx = I, \tag{2.8}$$

where I is the total current through the sample and F is some nonlinear function that represents the circuit. For the current source regime, the function F is simply the current density: $F \equiv j(x)$.

For the mathematical description of current filamentation kinetics, the nonstationary thermal conductivity equation is used:

$$\tau_T \frac{\partial T}{\partial t} = l_T^2 \frac{\partial^2 T}{\partial x^2} + \varphi\ (U,T), \tag{2.9}$$

where the coefficient τ_T represents a typical heating time for the particular structure.

For the analysis here $\tau_T = W^2/a^2$, where a^2 is the specific thermal conductivity coefficient: $a^2 = \kappa/\rho c$: κ, ρ, c are the thermal conductivity, the density, and the specific heat capacity of the semiconductor material, respectively.

The solution of (2.7) and (2.6) presents the S-shape I–V characteristic $I(U)$ that together with (2.9) provides the solution for a classical case of current filamentation.

2.2 Current Filamentation in Uniform Structures

In spite of a relatively simple view of (2.7), (2.6), and (2.9), the general solution and analysis is rather difficult due to the nonlinearity of the functions F and φ. However, a number of important conclusions can be reached by assumption of a uniform structure. Namely, this assumption is adopted in the theory of filamentation in semiconductors [15, 16]. The structure is considered uniform when all the parameters are constant along the direction of current filamentation. In this case, (2.7), (2.8) always have a uniform solution $T = $ const for given structure and fixed external circuit parameters. The classical methods of the stability theory can be applied to the initially uniform state.

2.2.1 Fluctuation Instability

With the assumption of small fluctuations of given T_0, U_0 parameters, the uniform distribution of temperature becomes slightly nonuniform $T = T_0 + \delta T(x)$ at a certain moment of time. Corresponding disturbance of voltage is $U = U_0 + \delta U$. Corresponding change in the structure state is determined by (2.8), (2.9). Since the steady-state solution T_0, U_0 is true for (2.7) and (2.8), the linearization for δT_0, δU_0 results in the equation

$$\tau_T \frac{\partial \delta T}{\partial t} = l_T^2 \frac{\partial^2 \delta T}{\partial x^2} + \varphi_T' \delta T + \varphi_U' \delta U, \qquad (2.10)$$

$$\int_0^{L_X} \left(F_T' \delta T + F_U' \delta U \right) dx = 0 \qquad (2.11)$$

with the boundary condition

$$\left. \frac{\partial \delta T}{\partial x} \right|_{X=0, L_X=0} = 0. \qquad (2.11a)$$

Conclusion about stability of the state T_0, U_0 can be done depending on the increase in time of an arbitrary disturbance $\delta T(x)$. The classical form of the stability analysis is presented below.

Using Laplace transformation of (2.10), the following equation can be obtained:

$$\tau_T \gamma \delta \ T(x,\gamma) - \tau_T \gamma \delta \ T(x,0) =$$
$$l_T^2 \frac{d^2 \delta \ T(x,\gamma)}{\partial \ x^2} + \varphi_T' \delta \ T(x,\gamma) + \varphi_U' \delta \ U(\gamma), \tag{2.12}$$

where $\delta \ T(x,\gamma) = \int_0^\infty e^{-\gamma t} \delta \ T(x,\gamma) \ dt$; $\delta \ T(x,0)$ is the initial disturbance fluctuation.

In this case, the solution $\delta T(x,\gamma)$ of (2.12) with the boundary condition (2.11a) is given by Fourier transformation:

$$\delta \ T(x,\gamma) = \sum_{m=0}^\infty C_m(\gamma) \cos \frac{\pi \ m}{L_X} x. \tag{2.13}$$

By substitution of (2.13) into (2.12), multiplying on $\cos \dfrac{\pi \ n}{L_X} x (n = 0, \ 1, \dots)$ and integrating from 0 to L_X a simple system of n linear equations can be obtained relatively to the coefficients C_n. Using δT the corresponding value $\delta U(\gamma)$ is given by the expression

$$\delta \ U = -\frac{\varphi_U' F_T'}{F_U'} C_0. \tag{2.14}$$

The determinant of the system is given by

$$Det = \left(\tau_T \gamma - \varphi_T' - \frac{\varphi_U' F_T'}{F_U'} \right) \prod_{n=1}^\infty \left(\tau_T \gamma - \varphi_T' + l_T^2 \left(\frac{\pi n}{L_X} \right)^2 \right)$$

and the roots of the equation at $Det = 0$ are given by

$$\gamma_0 = \frac{1}{\tau_T} \left(\varphi_T' + \frac{\varphi_U' F_T'}{F_U'} \right); \ \gamma_i = \frac{1}{\tau_T} \left(\varphi_T' - \left(\frac{l_T \pi n}{L_X} \right)^2 \right), i = 1, 2, \dots. \tag{2.15}$$

The physical sense of the analysis is that at reverse Laplace transformation [under transition $\delta T(x,\gamma) \to \delta T(x,t)$] in the equation for $\delta T(x,t)$ the terms proportional to $\exp(\gamma_i, t)$ appear. Therefore, when one of the roots γ_i is positive, the disturbance $\delta T(x,t)$ will grow, i.e., the state T_0, U_0 is unstable.

When $\gamma_0 > 0$ and all $\gamma_i < 0$, only a uniform disturbance will be increased. The thermal instability results in current filamentation when at least $\gamma_1 > 0$.

From (2.15), the filamentation criterion in the uniform structure is given by

$$\varphi'_T > \left(\frac{l_T \pi}{L_X}\right)^2. \tag{2.16}$$

The criterion (2.16) can be found in an even simpler way using substitution $\delta T \sim e^{\gamma t} \cos \dfrac{\pi \, n}{L_X} x$ in (2.10), (2.11) and after simplification of (2.15) for $\gamma_i \ (i = 1,2,...)$.

It is important that the criterion for instability (2.15) is always automatically true in the case of S-shape I–V characteristics for the negative differential resistance region $R_d < 0$.

As an example, the case of thermal breakdown of reverse-biased p-n junction can be analyzed. For an avalanche diode structure with the p-n junction at the top side of the structure, the perpendicular current flows in the OZ direction (Fig. 2.1); thermal generation due to Joule heating in the space charge region of p-n junction with thickness is much less than the chip thickness; the function φ in (2.7) and (2.9) is given by

$$\varphi = \frac{W}{\kappa} j_s U - (T - T_0), \tag{2.17}$$

where

$$j_S \sim \exp\left(-\frac{E_g}{kT}\right). \tag{2.18}$$

External circuit equation (2.8) is given by the continuity of the total current:

$$\int_0^{L_X} j_S dx = \frac{I}{L_Y} = \text{const}.$$

In this case from (2.8), (2.11), and (2.18):

$$\delta U = 0, \quad \int_0^{L_X} \delta j_S dx = \int_0^{L_X} \delta T dx = 0.$$

Therefore, filamentation criterion coincides with the condition (2.16) that corresponds to increase of the first harmonic of $\delta T(x)$ fluctuation, and is given by

$$\frac{WE_g}{\kappa \, kT^2} j_s U - 1 > \left(\frac{W\pi}{L_X}\right)^2. \qquad (2.19)$$

In the state of equilibrium, the ratio between U, I, and T is determined by the condition $\varphi = 0$, i.e.,

$$\frac{UI}{L_X L_Y} = \frac{\kappa}{W}(T - T_0). \qquad (2.20)$$

Using (2.20), the criterion (2.19) can be rewritten in a simpler form:

$$\frac{E_g(T - T_0)}{kT^2} > \left(\frac{W\pi}{L_X}\right)^2 + 1. \qquad (2.21)$$

The p-n junction temperature can be determined from (2.20) under given U, I, and T_0. After the substitution in (2.21), one can determine the possibility of current filamentation. Apparently in a long structure with $L_X \gg W$, the critical condition for current filamentation is realized at much lower dissipated power, while in the device with relatively small length $L_X \approx W$, the filamentation condition is hardly realized for temperatures of p-n junction below 700°C.

If the conditions of increase are true not only for the first, but for another $\gamma_i > 0$ harmonic in (2.15), the analysis of stability is necessary. It can be found that the first harmonic $\cos(\pi x / L_X)$ corresponds to the single filament state near the edge and the second harmonic of $\cos(2\pi x / L_X)$ corresponds to the single filament in the structure center. However, from the stability analysis, it can be found that only asymmetric filament at the edge is stable.

2.2.2 Bifurcation, Soft and Abrupt Filamentation

The condition of fluctuation instability in uniform structures was presented in the previous section. This idealized criterion allows evaluation of the instability in real structures too including the effect of physical parameters.

From the uniform structure criterion (2.16) the equality $\varphi'_T = \left(\dfrac{l_T \pi}{L_X}\right)^2$ and the stationary state equation $\phi = 0$ provide the S-shape I–V characteristic and the critical point M: (U_{CR}, I_{CR}). After transition through this critical point, the uniform state becomes unstable relatively longwave fluctuation disturbances $\cos(\pi x / L_X)$. The point M (Fig. 2.2) is called a bifurcation point, because at this point the uniform solution for $T(x)$ is split into two different solutions: spatially uniform and spatially nonuniform. The nonuniform solution represents the state with the current filament and corresponding nonuniform temperature distribution.

The splitting mode may be "abrupt" (Fig. 2.2a) or "soft" (Fig. 2.2b). In the abrupt mode, the splitting corresponds to a sharp or a jump-type filament excitation. For example, in current control mode a small fluctuation results in formation of a high-amplitude current filament during the transition through point M (i.e., at $I > I_{CR}$, Fig. 2.2a). The filament provides a high ratio of the current and temperature inside the filament to the same parameters farther away from the filament. The transition to the filament state MM' is accompanied by voltage decrease. The return to the uniform state is accompanied by a jump too, but at $I < I_{CR}$. This results in appearance of a hysteresis loop in the I–V characteristic.

In contrast, in the case of soft splitting at $I > I_{CR}$ the filament amplitude increases gradually with the current increase. Stable slightly nonuniform states always exist near point M too, unlike the case of abrupt splitting (Fig. 2.2a) where slightly nonuniform states near point M also exist (in the region KM), but are not stable and therefore cannot be observed experimentally.

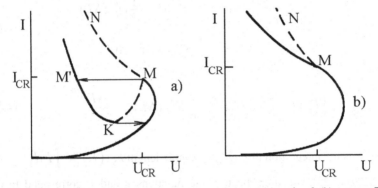

Fig. 2.2 S-shape I–V characteristics for the cases of abrupt (a) and soft (b) current filamentation scenario

The criterion for fluctuation instability (2.16), the critical point, and the different splitting modes are the direct consequence of mathematical representation of the current filamentation phenomena. In real models, this "theoretical" output requires some corrections.

The fluctuation instability criterion (2.16) is the main result of the linear theory of current filamentation. A nonlinear analysis is considerably complex, but allows approaching other more complex problems, for example the problem of soft or abrupt mode of current filament excitation. The method of nonlinear analysis is not so universal and its application requires a particular approach without a guarantee of success.

In the discussed case, we have an integration-differential system of nonlinear equations. To derive conclusions, it is necessary to make some simplifications. However, the general analysis of a similar equation has been widely used for the solution of various problems of catastrophic phenomena and self-organization in various nonlinear systems [20–22].

To distinguish between soft or abrupt splitting mode, the stationary states and their amount of the structure near the bifurcation point M (Fig. 2.2) in particular should be analyzed. To demonstrate this problem solution, the stationary equations (2.7) and (2.8) will be further analyzed near the critical point M under conditions of a small current increment ΔI: $\Delta I = I - I_{CR}$ in order to determine small increments ΔT and ΔU of the initial state T_0, U_0 with respect to the current increment ΔI.

An adequate approximation of S-shape dependence in various regions can be represented by a cubic-parabolic dependence. Therefore, in expansion of the functions ϕ and F in the vicinity of point M, the terms up to the third order should be taken into account. Respectively, (2.7) and (2.8) can be transformed into

$$
l_T^2 \frac{\partial^2 \Delta T}{\partial x^2} + \varphi_T' \Delta T + \varphi_U' \Delta U + \frac{1}{2} \varphi_{TT}'' (\Delta T)^2
$$

$$
+ \frac{1}{2} \varphi_{TU}'' \Delta T \Delta U + \frac{1}{6} \varphi_{TTT}''' (\Delta T)^3 = 0,
\tag{2.22}
$$

$$
\left\langle F_T' \Delta T + \frac{1}{2} F_{TT}'' (\Delta T)^2 + \frac{1}{6} F_{TTT}''' (\Delta T)^3 \right\rangle = \frac{\Delta T}{L_X},
\tag{2.23}
$$

where $<...>$ is the integration through the sample length L_x : $\langle F \rangle = \int_0^{L_X} F dx$.

In (2.22), (2.23), for simplification the functions ϕ and F correspond to the above-discussed example of thermal breakdown of p-n junction. Then ϕ''_{UU} and the following derivatives are calculated at the splitting point M where $\phi'_T = (\pi l_T / L_X)^2$.

As previously stated, the solution $\Delta T(x)$ will be found as a Fourier transformation:

$$
\Delta T(x) = \sum_{\ell=0}^{\infty} C_K \cos \frac{\pi k}{L_X} x,
\tag{2.24}
$$

where C_K are the expansion coefficients that depend on ΔT; $C_K \to 0$ at $\Delta I_K \to 0$. Substituting (2.24) in (2.22) and (2.23), multiplying (2.22) on $\cos(\pi n x / L_X)$ ($n = 0$, 1, 2,...) and integrating from 0 to L_X, the equation system for $(n + 2)$ equations for C_K and ΔU value is

$$\begin{cases} C_0\phi_T' + \Delta U\phi_U' + \dfrac{1}{2}\varphi_{TT}''\left(C_0^2 + \dfrac{1}{2}\sum_{'=1}^{\infty}C_K^2\right) + \varphi_{TU}''C_0\Delta U + S_0^3 = 0 \\[2ex] C_n\left[\phi_T' - \left(\dfrac{\pi d_T}{L_x}\right)^2 n^2\right] + \dfrac{1}{2}\varphi_{TT}''\sum_{|\pm m|=n}^{\infty}C_kC_m + \varphi_{TU}''C_n\Delta U + S_n^3 = 0, n \geq 1, \quad (2.25) \\[2ex] F_T'C_0 + \dfrac{1}{2}F_{TT}''\left(C_0^2 + \dfrac{1}{2}\sum_{'=1}^{\infty}C_K^2\right) + S_{n+1}^3 = \dfrac{\Delta I}{L_X} \end{cases}$$

where S^3 are the sums containing third-order terms.

Without altering the basic problem, the following approximations can be used for (2.25) simplification:

1. Two-mode approximation [i.e., in the expansion (2.24), only two first harmonics are used]. This is true near the first splitting point [23] at high attenuation of higher order harmonics.
2. In each equation of (2.26), only major terms for each variables are used. Finally:

$$\begin{cases} C_0\phi_T' + \Delta U\phi_U' + \dfrac{1}{4}\varphi_{TT}''C_1^2 = 0 \\[2ex] \varphi_{TT}''C_0C_1 + \dfrac{1}{2}\varphi_{TT}''C_1C_2 + \varphi_{TU}''C_1\Delta U + \dfrac{1}{8}\varphi_{TTT}'''C_1^3 = 0. \\[2ex] -3C_2\phi_T' + \dfrac{1}{8}\varphi_{TT}''C_1^2 = 0 \\[2ex] C_0(\phi_T' + 1) + \dfrac{1}{4}\varphi_{TT}''C_1^2 = \dfrac{\Delta I}{L_X}\dfrac{W}{\kappa}U \end{cases} \qquad (2.26)$$

In (2.26), it is considered that according to (2.17) and (2.19),

$$\phi_T' = \dfrac{W}{\kappa}UF_T' - 1; \quad \varphi_{TT}'' = \dfrac{W}{\kappa}UF_{TT}''.$$

Through the solution of (2.26) for C_1 variable, the following equation can be obtained:

$$C_1^3\left[\dfrac{1}{8}\varphi_{TTT}''' + \dfrac{(\varphi_{TT}'')^2}{48\phi_T'} - \dfrac{(\varphi_{TT}'')^2}{4(\phi_T'+1)} - \dfrac{\varphi_{TT}''\varphi_{TU}''}{4\phi_U'(\phi_T'+1)}\right]$$

$$+ C_1\left[\dfrac{WU\Delta I(\varphi_{TT}''\phi_U' - \phi_T'\varphi_{TU}'')}{\kappa L_x\phi_U'(\phi_T'+1)}\right] = 0. \qquad (2.27)$$

It can be further rewritten in the form

$$C_1\left(PC_1^2 + R\right) = 0, \tag{2.28}$$

where P and R are the coefficients corresponding to (2.27). The sign of R is determined by the sign of ΔI since $\varphi_{TT}'' \phi' - \varphi_{TU}'' \phi_T' \cong 1 > 0$. The coefficient P depends on the structure parameters and can change the sign.

The following analysis can be made for real roots of (2.28) relatively to the sign of ΔI and P. Here the root $C_1 = 0$ always exists that obviously corresponds to the uniform state: $C_0 \neq 0$, $C_1 = C_2 = 0$.

The case P > 0

At $R > 0$ ($\Delta I > 0$) $C_1 = 0$, there is only one real root. At $R < 0$ ($\Delta I < 0$), two additional roots exist: $C_1 = \pm(R/P)^{1/2}$. This case corresponds to the abrupt splitting and branch formation on the S-shape I–V characteristic (Fig. 2.2a). In fact, at $\Delta I > 0$ only the uniform state exists. At $\Delta I < 0$, besides uniform two slightly nonuniform states with temperature maximum near the left or right device edge appear. Since both of these states are equivalent with different sign at C_1, they correspond to the same branch "ML" on the I–V characteristic.

The case P < 0

By repeating the above procedure, it is easy to establish that this case corresponds to the soft splitting (Fig. 2.2b).

Now the conditions for soft and abrupt regimes of filament excitation can be obtained for the above-discussed case of thermal breakdown of p-n junction. Considering $E_G/kT \gg 1$, from (2.17),

$$\varphi_{TTT}'''\left(\phi_T' + 1\right) = \left(\varphi_{TT}''\right)^2;\ \varphi_{TU}''/\phi_U' = E_G/kT^2,$$
$$\varphi_{TT}'' = \left(\phi_T' + 1\right)E_G/kT^2$$

and
$$\tag{2.29}$$

$$P = \left(\varphi_{TT}''\right)^2\left[\frac{1}{12\phi_T'} - \frac{3}{2\left(\phi_T' + 1\right)} + \frac{\phi_T'}{\phi_T' + 1}\right].$$

It is easy to see that the condition $P = 0$, which divides the cases of abrupt and soft filament excitation, corresponds to $\phi_T' = 17/12$, as follows from (2.29). From $\phi_T' = \pi^2 l^2/L_X^2; l_T = W$.

Therefore, the hysteresis on S-shape I–V characteristic and abrupt filament excitation will be observed at $L_X > \sqrt{7}W$, i.e., when maximum chip dimension exceeds its thickness by at least 2.5–3 times.

Thus, by complex calculations a simple ratio between the geometrical sample dimensions is found. Of course, this ratio is obtained using serious simplifications that include at least the transition from three-dimensional to one-dimensional model. Therefore, quantitative coincidence with the experimental results can only be casual. However, it is important that by means of the above analysis, the effect that provides a major impact on the filamentation process becomes clear. The set of active parameters is less than, for example, that for the fluctuation instability criteria. In particular, the thermal conductivity κ, the heat sink temperature of T_0, and the parameters of semiconductors are not involved.

2.3 Current Filamentation in Nonuniform Structures

Assumption of semiconductor structure uniformity is the most significant simplification in the filamentation theory for mathematical analysis.

In reality, transistor structure can hardly be identified as a uniform object in the case of any discussed filamentation mechanism. A nonuniform current distribution in any time moment is caused by the design structure of transistor in principle for example due to the complex topology and structure of active regions. In addition, casual inhomogeneity is added by various uncontrolled technology defects. The source of inhomogeneity is natural defects in used materials and processes. Among them are lattice structure defects of thesemiconductor itself, metallization grains and contact boundary, mechanical strain in multilayer structures, regions of accumulated charge in the dielectric–semiconductor interface, photolithography alignment mismatch, and many others.

Thus, it is clear that all physical structure parameters incorporated in (2.1)–(2.9) can be varied in the current filamentation direction to some extent.

A logical question is: what is the impact of inhomogeneity, local structure defects, and other deviations from the ideal device structure on the instability criteria, derived in the previous section, and on the scenario of filamentation process itself?

To solve this problem analytically, a small local inhomogeneity $\Delta A(x)$ of some arbitrary physical parameter $A(x) = A_0 + \Delta A(x)$ is introduced, where A_0 is the value of the parameter in a cylindrical structure. The following statements will be proven below.

At some critical level of inhomogeneity amplitude, the deviation in the initial parameters might be larger than the deviation caused by fluctuations.

1. At some critical amplitude of inhomogeneity in a semiconductor structure, the notion of critical point is invalid. The bifurcation points disappear and split of the I–V characteristic is observed into a number of isolated branches.
2. With gradual quasi-static change of the structure operation regime, the current state always corresponds to the branch of the I–V characteristic for initial

slightly inhomogeneous and filament states. Thus, current filamentation in the structure always takes place deterministically and does not depend on small internal fluctuations. The initial cause of filamentation is stationary introduced inhomogeneities of electrophysical structure parameters.

3. In a nonuniform sample, the filament is localized on inhomogeneity. Edge of the structure presents an inhomogeneity by itself. The filament near the structure edge is stable, as well as the solitary filaments that are localized in inhomogeneity far from lateral structure boundary.

4. The critical regime of current filamentation is strongly influenced by the inhomogeneity parameters.

2.3.1 Splitting of I–V Characteristic

For simplification, it is assumed that the spatially dependent parameter $A(x)$ is included in the nonlinear function ϕ in (2.7) and (2.9). In this case, Taylor expansion ϕ must be completed near the point T_0, U_0 and in the left part of (2.22) the terms $\phi'_A \Delta A, \phi''_{TA} \Delta T \Delta A$ will appear. Respectively, in (2.25), terms like $\phi'_A \Delta A_n, \phi''_{TA} C_K \Delta A_n$ are added, where

$$\Delta A_n = \left\langle \Delta A(x) \cos \frac{\pi n}{L_x} x \right\rangle. \tag{2.30}$$

Since $\Delta A(x)$ is a given value, then without loss of generality all bilinear terms with ΔA can be neglected for an adequately small ΔA. Only the terms of $\phi'_A \Delta A$ type are preserved. For simplification, $\Delta A_0 = 0$ is assumed since the constant component of inhomogeneity results in a small shift of I–V characteristic without any qualitative changes. Besides, for further clear demonstration, the edge inhomogeneity is assumed $\Delta A_1 \gg \Delta A_n (n > 1)$. All mentioned assumptions result in change of only the second equation in system (2.26), due to appearance of free term $\phi'_A \Delta A_1$. Respectively, (2.26) can be transformed to

$$C_1 \left(P C_1^2 + R \right) = -\phi'_A \Delta A_1. \tag{2.31}$$

Assuming $\phi'_A \Delta A_1 > 0$, (2.31) can be analyzed graphically for the abrupt filament excitation regime $(P > 0)$ [24]. At $R > 0$ $(\Delta I > 0)$, (2.31) has the single real negative root (Fig. 2.3).

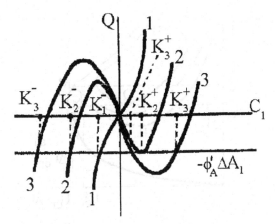

Fig. 2.3 Graphical solution of (2.31): $Q \equiv C_1\left(PC_1^2 + R\right)$; 1, $\Delta I > 0$; 2, $|\Delta I| = \Delta I_{CR}$; 3, $|\Delta I| > \Delta I_{CR}$; where $K_i^{\pm}$ are the positive and negative roots

At $R < 0$ ($\Delta I < 0$) the negative root always exists. Its modulus increases with decreasing R. Beginning from some $R_{CR} < 0$ ($\Delta I = -\Delta I_{CR}$), positive roots appear: the first root at $|\Delta I| = \Delta I_{CR}$ and two positive roots at $|\Delta I| > \Delta I_{CR}$. With increase in ΔI, the module of one of the roots is decreased, but the others are increased. From (2.26) according to all root dependencies on ΔU increment, the I–V characteristic can be easily built near the critical point (Fig. 2.4a).

For soft excitation regime, the I–V characteristic can be obtained using this approach (Fig. 2.4b).

As follows from Fig. 2.4, the I–V characteristic with inhomogeneity has no critical points and is splitting into isolated branches. The branch "OK" corresponds to "hot" states with the maximum temperature in the inhomogeneity region. Branch "BDE" represents "cool" states with the maximum temperature located far from the inhomogeneity region.

In the discussed case under the conditions of quasi-static (gradual) regime change, the states corresponding to the branch "BDE" should be excluded. This branch is absolutely isolated and has no common points with the branch "OKPT." However, the filaments in the region "DE" are stable with the temperature peak location is at the opposite boundary relatively the inhomogeneity. These states are slightly different from the filaments that correspond to the region "PT" due to small amplitude of the inhomogeneity. However, such filaments can be realized in the structure only in the pulsed regime or at rather high amplitude of additional disturbance.

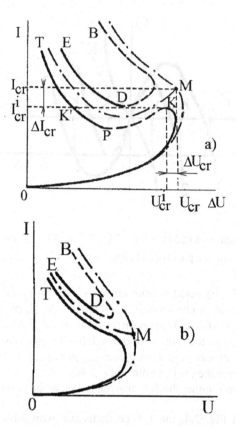

Fig. 2.4 Qualitative view of *I–V* characteristic in the presence of small inhomogeneity for abrupt (a) and soft (b) filamentation regimes. Dashed lines represent the unstable regions. Dashed-dotted lines represent *I–V* characteristics of corresponding uniform structure

2.3.2 Critical Regime

According to the shape *I–V* characteristic in the case of abrupt filament excitation (Fig. 2.4a), the critical regime for the structure with inhomogeneity corresponds to the common point area between the load characteristics and "OP" region of the *I–V* characteristic [25]. For assumed current source regime, the common point is the point "K" in the *I–V* characteristic. It corresponds to the current $I_{CRI} = I_{CR} - \Delta I_{CR}$.

It is important to underline that unlike in uniform structures, at $I > I_{CRI}$ the filament formation does not correspond to increase in the nonuniform fluctuations δT and δj, anymore. In contrast, the filamentation is the consequence of a deterministic transition process. The reason for this is the absence of stationary states near the point "K" in the current range that is slightly over the critical level I_{CRI}.

At $I > I_{CRI}$ in the structure the only possible state with filament is formed corresponding to the region $K'T$ for the current source load type.

In comparison with uniform case at abrupt filament excitation, the filamentation takes place at current level that is less by ΔI_{CR} value at least. Obviously, ΔI_{CR} is determined by inhomogeneity parameters. For ΔI_{CR} estimation, the condition of existence in (2.31) of the single positive root—the condition of "contact"—should be found. By differentiating (2.31),

$$3PC_1^2 + R = 0; C_1 = \sqrt{|R|/3P}$$

($R < 0$, since $\Delta I < 0$).

By substituting C_1 in (2.31) and taking into account (2.27),

$$\left[\frac{WU\Delta I_{CR}}{\kappa\varphi_U'(\varphi_T'+1)L_X}\right]^{\frac{3}{2}} \cong 3\sqrt{P}\varphi_A'\Delta A_1.$$

From the condition of abrupt filament excitation ($P > 0$) and for adequately long structures, $\varphi_A' \ll 1$ follows. Therefore, taking into account

$$\varphi_U' = \frac{W}{\kappa}U\frac{I_{CR}}{L_X},$$

$$\frac{\Delta I_{CR}}{I} \cong \sqrt[3]{P}\left[\varphi_A'\Delta A_1\right]^{\frac{3}{2}}. \tag{2.32}$$

The corresponding value ΔU_{CR} can be obtained from (2.26). Thus, an integral characteristic depends on the amplitude, dimension, and localization of inhomogeneity. The higher the relative critical regime deviation is, the stronger the filament excitation in comparison with uniform structure.

Since $P \sim \left(\varphi_T'\right)^{-1}$ (2.27), therefore for long structures with $L_X \gg W$ the rate of critical regime decrease with inhomogeneity increase $[d(\Delta I_{CR})/d\Delta A_1)]$ is significant. It may finally result in complete disappearance of the hysteresis in I–V characteristic.

One of the practically useful conclusions can be derived from (2.32) directly. Using the reversible measurements of the critical regime parameters (Chapter 5), at least high-amplitude structural defects and inhomogeneity can be detected.

2.3.3 Edge and Ordinary Filaments

The above results can be transformed toward a general case of arbitrary inhomogeneity model as well as for bifurcation points of higher order. It is clear that the demonstrated splitting of *I–V* characteristic takes place near any bifurcation point. A measure of splitting is the range of separation between the isolated branches. It is determined by an amplitude of the Fourier factors $\Delta A_n = \int \Delta A(x) \cos \frac{\pi n}{L_x} x \, dx$, where n is the bifurcation point number. A possible view of *I–V* characteristic for a general case of inhomogeneity localization is presented in Fig. 2.5 for the example of a second bifurcation point taken into account.

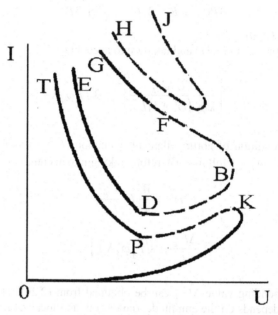

Fig. 2.5 Qualitative view of *I–V* characteristic in the presence of inhomogeneity

The interest in the second bifurcation point reflects the fact that this point corresponds to a solitary filament. This state can be stable in structures with inhomogeneity. This is reflected by the region "FG" (Fig. 2.5, solid line). A stable solitary filament is localized on inhomogeneity. Unlike the region "DE" for edge filaments, this filament can be realized without any additional disturbance for example in the pulse regime. The only necessary condition is that the inhomogeneity be sufficiently remote from the sample boundaries. The regions "OK," "PT," "DE," and "FG" represent all possible stable states in the structures with inhomogeneity under some thermal filamentation mechanism. Within the discussed model of (2.1)–(2.8), the multiple filament states, which might originally be localized in several inhomogeneities, are usually unstable relatively "pumping-over" into a single filament state process.

Before further discussion of some important problems related to the solitary filaments, it should be emphasized that solitary ordinary filaments are observed experimentally in real transistor structures. Moreover, edge filaments having the classical shape with vertical "wall" can exist only in the case of mirror boundary conditions. Unlike as assumed in numerical simulation problems, in nonideal real device structures the boundary conditions $\dfrac{\partial T}{\partial x}\bigg|_{x=0,Lx} = 0$ cannot be provided at the chip edge. Even in the case of silicon on oxide (SOI) process, lateral thermal flow always exists. Besides, in field-effect transistors the heat source is not at the chip edge.

Similarly to an isothermal filamentation mechanism for edge filament existence, the condition of the diffusion no-flux condition $\dfrac{\partial n}{\partial x} = 0$ is equivalent to the condition $\dfrac{\partial T}{\partial x}\bigg|_{x=0,Lx} = 0$. Obviously, the carrier density n is not reaching the maximum at the boundary.

At the same time, the structure with mirror boundary conditions in essence physically represents a case of a half-size of ideal structure. In this case, the edge filament simply corresponds to a filament in the middle of the full ideal structure. This approach allows one to study filament state as a state of an ideal nonlinear system independently from the particular boundary conditions.

For consistency, the analysis for edge filaments of the first order is completed above in spite of the fact that they are not observed under real experimental conditions. However, similar analysis and results are correct for solitary filaments and the second bifurcation point. The instability criterion (2.16) must be derived for the corresponding second harmonics with $n = 2$ using ΔA_2 according to (2.30) without factor ΔA_1.

The only possible stable nonuniform state corresponds to the transformation I–V characteristic in real structures to the transition into solitary filaments: the branches "FG" and "OK" (Fig. 2.5) are merged and form a continuous branch that corresponds to the solitary filaments. At the same time, the branch of stable edge filaments "PT" is realized.

2.3.4 Localization and Stability of Solitary Filaments

The conditions of formation and stability for filaments in the nonuniform structures can be analyzed rather accurately. Equations (2.7) and (2.8) can be modified by introducing a small inhomogeneity A, $A = A_0 + \Delta A(x)$ [26]:

$$l_T^2 \frac{d^2T}{dx^2} + \varphi(T,U,A) = 0, \tag{2.33}$$

$$\int_0^{Lx} F(T,U,A)dx = \text{const}. \tag{2.34}$$

A logical assumption is: before filament formation, the spatially uniform stationary states $T(x)$ and U in the structure with small inhomogeneity are only slightly different from the corresponding uniform states $T_0(x)$ and U_0 of the uniform structure. Then the solution $T(x)$ and U can be found as $T_0(x) + \Delta T(x)$ and $U_0 + \Delta U$, respectively. By linearization of (2.33), (2.34) in the vicinity of T_0, U_0, A_0 for ΔT and ΔU_0, we obtain

$$l_T^2 \frac{d^2 T}{dx^2} + \varphi_T' \Delta T + \varphi_U' \Delta U + \varphi_A' \Delta A = 0, \tag{2.35}$$

$$\langle F_T' \Delta T + F_U' \Delta U + F_A' \Delta A \rangle = 0 \tag{2.36}$$

at the boundary conditions: $\left. \dfrac{\partial T}{\partial x} \right|_{x=0,Lx} = 0$, where all partial derivatives are taken in the point T_0, U_0, A_0.

According to the classical method [16], $\Delta T(x)$ can be found through serious expansion in fundamental functions of operator $\hat{H}_0 \equiv d^2/dx^2 + \varphi_T'$ of equation. $\hat{H}_0 \varphi_m = \lambda_m \varphi_m$ at the boundary conditions:

$$\Delta T = \sum_{m=0}^{\infty} C_m \varphi_m. \tag{2.37}$$

The variable ΔU is derived from (2.36) and is substituted in (2.35) taking into account (2.37). Then by multiplying (2.35) on ψ_n and by integration along the sample for $n = 0, 1, 2, \ldots,$

$$\sum_{m=0}^{\infty} C_m \left(\lambda_m \delta_{m,n} - \frac{D_n \beta_m}{b} \right) = \frac{D_n}{b} \langle F_A' \Delta A \rangle + \langle \psi_n \varphi_A' \Delta A \rangle, \tag{2.38}$$

where $\delta_{m,n}$ is Kronecker's symbol and

$$b = \langle F_U' \rangle; \ D_n = \langle \psi_n \varphi_U' \rangle; \ \beta_m = \langle \psi_m F_T' \rangle. \tag{2.39}$$

Linear equation system (2.38) has the only solution $C_n = \det^{(n)}/\det$ if the determinant of system (2.38) is not equal to zero. $\det^{(n)}$ is the determinant array with the column number n replaced by the corresponding column of free terms in (2.38).

Apparently, if $\Delta A \to 0$, then $C \to 0$ since $\det^{(n)} \to 0$ and ΔT is a small value.

For the solitary filaments $T(x)$, the solitary filaments $T_0(x)$ are initial solution too. For filaments, fundamental functions of operator H_0 cannot be derived, since φ_T' is a function of coordinate. However, as known from [15, 16], dT_0/dx coincides or just slightly differs from fundamental function with fundamental value $\lambda = 0$. This case is easy to prove using differentiation of (2.33) on x.

Since dT_0/dx has only one zero value inside the interval $(0, L)$, this is the fundamental function with index "1": $dT_0/dx = \psi_1$. Therefore, $\lambda_1 = 0$. Due to the oscillation theorem [16], $\lambda_0 > 0, \lambda_m < 0 \ (m > 1)$.

Determinant of (2.38) in the case of filament is equal to [16]

$$\det = \prod_n \lambda_n \left(b - \sum_{m=0}^{\infty} \frac{D_n \beta_n}{\lambda_n} \right), \tag{2.40}$$

if $\lambda_n \neq 0$.

In a long structure $(L \gg l_T)$, the impact of boundaries can be neglected. For a solitary filament localized farther away from the boundary, it remains in equilibrium relatively fluctuations of shift. Thus, such filament can be localized in any arbitrary point of the structure. This is a direct result of $\lambda_1 = 0$. So, corresponding fundamental function $\psi_1 = dT_0/dx$ is the shift function for the filament. Since $\lambda_1 = 0$ in calculation of det (2.40), nothing can be stated about obtained expansion coefficients C_m yet.

In order to analyze (2.38) for $n = 1$, one can consider that $D_1 = 0$ due to parity $T_0(x + M)$, Fig. 2.6a, and $\lambda_1 = 0$. The following indispensable condition can be obtained:

$$\left\langle \varphi_A' \Delta A(x) \frac{dT_0}{dx} \right\rangle = 0. \tag{2.41}$$

It is easy to demonstrate that if (2.41) is true, then system (2.38) has only one solution $C_n \ (n \neq 1)$ and $C_n \to 0$ at $\Delta A \to 0$. Considering that the equation for $n = 1$ is disappearing from (2.36) as well as the terms C_1 and $B_1 = 0$, the determinant of the final system does not equal zero since all $\lambda_n \neq 0$.

Thus, in the structure with small inhomogeneity $\Delta A(x)$, the stationary solitary filaments exist. At the same time, the final $T(x)$ distribution is only slightly different from the filaments $T_0(x)$ of spatially uniform structure thus satisfying condition (2.41). Therefore, (2.41) approximately provides possible positions of solitary filaments and according to Fig. 2.6c these positions are localized near the extremes of function $\Delta A(x)$.

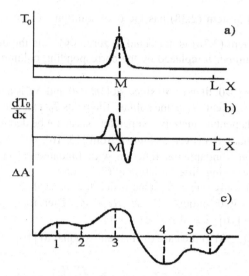

Fig. 2.6 Qualitative view for the solitary filament (a) and its derivative (b) in a uniform sample; (c) possible positions of corresponding filament (1–6) in the presence of inhomogeneity $\Delta A(x)$

The number of these positions depends on $\Delta A(x)$ and a typical filament dimension in given spatially uniform structure.

A nonstationary equation should be analyzed in order to identify stable states among possible states of (2.41):

$$\tau\frac{dT}{dt} = l_T^2\frac{d^2T}{dx^2} + \varphi(T,U,A),\tag{2.42}$$

where τ is the characteristic time.

For simplification, it is further assumed that t is measured in τ units, and x is measured in units of l_T. By linearization of (2.42) and (2.34), relatively small fluctuations $\delta T(x,t)$ and δU near the stationary state after Laplace transformation (2.43) can be presented in the form

$$A\delta T(x,\gamma) + \frac{\partial\varphi}{\partial U}\bigg|_{T,U,A}\delta U(\gamma) = \gamma\delta T(x,\gamma) - \gamma\delta T(x,0),\tag{2.43}$$

where $\delta T(x,\gamma) = \gamma\int_0^\infty e^{-\gamma t}\delta T(x,t)dt; \delta T(x,0)$ is the initial disturbance;

$$\hat{H} \equiv \frac{d^2}{dx^2} + \frac{\partial\varphi}{\partial t}\bigg|_{T,V,A}.$$

By presenting $\delta T(x, \gamma)$ as expansion in fundamental functions φ'_m of operator $\hat{H}$, and by making corresponding transformations and transiting to the equation system for expansion coefficients C'_m:

$$\sum_{m=0}^{\infty} C'_m \left((\gamma - \lambda'_m) \delta_{m,n} + \frac{D'_n \beta'_m}{b'} \right) = \gamma \langle \delta T(x,0) \varphi'_n \rangle, \qquad (2.44)$$

where λ'_m are fundamental values of $\hat{H}$, and D'_m, β'_m, b are calculated using formulas similar to (2.39). Fundamental values λ'_m can be estimated by using the methodology of disturbance theory.

The view of equation $\hat{H}\psi = \lambda\psi$ is similar to the Schrödinger equation where $(-\lambda)$ is the analog of energy and $(-\partial\psi/\partial T)$ is the analog of potential [16]. Operator $\hat{H}$ can be presented as a sum $(\hat{H}_0 + \hat{V})$ of the operator in spatially uniform structure $\hat{H}_0$ and is the disturbance operator

$$\hat{V} = \varphi''_{TT}\Delta T + \varphi''_{TU}\Delta U + \varphi''_{TA}\Delta A. \qquad (2.45)$$

Correction of the first approximation of λ_1 can be written [27] as

$$\lambda_1^{(1)} = \left\langle \frac{dT_0}{dx} \hat{V} \frac{dT_0}{dx} \right\rangle. \qquad (2.46)$$

Since $\lambda_1 = 0$, then $\lambda'_1 = \lambda_1^{(1)}$.

Stability of filament $T(x)$ can be determined by the sign of the roots γ_n of characteristic equation to the system (2.45):

$$\prod_n (\gamma - \lambda'_n) \left(b + \sum_{m=0}^{\infty} \frac{D'_n \beta'_n}{\gamma - \lambda'_n} \right) = 0. \qquad (2.47)$$

In a given current regime, the basic information about stability $T(x)$ is provided by the sign of γ_1. In uniform structure, $\gamma_1 = 0$, and in the rest, $\gamma_n < 0$. Small inhomogeneity brings only small disturbances in the γ_n spectrum; therefore, all γ_n besides γ_1 preserve its sign. Thus, solitary filament is stable if $\gamma_1 < 0$. The sign γ_1 coincides with the sign of λ'_1 and therefore with the sign of $\lambda_1^{(1)}$ that is given by [26]

$$\lambda_1^{(1)} = -\left\langle \varphi'_A \frac{d\Delta A}{dx} \frac{dT_0}{dx} \right\rangle. \qquad (2.48)$$

According to (2.49), at $\partial\varphi/\partial A > 0$ and narrow filaments with respect to distribution T0(x) versus $\Delta A(x)$ (Fig. 2.6), the filaments localized in the "ridge" of inhomogeneity are stable, but unstable in the "valley." If $\partial\varphi/\partial A < 0$, the conclusion is reversed. With relatively wide filaments, the inhomogeneity is adopted integrally and for detection of the filament stability direct calculation of integral (2.48) is required. Detection of the sign of $\gamma 1$ with the help of (2.48) allows analysis of the case of spatial interference of inhomogeneities on independent parameters.

Contribution in $\lambda_1^{(1)}$ from each inhomogeneity is determined by integral (2.48) and the sign of this contribution sum determines conclusions about the state stability.

For the state with solitary filament, a change in structure parameters can cause a shift in filament position due to both the stability lost and change in the possible state number. For example, a stability loss might occur with a change in external circuit parameters. The simplistic case is an increase of filament dimension with current increase. Overlap of the filament wall with other inhomogeneities can result in appearance of a positive sign of $\lambda_1^{(1)}$ in (2.48) and the filament may migrate from one inhomogeneity to another. A similar instability can appear with the filament dimension increase when the filament "feels" the sample boundary, which adds in λ_1 the positive component $\exp\left(-\dfrac{L_X}{l_T^{fil}}\right)$.

Therefore, for relatively short samples the criterion of filament stability on inhomogeneity is given by

$$\exp\left(-\frac{L_X}{l_T^{fil}}\right) - \left\langle \varphi_A' \frac{d\Delta A}{dx} \frac{dT_0}{dx} \right\rangle \leq 0. \tag{2.49}$$

If (2.50) is not true, the solitary filament can move from the inhomogeneity position and migrate toward the edge of the sample.

2.3.5 Current Filamentation at Positive Differential Conductivity

A classical case of current filamentation is accompanied by S-shape I–V characteristic. The critical point (Fig. 2.2) corresponds to a current-controlled negative differential conductivity region (NDC). However, in some transistors and semiconductor structures, filamentation could be observed without S-shape formation, at $dI/dU > 0$. According to [28, 29], the current filamentation at positive differential conductivity (PDC) is related to self-organization phenomena of dissipative systems [18, 28, 30]. The process of filamentation in such systems is the result of simultaneous redistribution not of one but both distributed parameters: $\theta(x)$ and $\eta(x)$. In the filament region, these parameters can be spatially dependent. For example, in bipolar transistors at emitter current filamentation [29], temperature acts as the activator parameter θ and potential of the emitter junction acts as the inhibitor parameter η.

At given external regime for total current I and voltage U, the values θ and η can be described by differential equations similar to (2.7):

$$l_\theta^2 \frac{d^2\theta}{dx^2} + \varphi_1(\theta,\eta,I,U) = 0, \tag{2.50}$$

$$l_\eta^2 \frac{d^2\eta}{dx^2} + \varphi_2(\theta,\eta,I,U) = 0, \tag{2.51}$$

where ϕ_1 and ϕ_2 are the nonlinear functions in general, and l_θ and l_η are the corresponding typical dimensions for both parameters. Usually in this problem statement, stratification, not filamentation, is discussed since the system has stable periodic solutions for $\theta_{(x)}$ and $\eta_{(x)}$. These solutions are called strata [28] or multiple filament states.

In transistors considering thermal instability mechanisms, as a rule the simplest case is realized: $l_\theta \gg l_\eta$ and $l_\eta \gg L$, where L is the total structure dimension. Therefore, variation of η along the direction of θ redistribution can be neglected, i.e., η = const.

In this case, the description is practically similar to (2.7), but η is now a spatially independent value and θ belongs to the external circuit parameters as well as the current I:

$$l_\theta^2 \frac{d^2\theta}{dx^2} + \varphi_1(\theta,\eta,U) = 0, \tag{2.52}$$

$$\varphi_2(\theta,\eta,I,U) = 0, \tag{2.53}$$

$$\int_0^L F(\theta,\eta,U)dx = I, \tag{2.54}$$

where (2.50) is the analog of (2.8).

At given I or U, the solution of (2.52)–(2.54) gives all variables $\theta(x)$, η, U, the I–V characteristic $I(U)$ and other dependencies: $\eta(U)$, $\eta(I)$. $\eta(U)$, $\eta(I)$ have no analogy with the case of classical filamentation mechanisms. Due to "self-tuning" of the parameter η to the distribution $\theta(x)$, the I–V characteristic $I(U)$ can maintain positive slope ($dI/dU > 0$) even in the case of nonuniform states or filament formation. Thus, a nonuniform distribution $\theta(x)$ might not be visible through peculiarities of I–V characteristics.

Analysis of the critical condition in uniform structures using (2.52)–(2.54) corresponds to Section 2.2. Equations (2.52) and (2.53) can be transformed into the view (2.7). The same methods can be used for analysis of soft and abrupt filament excitation as well as for analysis of small inhomogeneity influence. As a bifurcation parameter, both current I and voltage U can be accepted. The bifurcation characteristics appear as dependencies of the parameter η on voltage and current at fixed current and voltage, respectively: $\eta(U)|_I$ and $\eta(I)|_U$. The properties of these characteristics are similar to $I(U)$ in the case of classical filamentation discussed above. In particular, splitting of $\eta(U)|_I$ into different branches is observed in nonuniform structure with inhomogeneity. Similarly, relative change in critical regime sublinearly depends on the inhomogeneity:

$$\frac{\Delta U_{CR}}{U_{CR}^0} \sim \left(\Delta A_1\right)^{\frac{2}{3}}. \tag{2.55}$$

In the discussed case, the standard output I–V characteristic $I(U)$ is not a bifurcation characteristic. It exhibits a discontinuity and remains ordinary even at $U > U_{CR}$. However, if in this system the additional "degree of freedom" is fixed due to variable η, then "particular" I–V characteristic $I(U)|_{\eta=\text{const}}$ has S-shape where its differential conductivity changes the sign in the critical point:

$$\frac{dI}{dU}\bigg|_{\eta=const} \leq 0. \tag{2.56}$$

2.4 Real Inhomogeneity and Current Filamentation

In spite of the results presented above, the full mathematical description of current filamentation in transistors is rather complex and is not as yet developed.

By introduction of small inhomogeneity $\Delta A(x)$ into idealized uniform model (2.7)–(2.9), a number of new important results were obtained toward clarification of real filamentation process. However, quantitative problems of reliability of direct interest for device and circuit design are not as yet solved. The problem of critical dissipated power for given transistor structure with an appropriate accuracy can be solved only experimentally. This conclusion is motivated not only by simplification of the mathematical model. Systematic inhomogeneities subdivide power transistors into a number of parts. This excludes the possibility of common analytical solutions.

One of the reliable alternative ways to address the problem is numerical simulation. Efficiency of numerical simulation will be demonstrated in Chapters 3, 4, 5, and 7 using a number of particular examples. However, it is necessary to take into account the applicable range for the used mathematical model. On the other hand, a consideration of structure inhomogeneities in numerical models is in reality limited so far by the 2D case due to limitation of the solution stability and computer power in the case of 3D numerical analysis.

Systematic inhomogeneities related to the transistor design itself are added to the casual structural defects with unknown parameters. This results in some errors of the used numerical model. Finally, the exact analysis is limited by assumption of small or "primary" inhomogeneity resulting in small deviation of temperature $\Delta T(x)$. If the inhomogeneity ΔA of the parameter A_0 is not a small value, the analysis of the filamentation conditions is apparently difficult too. In some cases, it is possible to solve the equations inside the inhomogeneity region and far from it followed by stitching the solution at the interface [22]. To specify the application area of the theoretical approach in Section 2.3 and formula (2.32) in particular, it is necessary to clarify the physical sense of "primary" inhomogeneity. It is easy to ensure that the smallest of $\Delta T(x)$ [i.e., expansion coefficients C_n in (2.24)] is provided by the smallest term in the integral factor $\quad \Delta A_n = \left\langle \Delta A(x) \cos \dfrac{\pi\, n}{L_x} x \right\rangle$.

Therefore, the definition of a small primary inhomogeneity in most cases can be treated as follows. The inhomogeneity is small if all coefficients are small in its Fourier expansion. This means that a large spatial dimension inhomogeneity, but small amplitude, as well as high-amplitude inhomogeneity of small dimension could both satisfy the small inhomogeneity definition.

The degree of defect impact on the critical filamentation conditions depends on the ratio of the dimension l_T to the dimension of the defect itself l_A. At $l_A \ll l_T$, the defect is only slightly sensible, especially if inhomogeneity of $\sim l_T$ is already present in the structure design itself. At $l_A \gg l_T$, the transition is made to the structure with a new value of the parameter A. The last differs from the initial value by the value $\Delta A_0 = \langle \Delta A \rangle$. Inhomogeneity of this parameter on the background of its median variation can be nonsignificant. Inhomogeneities with $l_A \approx l_T$ besides the significant change in the critical regime can provide power localization. According to (2.49), the filaments at this inhomogeneity can be much more stable.

2.5 Summary

In this chapter, the semiconductor device is treated from a physical point of view as an open, distributed, and dissipative system. In the active regime, this system operates under permanent exchange of energy and matter between the transistor and the ambient space. The basic equations are provided for the mathematical description of the device by the following distributed physical parameters: the temperature, the electric field, and the electron–hole current density. Examples of the analysis are then provided. The general results of this chapter are discussed in greater detail for particular thermal and electrical instabilities in Chapters 3–5 and 7.

Chapter 3
Thermal Instability Mechanism

The following statement should be given prior to the feature presentation of the chapters devoted to transistor failure mechanisms. The purpose of this book is not to serve as e some reference manual with a classified set of external features and peculiarities of transistor failures. On the contrary, the purpose is to establish and clarify major regularities and basic mechanisms in "purified" form in order to give a modern physical interpretation and to enable creation of correspondent physical and mathematical tools for particular analysis. Application experience demonstrates that in most real cases, multiple forms of similar mechanisms are involved rather than multiple mechanisms themselves due to multiple design and technological peculiarities of real devices.

3.1 Emitter Current Filamentation in Bipolar Transistors

Among the phenomena related to thermal instability in transistors, the phenomenon discussed in this section is most studied both experimentally and theoretically in detail. In published studies, the term "thermal instability" is usually used to define the phenomenon mentioned above. We will use the more generic broad term "thermal filamentation of emitter current" in order to separate this phenomenon from a similar process that is observed in other types of semiconductor devices, for example in field-effect transistors.

3.1.1 Some Experimental Results

The phenomenon of emitter current thermal filamentation was first observed in power bipolar transistors due to the design peculiarities discussed in Chapter 1. Attention to this phenomenon stimulated a number of studies in the 1960s and 1970s. The reason for such attention was that the phenomenon had been observed in the active operation regime (forward emitter bias) and preceded the catastrophic failure of the transistor followed by so-called secondary breakdown [31–33].

An abrupt reduction of the collector-base voltage, filamentation of the collector current, and increase of the local lattice temperature accompany secondary breakdown. This set of thermoelectrical instability phenomena will be discussed below.

It is important to emphasize that secondary breakdown is an irreversible process under most real application conditions. At the same time, the preceded current thermal filamentation (for simplicity, "emitter" is not used here) in a number of cases is reversible. It could be observed experimentally under quasi-static conditions.

Thermal current filamentation is observed in distributed transistor structures (typically in power devices) in both static and dynamic operation regimes. In the case of low avalanche multiplication, the phenomenon practically does not depend on the used circuit bias configuration [common-base (CB) or common-emitter (CE)]. It is accompanied by the current localization in some part of the transistor structure.

In the DC operation regimes, a typical indication of thermal current filamentation is formation of a "hot spot" region on the structure surface. A hot spot presents itself as a region of local overheating. It is experimentally observed by scanning IR microscope (micro pyrometer) [31], liquid crystals (Fig. 3.1), potential scanning method [32], electroluminescence measurement [34], or more advanced methods. In contrast to secondary breakdown, which corresponds to the S-shape region in I–V characteristics, the emitter current filamentation is observed at positive differential conductivity (PDC) without any significant changes in the I–V characteristic, $I_C(U_{CE})|I_B = $ const. This effect brings the emitter current filamentation into a special class of fundamental instabilities that are not accompanied by the negative differential resistance in the I–V characteristics [35].

Certain critical voltage current and temperature regime or, in general, safe operating area (SOA) represents the critical conditions for thermal current filamentation, similarly to other known instabilities. Increase of the device parameters above this critical regime (the boundary of the safe operation area) results in filament formation. In the DC regime, the critical power $P_{CR} = I_C U_{CB}$ can be used (if only the current and the voltage are varied) or the critical voltage U_{CR} (if the collector current is fixed). At transition of the operation regime over the SOA boundary, the filament can arise both spasmodically, abrupt (rigid) and gradually (soft). The mode of filament excitation depends on both the transistor structure parameters and the operation regime.

The event of stationary filament formation in a static regime can be found by an indirect method using the dependence of the emitter-base voltage upon the collector-base voltage at constant emitter current $U_{EB}(U_{CB})|I_E = $ const. A peculiarity in form—a break or a jump—of this characteristic corresponds to "soft" or "abrupt" filamentation process, respectively.

Regarding emitter current filamentation, a region of stable hot spot formation is typical in the reversible DC operation regimes. The region increases with collector current increase. At low collector current, this stable hot spot formation region degenerates and the thermal filamentation boundary practically coincides with the boundary regime for secondary breakdown. In this case, the filamentation may result in catastrophic failure.

Several regularities can be derived from the pulsed measurements. With pulsed overstress over the critical regime, the filament is formed after some delay time τ_D. τ_D does not always correlate to the thermal time constant of the chip overheating and can vary in a rather wide time domain. The delay process of thermal current filamentation was observed in [32] on the oscillogram of the characteristic

$U_{EB}(t)$ measured at the transistor heating by power pulse in the circuit for thermal resistance measurement. Regarding transistor structure heating, the value $|U_{CE}|$ is decreased (n-p-n transistor in CB circuit), since at higher temperature the fixed current value is provided by lower emitter junction voltage. If pulsed power exceeds some critical value, an additional typical slump ("step") is observed in the $U_{EB}(t)$ characteristic. In [32] this was identified as a sign of current pinching in some small part of the device, i.e., filamentation. With the collector voltage increase, the step appears earlier. The following increase in U_{CB} results in the transistor thermal runaway [32].

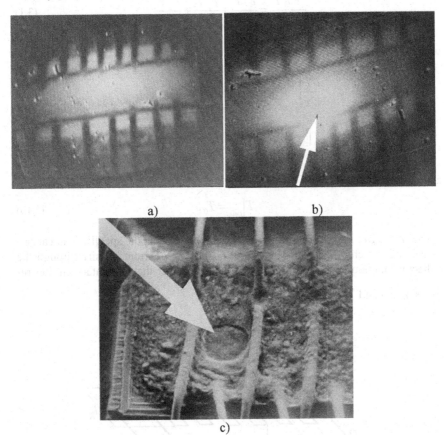

a)   b)

c)

Fig. 3.1 Obtained by IR-microscope for uniform (a) and hot spot current distribution (b). (c) Hot spot detection using liquid thermal indicator

3.1.2 Thermoelectrical Model and Stability Criteria

The model of power bipolar transistor is similar to that discussed in Section 2.1 and [33, 36, 37]. Typical structure of a planar epitaxial high-frequency transistor is a semiconductor chip of dimension $L_X \times L_Y \times W$ with the emitter region of dimension $l_X \times l_Y$ (Fig. 3.2). The emitter region can have rather complex configuration. In this case, it consists of multiple parallel emitter regions of length l_X and width a at a distance b from each other; in addition, $a \approx b$, $(a + b) \ll l_Y$. The bottom plane of the chip $Z = W$ is connected to an ideal heat sink with the maintained constant temperature T_0. The heat source (collector junction) is assumed to be flat at the chip surface $Z = 0$ and coincident with the emitter region topology. Heat propagation in the chip is described by the following nonstationary equation for thermal conductivity:

$$\frac{1}{a^2}\frac{\partial T}{\partial t} = \frac{\partial^2 T}{\partial x^2} + \frac{\partial^2 T}{\partial y^2} + \frac{\partial^2 T}{\partial z^2} \qquad (3.1)$$

with the boundary conditions

$$\frac{\partial T}{\partial x}\bigg|_{x=0,L_X} = \frac{\partial T}{\partial y}\bigg|_{y=0,L_Y} = 0, \qquad (3.1a)$$

$$T\big|_{z=W} = T_0; \frac{\partial T}{\partial z}\bigg|_{z=0} = -\frac{U_{CB}j_C(T)}{\kappa}$$

and the initial condition

$$T\big|_{t=0} = T_0, \qquad (3.1b)$$

where $a^2 = \kappa/\rho c$; κ, ρ, c are the specific heat conductance, the specific heat capacitance, and the chip material density; j_C, U_{CB} are the current density through the collector junction and the collector-base voltage. Over the elementary emitter regions, $j_C = 0$ and $\dfrac{\partial T}{\partial z}\bigg|_{z=0} = 0$.

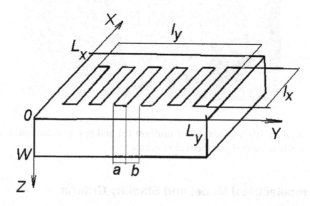

Fig. 3.2 Thermal model structure of the epitaxial high-frequency transistor

The electrical model of transistor is based on the classical Ebers-Moll mathematical model [38]. In the absence of avalanche multiplication in the collector p-n junction region, the current density is given by

$$j_C = \alpha j_{SE}\left(\exp\frac{qV_E}{kT} - 1\right) + j_{SC},$$ (3.2)

where V_E is the voltage on the emitter p-n junction, j_{SE} and j_{SC} are the thermal current densities through the emitter and collector p-n junctions, and α is the current gain.

In the general case, V_E is a distributed value. It obeys the current continuity equation in the transistor base [36]. However, in microwave transistors with heavily doped active base, the electrical connection between the parts of the emitter p-n junction is rather close to each other and the expression for V_E can be derived in simplified form:

$$V_E = U_{EB} - j_E r_{Eeff} - j_B r_{Beff},$$ (3.3)

where U_{EB} is the voltage between the emitter and base terminals, j_E and j_B are the current densities in the element of the emitter and base, and r_{Eeff} and r_{Beff} are the effective resistances of the emitter and base. The physical sense of r_{Eeff} and r_{Beff} can be illustrated with the example of a typical self-aligned transistor (Fig. 3.3).

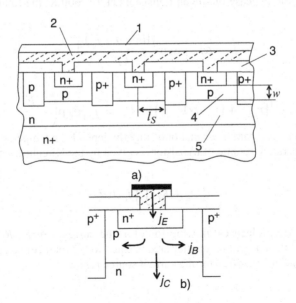

Fig. 3.3 Simplified cross section of the self-aligned BJT structure (a) and its emitter region (b): 1, emitter metallization; 2, polysilicon ballasting emitter; 3, oxide; 4, base; 5, collector

The potential difference U_{EB} is the sum of the voltage drop on the emitter ballasting resistors (and emitter metallization), on the emitter p-n junction, and on the active region of the p-base. Usually, the external p^+-base is heavily doped to enhance the microwave properties of the transistor. Effective resistance of the emitters in the given structure is distributed across the area rather uniformly and can be roughly estimated using the expression

$$r_{Eeff} \approx R_E S_E,$$ (3.4)

where R_E, S_E are the total resistance and area of emitter regions.

Neglecting the effect of current displacement to the emitter edge, r_{Beff} can be estimated by the expression

$$r_{Beff} = \frac{l_B}{\sigma L_E w} S_E,$$ (3.5)

where σ, w, and l_B are the conductivity, thickness, and length of the p-base, respectively (Fig. 3.3) and L_E is the total perimeter of the emitter. Experimental evaluation of the $(r_{Beff} + r_{Beff})/S_E$ value can be obtained by measurements of the resistance between the emitter and base terminals of transistor under forward bias of the emitter junction under the conditions of floating collector.

Considering that $j_E = \alpha^{-1} j_C$ and $j_B = \alpha^{-1}(1-\alpha)j_C$ under negligible values of j_{SC} and j_{SE} in (3.2) in the case of silicon transistors at $T < 600$ K, the final expression for j_C is

$$j_C = \alpha j_{SE} \left(\exp \frac{q(V_E - r_{eff} j_C)}{kT} \right),$$ (3.6)

where $r_{eff} = \alpha^{-1}\left(r_{Eeff} + (1-\alpha)r_{Beff} \right)$; $j_{SE} = j_{S0} \exp\left(-\frac{E_G}{kT} \right)$, E_G is the band gap, and j_{S0} is a parameter that only slightly depends on temperature.

External operation regime of transistors is given by the Kirchhoff equation:

$$E = U_{CB} + R_L \int_0^{L_X} \int_0^{L_Y} j_C dx dy,$$ (3.7)

where E is E.M.F. (electrical moving force) of the current source, R_L is the load resistance, and the integral on emitter area is equal to the total collector current I_C. Equation (3.7) can be written for the current source regime:

$$\int_0^{L_X} \int_0^{L_Y} j_C dx dy = I_C.$$ (3.7a)

Equations (3.1), (3.6), (3.7) are fundamental for mathematical description of emitter current thermal filamentation in bipolar transistors.

In order to obtain the analytical stability criterion for the critical regime of current filamentation, as mentioned in Section 2.2, uniform structure should be analyzed first assuming that the heat source (collector junction) is distributed uniformly at least along one of the directions (OX or OY) on the chip surface (Fig. 3.2). The criterion in the simplest one-dimensional case for the current source regime (I_C = const) is given by expression (2.16), where

$$\varphi = \frac{W}{\kappa} U_{CB} j_C - \left(T - T_0 \right).$$ (3.8)

According to (2.16), the uniform current and temperature distribution becomes unstable relative to the longwave fluctuation of $\cos(\pi x / L_X)$, if

$$\frac{W}{\kappa} U_{CB} \frac{\partial j_C}{\partial T} > 1 + \left(\frac{l_T \pi}{L_X} \right)^2$$ (3.9)

is true, where $l_T \sim W$.

The criterion (3.9) is practically obtained using the approach [33], and at $L_X \gg l_T$ it transforms into the instability condition [32]. Evolution of the instability must be accompanied by an increase in total current. The last condition is frequently written in the form

$$R_T \frac{dP}{dT} > 1,$$ (3.10)

where R_T is the thermal resistance and P is the dissipated power ($R_T \sim W/\kappa$).

In a similar form, the criterion (3.9) is given by

$$\widetilde{R}_T \frac{dP}{dT} > 1,$$ (3.11)

where $\widetilde{R}_T$ is the thermal resistance for longwave disturbances:

$$\widetilde{R}_T = R_T \bigg/ \left[1 + \left(\pi l_T / L_X \right)^2 \right];$$
$$\frac{dP}{dT} = U_{CB} \left(\frac{\partial j_C}{\partial T} \right) \bigg/ \left(1 + r_{eff} \frac{\partial j_C}{\partial V_E} \right).$$ (3.11a)

The detailed theoretical study of thermal filamentation conditions can be found in [36, 39]. For this chapter, the criterion (3.9) in the simplistic form is quite adequate for both qualitative and quantitative analysis. By differentiating (3.6) at consideration (3.11):

$$\frac{\partial j_C}{\partial T}\left(1+\frac{qr_{eff}j_C}{kT}\right)=$$

$$-\frac{[q(U_{EB}-r_{eff}j_C)-E_G]j_C}{kT^2}=\frac{j_C}{T}\ln\frac{\alpha j_{S0}}{j_C}.$$

Substituting $\frac{\partial j_C}{\partial T}$ in (3.9), the final instability criterion is given by

$$\frac{WU_{CB}j_C}{\kappa T}\ln\frac{j_C}{\alpha j_{S0}}\left(1+\frac{qr_{eff}j_C}{kT}\right)^{-1}\geq 1+\left(\frac{l_T\pi}{L_X}\right)^2. \tag{3.12}$$

At given collector current density j_C, the critical power $P_{cr}=U_{CBcr}I_C$ and critical temperature T_{cr} can be calculated from (3.8) using $\phi=0$ and criterion (3.12).

It is important to emphasize that criterion (3.12) is opposite to the "classical" criterion (2.16). It is true for the region of I–V characteristic with $R_D>0$ and, therefore, the emitter current filamentation can take place even under the conditions of positive differential resistance of transistor. This effect is the result of the explicit dependence of the collector current density j_C (3.6) on both T and U_{EB} values. Redistribution of the first value is compensated by decrease of the second value. In principle, this physical effect yields the possibility of filamentation under the conditions of unchanged I_C and U_{CB}.

It is easy to see that at emitter current filamentation, the instability boundary $U_{CBcr}(I_C)$ does not coincide with the typical line for constant dissipated power $U_{CB}I_C=$ const, which presents a boundary for some critical temperature limitation T_{lim}.

With the help of (3.12), it is easy to demonstrate that P_{cr} decreases with increasing U_{CB}: $P_{cr}\sim(A-U_{CB}^{-1})^{-1}$. It is considered that $\ln j_C$ is only a slightly variable function.

Therefore, the line of constant power and the boundary of filamentation always will have a point of intersection. This understanding explains the typical SOA shape for power bipolar transistors (Fig. 3.4). Similarly, it is easy to conclude that with increase of the collector current j_C the critical power P_{cr} increases obeying almost a linear law. The last conclusion is confirmed by the experimental data (Fig. 3.5). In principle, the linear dependence $P_{cr}(c_j)$ allows detecting the stability boundary for real transistors using only two measurements, i.e., two boundary points in the value region of j_C where the measurements are reversible.

It is necessary to discuss a practical value of criterion (3.12). As was emphasized in Chapter 2, the transistor structure uniformity assumed for this fluctuation instability criterion is far from reality. In real structures, enriched by various inhomogeneities or the material and contacts, as well as local structural defects, the idea of fluctuation instability is diminished. Is it possible to use criterion (3.12) that has been obtained in one-dimensional approximation? Nevertheless, according to real application examples it has been demonstrated that given qualitative regularities by the criterion are useful.

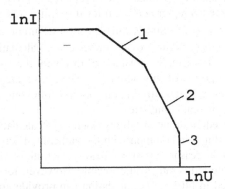

Fig. 3.4 Qualitative view of SOA for power bipolar transistors: 1, line of constant power; 2, boundary of the emitter current filamentation; 3, isothermal filamentation

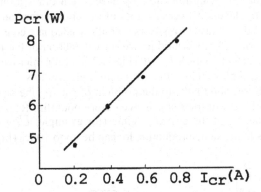

Fig. 3.5 Experimental dependence of the critical power on current

For example, the experimentally observed weak dependence of P_{cr} on ambient temperature T_0 can be derived from (3.12) directly. Indeed, at constant j_C increase of T_0 is compensated by the corresponding emitter-base voltage drop U_{EB}. The temperature dependencies of functions $\alpha(T), j\,T_{s0}(T),\ \kappa(T),\ r_{eff}(T)$ are only slightly varied in comparison with the exponent dependencies on T on the left side of (3.12).

From (3.12), it follows that in the distributed structures with high collector junction area ($L_X \gg W$) the current filamentation begins under the conditions of lower specific power in comparison with the structures with low junction area and high thermal feedback ($L_X \geq W$) at the same equal conditions. A positive effect of the ballasting emitter resistors becomes apparent. Increase in r_{eff} value results in corresponding increase in the critical regime for filamentation.

As for the quantitative estimation of the critical regime, it is not clear that satisfactory agreement with experiment has ever been demonstrated for any real type of transistor. Apparently, a very good coincidence of the theoretical calculations with results of stability boundary measurements is possible. This was demonstrated in [40] where a simpler condition for instability was used [32] related to condition (3.12) at $L_X \gg W$. Successful selection of numerical parameters for (3.12) α, j_{S0}, κ, l_T, r_{eff} might provide rather exact calculation of the stability boundary. Moreover, it is usually satisfactory to obtain the coincidence with the experiment only two points of this boundary. The above-mentioned linearity of the $P_{cr}(j_C)$ dependence practically guarantees the following coincidence of the theoretical and experimental boundaries.

The discussed procedure of the instability boundary "calculation" for given transistor structure has nothing in common with the authentic theoretical calculation of the boundary. Due to the reasons discussed below, the truth of the calculation using (3.12) or a similar criterion is rather questionable. However, for a number of structures, as demonstrated in [40], such calculation can provide good results. Apparently, it must be the structures with several specific features: adequate uniform elementary emitter distribution on the structure surface; linear dimensions of emitter region exceeding the chip thickness by at least ten times; minimized amount and thickness of the layers between bottom of the chip surface and heat sink and with the design of the ballasting resistors that allow adequate consideration of parameter r_{eff}. Criterion (3.12) is hard to use for calculation of the filamentation in medium-power structures with $l_X, l_X \sim W$ (Fig. 3.2) and for microwave power transistors due to the complex architecture of the multiple emitters.

Application of criterion (3.12) to the problem of finding the current filamentation conditions for a given transistor is even more doubtful. Real examples demonstrate that scattering of the critical regime (for example, U_{CBcr} at given $I_C = $ const) for devices in the same production lot can be up to 100% (Fig. 3.6).

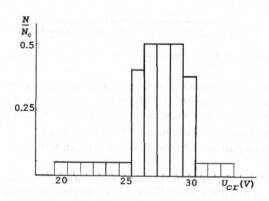

Fig. 3.6 Histogram plot for the experimental bipolar transistor distribution versus parameter U_{cr}

This fact is related not only to a nonuniform variation of the intrinsic parameters, which cannot always be measured directly, but mainly to the presence of practically uncontrollable structural inhomogeneities and defects in the semiconductor device. Since it is hard to provide either an adequate quantitative description of the majority of technological and design defects or extraction of the defect parameters on the indirect experimental measurement results, the problem of individual estimation of the critical regime for given semiconductor structure could hardly be solved even using numerical simulation methods. In this case, criterion (3.12) can give only rough estimation of the critical regime for the ideal structure without defects. Such an estimation as a rule is a significantly excessive value.

3.1.3 Informative Characteristic

Apparently, the direct measurement methods for the case of thermal current filamentation are the most informative. The methods allow one to measure hot spots or local regions of electroluminescence using IR microscopy or other scanning methods. However, in most practical cases the surface of the transistor structure is not always available for experimental observation due to the specifics of packaging, backend metallization layer design, passivation layers, or even flip chip or stack die attachment. If application of the direct methods is either impossible or unreasonable, the indirect methods can be used. These methods are developed and based on the measurements of those external electrical characteristics that are sensitive to the current filamentation effect.

Usually, output I–V characteristics of the transistor are poorly informative for this problem solution. Of course, some peculiarities of I–V characteristics can be observed at emitter current filamentation, but this "response" is not simple. Some of the typical indicators of nonlinear current distribution behavior are S-shape, N-shape, and kink features on the output I–V characteristics [39]. In addition, the I–V characteristic exhibits a secondary phenomenon that results in I_{CB0} increase or local gain α decrease in the hot spot region. These processes can produce an effect on the collector current I_C in the opposite directions and, thus, can compensate each other in general.

Apparently, the emitter current localization must be directly connected with the emitter junction potential and therefore directly available to the measurements of the potential difference between the emitter and base terminals U_{EB}. In the simplified transistor model for (3.12) $U_{CB} =$ const, in the absence of multiplication on collector and neglecting I_{CB0}, the electrical conditions on the collector junction are external relative to the emitter junction. In this case, the I–V characteristic of the emitter junction $I_E(U_{EB})$ exhibits typical properties for S-shape dependence at emitter current filamentation. It can be demonstrated that at the point of current instability, criterion (3.12) is true for uniform disturbances ($R_T U_{CB} \frac{\partial j_C}{\partial T} = 1$), the differential resistance $\frac{dU_{CB}}{dI_E}$ is equal to zero, and $\frac{dU_{CB}}{dI_E} < 0$.

For a more general model, which takes into account the processes in the collector junction [36, 39], the emitter current filamentation is formed under change in the sign of the derivative $\partial U_{CB}/\partial I_E$. This effect essentially means a hidden negative differential resistance [39].

Under real conditions, the emitter current I_E and emitter-base voltage U_{EB} are the basic external electrical parameters of transistor operation regime. U_{EB} is self-adjusted according to the stationary structure state. In bipolar transistors, U_{EB} plays the role of the parameter η (Section 2.3.5); bifurcation characteristic is the dependence $U_{EB}(U_{CB})$ at I_E = const [or $U_{EB}(I_E)$ at U_{CB} = const]. For the first time, static characteristics $U_{EB}(U_{CB})|I_E$ as informative characteristics of the device have been used in [41]. Independently, the same characteristics were suggested and analyzed in [39, 42].

Formation of the emitter current filament is inevitably accompanied by the appearance of a nonmonotonic dependence of the U_{EB} characteristic in theform of a break, a kink, or a jump. This feature allows detection of both the critical regime (I_{Ecr}, U_{CBcr}) and the stability boundary $I_{cr}(U_{cr})$. Simultaneous measurements of the

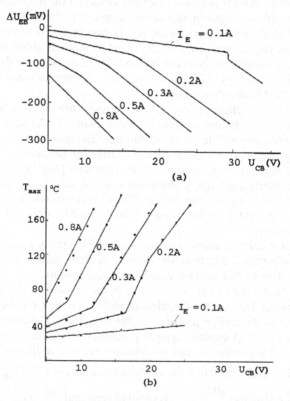

Fig. 3.7 Experimental dependencies of the $U_{EB}(U_{CB})|_{IE}$ (a) and maximum temperature (b) for different emitter current levels

$U_{EB}(U_{CB})$ characteristics and the maximum temperature T_{max} in the transistor structure have confirmed proper understanding of these peculiarities. The break or change of the slope is evidence of a gradual (soft) filament formation; the maximum temperature increases continuously with U_{CB} increase. The speed of increase dT_{max}/dU_{CB} can increase by a few times (Fig. 3.7b).

The jump in the characteristic $U_{EB}(U_{CB})|I_E$ is evidence of abrupt filament excitation. In this case, the maximum temperature in the filament region T_{max} under this process can be suddenly (with the typical heat dissipation time of the device) increased a few tens of degrees. Often, this may result in irreversible changes in the transistor structure and failure. In this case, the method of filamentation regime measurements with the help of DC UEB(UCB) characteristic might not be reversible.

Using the approach from Section 2.2.2 for the uniform transistor model (3.6), (3.8), the filament excitation mode can be evaluated in terms of abrupt or soft scenario. Indeed, (2.29) at $P = 0$ can be transformed into the approximate view:

$$\left(\frac{L\sqrt{3}}{W\pi}\right)^2 = 12\left(1 - \frac{2kT}{E'_G}\right)^{-1} - 7,$$

where $E'_G = \dfrac{E_G - qU_{Ep-n}}{1 + r_{eff}(q/kT)j_C}$.

Taking into account $U_{Ep-n} \approx 0.6$ V and $r_{eff}(q/kT)j_C \gg 1$, the condition for abrupt filamentation mode is

$$\frac{L}{2\pi W}\sqrt{1 - \frac{4r_{eff}j_Cq}{E_G}} > 1. \qquad (3.13)$$

Not only the chip dimension, as in the discussed case of classical thermal filamentation in Section 2.2.2, but several transistor structure parameters provide the impact on mode of hot spot appearance, among them: the value of emitter ballasting resistors and the operation regime. In reality, the increase of given total current always results in "softening" of the filamentation and the smoothing or even disappearance of the jump feature in the experimental characteristics (Fig. 3.7a). The emitter ballasting region is usually distributed across the array or multifinger device and connected to every emitter finger [38, 43].

At the same time, introduction of the emitter ballasting resistors r_E according to (3.12) does not provide any shift in the instability boundary. According to (3.13), the possibility of instantaneous failure at the critical regime is even increasing. Nevertheless, real characteristic $U_{EB}(U_{CB})|I_E$ in the polyemitter balanced BJT structure has neither a jump nor a break. In the real device, the filamentation with U_{CB} increase is rather slow and it is rather hard to determine any definite critical voltage, while the slope dU_{EB}/dU_{CB} is changing gradually.

The characteristic $I_B(U_{CB})|I_E$ can be used for detection of the emitter current localization under the quasi-stationary DC conditions. It becomes possible due to the reduction of the local gain α with the emitter current density increase in the hot spot region. Since the major part of the emitter current I_E flows in this region, the total gain is reduced too. The total base current is

$$I_B = \int_0^{L_X} \left[1 - \alpha\left(j_E\right)\right] j_E \, dx. \tag{3.14}$$

A smooth or an abrupt increase of I_B can be observed for the characteristic $I_B(U_{CB})|I_E$[41]. The characteristic $I_B(U_{CB})$ is relatively poorly informative in comparison with $U_{EB}(U_{CB})$ due to unknown empirical function $\alpha(j_E)$. As follows from (3.14), the value I_B is mainly determined by the parameters of bell-shape dependence $\alpha(I_E)$. Therefore, it is not obvious that I_B will bring a good correlation with the emitter current density distribution $j_E(x)$. Using numerical simulation results, it has been demonstrated [44] that with respect to the ratio $I_E \big/ I_E^\alpha$, where I_E^α is the point of maximum of function $\alpha(I_E)$, the filamentation event can be observed in the $I_B(U_{CB})|I_E$ regime both with increase and with reduction in I_B under unchanged $U_{EB}(U_{CB})|I_E$ characteristic.

3.1.4 Local Defect and Inhomogeneity Impact

The fact that local inhomogeneities and structural defect or semiconductor material and contact regions in real transistor structures have a considerable impact on the process of thermal filamentation, is not in doubt. This conclusion is based on numerous experimental data for example [34, 45–47] and particularly demonstrated in Fig. 3.6. These studies deal with defect classification according to its influence on the filamentation [34, 46] or using the methods of statistical analysis of the measurement results [45, 47]. The direct experiment to consider the nature of the particular defect and its parameters requires a precise methodology. This methodology, however, still might interfere with device integrity. Several methods are described in Chapters 5 and 8.

Typically, published experimental studies provide rather limited results about parameters of structural defects. According to [45], the local structure crystal defects (dislocations, regions of mechanical strengths, and so on) have practically no influence on the thermal filamentation process. However, its accumulations and formation of macrodefect clusters, for example voids in the region of die attachment or unmatched value of the emitter ballasting resistors, may have a significant impact on the current filamentation scenario [34, 46–48]. In a multidie power device, one defect that may decrease the thermal stability is unmatched values of thermal resistance R_T for different dies [42, 48].

Application of analytical approach to the device with inhomogeneity is complicated. In this case, practically everything that is related to inhomogeneity becomes a problem that requires nonlinear analysis and therefore needs either significant simplification or computer numerical modeling. This section is presented in correlation with the analysis presented in Chapter 2 in order to demonstrate greater details in the previously obtained conclusions. On the other hand, computer modeling results are involved and widely used for thermal filamentation simulation [49–54].

There are several numerical models for nonisothermal processes in bipolar transistors. In the complex models [49–51], the heat equation and Poisson equation for electrical potential in base are solved taking into account the recombination processes and lattice temperature dependencies of various semiconductor parameters. However, sometimes, simple models may provide rather adequate agreement with the experimental data [40]. Previously, the type of model selection has been limited by the problem statement and available CPU power. The numerical simulation results are mainly used to demonstrate a number of important qualitative peculiarities, dependencies and regularities, rather than quantitative calculation of thermal filamentation characteristics. This is beyond the capabilities of analytical solution [53, 54].

The numerical simulation results are presented for the simple model of uniform transistor structure (Section 3.1). The nonstationary two-dimensional thermal problem (3.1) is solved numerically. The pulsed $U_{CB}(I_C)$ power source is simulated with the heat source located at the surface $Z = 0$ of two-dimensional chip model (Fig. 3.8a). Under given $j_C(x)$ distribution, the temperature distribution on the chip surface is provided by Fourier expansion in fundamental functions:

$$T(x,0,t) = T_0 + \frac{z}{LW}\sum_{n=1}^{\infty}\sum_{m=0}^{\infty}G_{mn}(t)\cos p_n x,$$

$$G_{mn}(t) = a^2 \int_0^t\int_0^L \frac{U_{CB}}{\kappa}\, j(x,t')\cos p_n x$$

$$\times \exp\left[a^2\left(p_n^2 + g_m^2\right)(t'-t)\right]dxdt',$$

$$p_n = \frac{\pi n}{L}, g_m = \frac{\pi(2m+1)}{2W}.$$

(3.15)

In order to find the $T(x,0,t)$, $j(x,t)$, and $U_{EB}(t)$ dependencies, (3.15) and (3.6) are solved in combination with the condition (3.7a) for total current constancy

$$I_C = \int_0^L j_C dx = \text{const}$$ using Newton's iteration method.

As mentioned above, in the uniform structure, thermal filamentation evolution is possible in the presence of nonuniform fluctuations of temperature and current density. Introduction of appropriate fluctuations in the initial time moment results in excitation of filamentation under $U_{CB} > U_{CBcr}$ and thereby enables calculation of the stationary state of the structure (Fig. 3.8c). In this case, the only possible stable states in over critical area of operation regimes are the filaments of current and temperature localized at the edge of the structure.

In uniform structures, the other types of filaments, including solitary filaments observed in real devices, are unstable relative to the shift toward the edge. In the nonstationary case, these filaments are "pumped" into the edge filament state. In the structure with inhomogeneity, the filament state with multiple filaments is unstable relative to mutual "pumping over" processes. However, the solitary filament on the inhomogeneity can be stable, as demonstrated in Section 2.3.4.

In the general case, nonuniform structure does not need an "excitation" event for filament formation. After exceeding a certain critical regime, the structure spontaneously makes rather deterministic transition into the filament state. These general conclusions are confirmed and demonstrated in detail by the numerical simulation results for emitter current thermal filamentation.

The inhomogeneity in transistor structure model is introduced by a coordinate dependence of the current density parameter j_{S0} in (3.6a):

$$j_{S0} = j_{S0}^0 (1 + \Theta(x)),$$ (3.16)

where the value j_{S0}^0 corresponds to uniform structure and $\Theta(x)$ is the arbitrary function of the variable x. For simplicity, $\Theta(x)$ is assumed as a step function: $\Theta(x) = \Theta_0 > 0$ in the inhomogeneity region of dimension v, that is positioned a distance χ from the center of the structure (Fig. 3.8b). Outside the inhomogeneity region, $\Theta(x)$ is equal to 0.

A number of generic results of numerical simulation can be listed prior to discussion of the details of the filamentation process with introduced inhomogeneity.

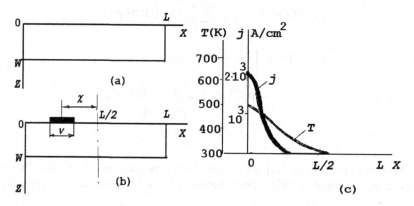

Fig. 3.8 Device cross section for two-dimensional thermal model of uniform (a) and nonuniform (b) structure. (c) Calculated quasi-static temperature and current density distributions

Depending on the structure parameters and operation regime, filamentation can have either "abrupt" or "soft" excitation mode. In the characteristic $U_{EB}(U_{CB})$, either the hysteresis (Fig. 3.9a, curves 1 and 2) or the break (curve 3) is observed. Increase in total current I_C or typical thermal structure dimension l_T ($l_T \sim W$) results in reduction and degeneration of the hysteresis (Fig. 3.9a). This result remains in agreement with both the experimental (Fig. 3.7) and the analytical (3.13) results. The numerical simulation demonstrates that with increase in $U_{CB} > U_{cr}$ the typical dimensions of the temperature and current filaments l_T^{fil} and l_j^{fil} are reduced. In addition, the dependencies $l^{fil}(U_{CB}^{-1})$ have a good approximation by the linear function in a wide U_{CB} range [53]. The physical meaning of l_T^{fil} and l_j^{fil} values are represented by relations $j(l_j^{fil}) = j_{max}/e$; $T(l_T^{fil}) = T_0 + (T_{max} - T_0)/e$.

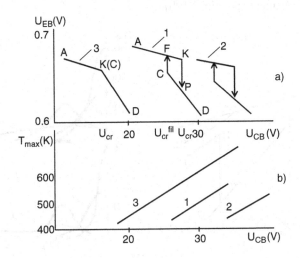

Fig. 3.9 Stationary characteristics $U_{EB}(U_{CB})$ at various I_E values (a) and corresponding dependencies $T_{max}(U_{CB})$ (b): 1, $j_M = 300$ A/cm^2; 2, $j_M = 200$ A/cm^2 ($W = 50$ μm, $L = 240$ μm); 3, $j_M = 300$ A/cm^2 ($W = 100$ μm, $L = 240$ μm)

It has been demonstrated that in spite of current pinching the dependence $T_{max}(U_{EB})$ is still close to linear (Fig. 3.9b). The regions of $U_{EB}(U_{CB})$ characteristics in the filament state (Fig. 3.9a) demonstrate the linearity too. The mentioned facts are confirmed experimentally (for example, Fig. 3.7) and can provide the basis for an indirect method of maximum temperature measurement inside the hot spots formed in the transistor structures [53]. Indeed, the linearity of $T_{max}(U_{EB})$ and $U_{EB}(U_{CB})$ provide constant value of the derivative dU_{EB}/dT_{max} in both regions of $U_{EB}(U_{CB})$ characteristics and therefore T_{max} can be determined by

$$T_{max} = \frac{\Delta U_{EB1}}{\left.\dfrac{dU_{EB}}{dT}\right|_{I=I_E}} + \frac{\Delta U_{EB2}}{\left.\dfrac{dU_{EB}}{dT}\right|_{I=I_E}} + T_0, \qquad (3.17)$$

where ΔU_{EB1} is the increment of U_{EB} with increase in U_{CB} from zero to U_{CBcr}, and ΔU_{EB2} is the increment with increase in U_{CB} from ΔU_{CBcr} up to the given value. What is the impact of an inhomogeneity or defect on the above conclusions in principle and in detail for the case of emitter current filamentation process?

Let us consider the structure with a small inhomogeneity (3.16) near the center of the structure at $\chi \ll L$ (Fig. 3.8b) as a general case. Stationary characteristic $U_{EB}(U_{CB})|I_E$ for this structure coincides in most details with similar characteristic of the initial uniform structure (Fig. 3.10a). The branch "AK" corresponds to slightly inhomogeneous distribution of the temperature and current density (curve "k" in Fig. 3.10b). A minor increase of U_{CB} over $U_{EB}{}^0$ results in the fast transition K→P into the stable state with current filament at the nearest or more heated structure wall (filament "p," Fig. 3.10b).

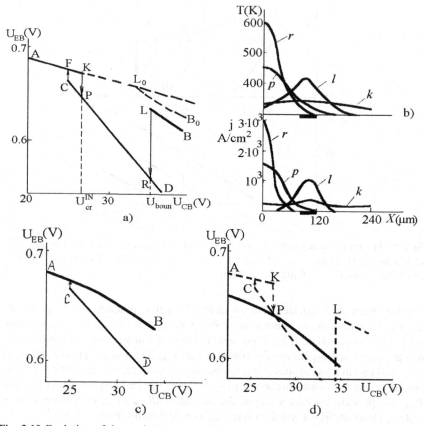

Fig. 3.10 Evolution of the stationary states in the structure with inhomogeneity. Emitter-collector $U_{EB}(U_{CB})$ characteristics (a) and corresponding distributions of the current density and temperature (b) at $\Theta = 0.2$. The characteristic of uniform structure is shown by a dashed line. Transformation of the $U_{EB}(U_{CB})$ characteristic at the inhomogeneity amplitude increase: $\Theta = 0.5$ (c); $\Theta = 2$ (d). Other parameters: $j_{cr} = 300$ A/cm^2; $v = 30$ µm; $\chi = 15$ µm

This transition, in contrast to the similar transition "KP" of uniform structure (Fig. 3.9a), takes place spontaneously without introduction of any fluctuation (2.13). The inhomogeneity itself (slightly nonuniform distribution near point "K")

creates all necessary preconditions for the transition into filament state. Branches "CD" of edge filaments in uniform and nonuniform structures are only slightly different depending on the amplitude of the inhomogeneity. In point "C" at $U_{CB} = U_{cr}^{fil}$, the reverse transition C→F to the branch "AK" takes place into a state with the only nonuniform distribution that is defined by the inhomogeneity itself.

The solitary branch "LB" corresponds to the filament formation on the inhomogeneity. It is formed below the branch "L_0B_0" that corresponds to the unstable solitary filaments. The transition into "LB" at monotonous U_{CB} voltage variation is impossible. Corresponding states, for example filament "l" (Fig. 3.10b), can be realized under pulsed voltage conditions $U_{CB} > U_{boun}$ if the pulse rise time is much smaller than the characteristic time of the whole structure heating.

The filament, formed on inhomogeneity, is stable relatively to additional disturbances. However, with gradual U_{CB} reduction down to point "L" the filament is losing stability and spontaneously moving to the nearest edge of structure position that corresponds to the transition L→R (Fig. 3.10a). This transition is accompanied by corresponding sharp increase in the maximum temperature (filaments "l" and "r" from Fig. 3.10b).

Existence of the solitary branch "LB" completely confirms the theoretical conclusions of Section 2.3.4. It represents solitary filaments localized on inhomogeneity. Indeed, with decrease in U_{CB} the filament becomes wider and its walls "feel" the sample boundary conditions. According to (2.49), the module of fundamental value λ_1, representing the filament stability relative to shift fluctuations, is decreased. At some $U_{CB} = U_{boun}$, the value λ_1 becomes equal to zero. Thus, criterion (2.49) becomes true and the filament is moving toward the nearest edge of the structure. In terms of Section 2.3.3, the branch "LB" represents a stable region of the isolated branch "BFG" (Fig. 2.5). This branch is formed at the splitting of the I–V characteristic near the second bifurcation point.

Consideration of the influence of inhomogeneity parameters on the shape of $U_{EB}(U_{CB})$ characteristic is presented below for the parameters of amplitude Θ_0, dimension v, and position χ. First, the change in the parameters results in localization of the branch "LB" due to strong dependence of the increment voltage ($U_{boun} - U_{cr}^0$) on Θ_0. With increase in Θ_0, both U_{boun} and U_{cr} are reduced, but U_{boun} reduction is more significant. This fact reflects elevation of the filament stability on inhomogeneity and results in elimination of the gap "KL" (Fig. 3.10c). Realization of edge filaments becomes impossible without additional disturbance of an adequate value. Beginning from some Θ_0, the filament is practically impossible to "separate" from the inhomogeneity, the branch of edge filaments "CD" disappears, and the characteristic $U_{EB}(U_{CB})$ becomes simple (Fig. 3.10d). On variation in dimension v, the regularities are the same, but the relative influence of v on the shape $U_{EB}(U_{CB})$ is much lower. For the discussed case of Fig. 3.10 with the inhomogeneity near the structure center, 15% increase in Θ_0 results in twofold reduction in the gap "KL while a 1.5-fold increase in v reduces the gap only 20%.

Shift of inhomogeneity from the center to the edge results in reduction of the gap "KL." Beginning from some value χ ($\chi \approx L/2 - W$) the heat feedback with the structure edge becomes so strong that the stationary filament state on inhomogeneity disappears and only edge filaments are formed. Obviously, it is possible to choose such minimal amplitude Θ_0 that will be independent of the shift in χ value. In this case, the filament will always be localized in inhomogeneity.

Comparison of numerical simulation results with the analytical evaluation in Section 2.3 is given below.

As was previously stated, the reduction of critical regime (in the given case U_{cr}) at local inhomogeneity insertion in the structure $\Theta(x)$ is proportional to factor $(\Delta\Theta_1)^{2/3}$, where $\Delta\Theta_1 = \int_0^{\check{A}} \Delta\Theta(x) \cos\frac{\pi}{L} x dx$. Apparently, for the inhomogeneity near the center of the structure, this factor is close to zero and thus has only minor impact on U_{cr}. For the edge inhomogeneity, $\Delta\Theta_1 \approx \Theta_0 v$ is true. Therefore, in this case both the dimension and amplitude of the inhomogeneity equally influence U_{cr}. In the case of shift of the inhomogeneity from the center to the edge, the factor $\Delta\Theta_1$ is increased from zero to $\Theta_0 v$. It results in appreciable reduction of U_{cr} and elimination of the hysteresis shape of the $U_{EB}(U_{CB})$ characteristics. Simulation results demonstrate that with increase in Θ_0 , the value U_{cr} can be extrapolated by a linear dependence $\Theta_0^{2/3}$. This is in agreement with (2.32).

The position of the branch "LB" and the value U_{cr} are determined by (2.48), the criterion of filament stability on inhomogeneity. Considering that for the discussed model $\phi = U_{CB} j_C R_T - (T-T_0)$, and inhomogeneity is contained in the parameter j_{S0} in (3.6) for j_C, the following can be obtained: $\partial\phi / \partial j_{S0} = U_{CB} j/j_{S0}$.

From the condition $\lambda_1 = 0$:

$$\int_0^L \frac{j(x)}{j_{S0}} U_{boun} \frac{d\Delta j_{S0}}{dx} \frac{dT}{dx} dx = \exp\left(-L / l_T^{fil}\right).$$

To detect the qualitative regularities, the derivatives can be roughly approximated by the maximum values of the function divided on the dimension of the region of significant change. Then dT/dx becomes a step function that is equal in modules, but different in sign from the values $2(T_{max}-T_0)/l_T^{fil}$ in the regions $l_T^{fil}/2$. Similarly,

$$\frac{d\Delta j_{S0}}{dx} \rightarrow \frac{\Delta j_{S0max}}{v/2} = \frac{2 j_{S0}\Theta_0}{v}.$$

Since the dimension v is less than l_f^{fil}, then in the vicinity of inhomogeneity the current density is changed slightly, and the integration is limited by the segment v, assuming that far from the inhomogeneity the distribution is uniform and

$$\frac{d\Delta j_{S0}}{dx} \approx 0.$$

Finally, the following expression can be obtained:

$$j_{max} U_{boun} \frac{\Theta_0}{\nu} \frac{T_{max} - T_0}{l_T^{fil}} \nu = \exp\left(- L \Big/ l_T^{fil}\right) . \qquad (3.18)$$

Thus, the dependence on the dimension ν is vanishing. This means that in the case of a real problem, this dependence is rather weak. From (3.18), it clearly follows that U_{boun} must decrease with the increase in inhomogeneity amplitude Θ_0.

The above-discussed inhomogeneity $\Theta(x)$ has a casual defect nature. However, in real transistor structures and semiconductor devices, there are a number of systematic inhomogeneities due to principal design peculiarities. The dominant design peculiarity is self-aligned emitter-base region topology and device architecture. One of the significant factors to take into account is the fact that the emitter region dimension is always significantly less than the corresponding chip dimension, i.e., $l_X < L$ (Fig. 3.2). In the case of power device array, the value of L could be larger than the chip thickness W. This difference can be treated as "negative" edge inhomogeneity in the sense that above the limits l_X the current does not flow and heat is not generated.

In such structures, the edge filaments cannot be formed, and the only possible filament state is solitary filament, localized within l_X limits. As will be shown in Chapter 5, this case can be simulated by replacing the mirror boundary conditions at the edge used above into conditions that represent isolation of the structure width.

The structures with $l_X \ll L$ do not allow a correct analytical evaluation. However, for the numerical simulation results a complex emitter region configuration does not present any serious obstacle [37, 51, 52]. From the simulation results, the filament is localized at the structure center, and its dimension and kinetics of formation are mainly determined by ratios between l_X, L, and W.

A question from a practical point of view is: does it make any sense to discuss the impact of small inhomogeneities and casual structural defects "on the background" of such significant inhomogeneities like the presence of the regions without heat generation on the chip surface? What is the practical value of the results that might be obtained using analytical analysis and numerical simulation of the filamentation in structures with small inhomogeneity?

Experimental studies and numerical analysis provide justification confirming that in power transistor structures with $l_X \gg W$ the temperature and current distribution remain uniform in a wide range $U_{CB} < U_{cr}$ far from the chip edge. This condition is true for practical application of the main results for conditions of thermal filamentation in the model of inhomogeneity structure shown in Fig. 3.8b. The only difference is the absence in real structures of the filaments formed in the chip edge. This role is played by the solitary filaments (second-order filaments). The value U_{cr} for such filaments must be calculated with a slight correction taking into account the difference between γ_1 and γ_2 (2.15). However, for structures with $l_X, L \gg W$, this difference is not significant. The same negligible correction requires the criterion of abrupt filamentation regime (3.13). Apparently, in the characteristic

$U_{EB}(U_{CB})$ the isolated branches (of "LB" type, Fig. 3.10a) are not formed, since all higher-order filaments above the second are unstable and the edge filaments are not realized. Due to absence of coincidence between l_X and L, the transition to continuous characteristic is observed similarly to Fig. 3.10d. Solitary filament localization is determined by the presence and value of defects in l_X limits, i.e., by comparison of the defects and design inhomogeneities.

Thus, the main conclusion that can be made for practical cases is: the transistor structure "defectiveness" can be detected by the characteristic $U_{EB}(U_{CB})|_{I_E}$, particularly in its quasi-linear region slope, where the value U_{CBcr} of jump, break, or changes in slope of characteristics can be indicated. This conclusion is confirmed both by numerical simulation results and by multiple experimental studies (Fig. 3.11).

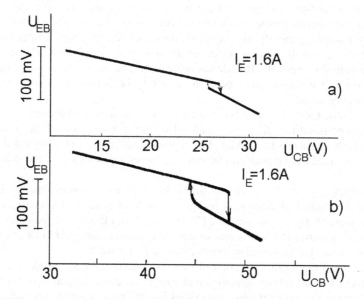

Fig. 3.11 Experimental $U_{EB}(U_{CB})$ characteristics at hot spot formation in the uniform structure (a) and in the structure with defect (b) at $U_{CB} \geq U_{cr}$

3.1.5 Delay Time

In the case of pulsed power overload above the critical regime for filamentation, redistribution of the current and temperature in the transistor structure requires a certain delay time [31, 32]. In general, thermal filamentation of emitter current involves two major parameters included in the physical mechanism of instability: the critical voltage $U_{CBcr}(I_E = \text{const})$ and the delay time τ_D (at $P > P_{cr}$). The methodology of U_{cr} estimation has been discussed in the previous sections.

The interest to τ_D is motivated by several application aspects: dynamic regime of bipolar transistor application, power pulsed circuits, reversible methods for current filamentation control and particularly the device operation in case of electrostatic discharge and electrical overstress conditions.

There are many methods of τ_D measurement [55]. One of the major methods is the measurement of transient characteristics $U_{EB}(t)|_{I_E}$. For this characteristic, abend is observed at the moment of filamentation beginning. The bend can be easily detected by additional measurements of the derivative dU_{EB}/dt [32, 41] (Fig. 3.12a).

One of the first measurements of τ_D [31] demonstrated that the delay time has a strong dependence on the transistor operation regime I_C, U_{CB} and its value has a significant scattered distribution for transistors of the same production lot.

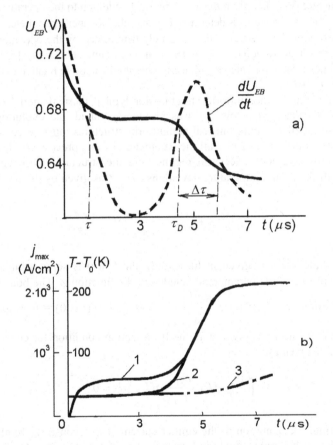

Fig. 3.12 Kinetics of the emitter current filamentation: (a) calculated characteristic $U_{EB}(t)$; (b) corresponding dependencies ($T_{max} - T_0$)—"1," j_{max}—"2," and $\delta j(0)\exp \gamma t$—"3"

The first theoretical analysis of the filamentation kinetics was obtained for uniform model approximation [56]. The delay time is determined by the rise time of the nonuniform fluctuation in current density and temperature up to the macroscopic values. Respectively, the rate of fluctuation rise is determined by the rise increment γ_1 that can be calculated using (3.12). Within the limitations of the fluctuation instability approach, it has been demonstrated that the instability at emitter current filamentation is initiated practically from the initial time moment without any significant delay time of the total structure heating [56]. Numerical simulation of the filamentation process in uniform structures [53] and the structures with small inhomogeneity [54] demonstrated similar filamentation kinetics up to a time when the increased value of disturbance remains small. In this case, the external characteristic $U_{EB}(t)$ practically does not "feel" the disturbances. In the time domain $0 \le t \le \tau_D$, it coincides with the dependence $U_{EB}(t)$ for the case of uniform heating of similar structure (Fig. 3.12b). In this case, $\tau_D > \tau$. In addition to the operation regime, the value τ_D in real structures is determined by the inhomogeneity parameters. Interval $\Delta\tau$ of macroscopic heat redistribution is mainly determined by the operation regime and the overvoltage value ($U_{CB}-U_{cr}$). The numerical simulation study [53] demonstrated that the value $(\Delta\tau)^{-1}$ increases linearly with the increase in relative overvoltage level $(U_{CB}-U_{cr})/U_{cr}$.

The delay time in nonuniform structure for typical thermal model of power transistors is calculated below. As was previously stated, the evolution of the thermal current filamentation in real nonuniform structures must be considered a deterministic transition process: all the parameters of this process are determined by the initial conditions, electrical regime, and the structural inhomogeneities. Evolution of the small temperature deviations $\delta T(x, t)$ is given by the linearization of (2.9):

$$\tau_T \frac{\partial \delta T}{\partial t} = l_T^2 \frac{\partial^2 \delta T}{\partial x^2} + \frac{\partial \varphi}{\partial T} \delta T + \frac{\partial \varphi}{\partial U_{EB}} \delta U_{EB} + \frac{\partial \phi}{\partial A} \Delta A, \qquad (3.19)$$

where the function φ is given in form (3.8) and $\Delta A(x)$ is some arbitrary inhomogeneity of one of the parameters included in the function φ. The boundary and initial conditions for (3.19) are $\dfrac{d\delta T}{\partial x}\bigg|_{x=0,L} = 0$ and $\delta T(t = 0) = 0$, respectively.

Equation (3.19) must be solved with the linearization condition for constant total current that is given by

$$\int_0^L \left(\frac{\delta j_C}{\partial T} \delta T + \frac{\partial j_C}{\partial U_{EB}} \delta U_{EB} \right) dx = 0. \qquad (3.20)$$

For thermal filamentation of the emitter current at $I_C =$ const, the total heating of the structure has rather negligible influence on the conditions of nonuniform temperature disturbance elevation [56]. Therefore, it is correct to use the method of "freezing" for variable coefficients. Assuming that the derivatives $\phi_T', \phi_{U_{EB}}', \phi_A'$ do not depend on average temperature, by transformations similar to (2.12) and (2.13) the analytical expression for nonuniform temperature component evolution $\delta T_N(x,t)$ can be obtained:

$$\delta T_N(x,t) = \sum_{m=1}^{\infty} \frac{\phi_A' \Delta A_m}{\gamma_m^N} \left(\exp\left(\gamma_m^N t / \tau_T \right) - 1 \right) \frac{2}{\sqrt{L}} \cos \frac{\pi m}{L} x, \qquad (3.21)$$

where ΔA_m are the expansion of coefficients on inhomogeneity in the Fourier expansion; γ_m^N are the fundamental values of the operator $\hat{H} \equiv \partial^2/\partial x^2 + \phi_T' + \phi_{TA}'' \Delta A$ that differs from operator $\hat{H}_0$ of uniform problem (2.1) by the presence of the disturbance operator $\hat{V} = \phi_{TA}'' \Delta A$. Respectively, the value γ_m^N differs from γ_m of uniform problem (2.15) in the value of small correction $\Delta \gamma_m$ that is created by the inhomogeneity:

$$\gamma_m^N = \gamma_m + \Delta \gamma_m = \phi_T' - \left(\frac{\pi m l_T}{L} \right)^2 + \int_0^L \hat{V} \cos^2 \frac{\pi m}{L} x dx. \qquad (3.22)$$

If the pulsed regime does not exceed the critical level ($\gamma_m^N < 0$), then a slightly nonuniform temperature distribution is stable in the structure (3.21). If $\gamma_1^N > 0$, but all remaining $\gamma_m^N < 0$, the longwave nonuniform disturbance with $m = 1$ is increasing. In this case, the time required to reach a certain amplitude δT_0 of nonuniform temperature deviation δT_{max} can be assumed as definition criterion for the delay time τ_D. For example, $\delta T_0 = 0.01 T_0$.

Then,
$$\tau_D \cong \frac{\tau_T}{\gamma_1^N} \ln \frac{\delta T_0 \sqrt{L} \gamma_1^N}{2 \phi_A' \Delta A_1}. \qquad (3.23)$$

Since $\phi_A' = U_{CB} j_T' R_T - 1$ from (3.23) it follows that the delay time decreases with the increase in $U_{CB} > U_{cr}^N$ and corresponding electrical power during the pulse. Experimentally observed scattering in τ_D parameter can be correlated either with the scattering in the thermal resistance parameter R_T or with the presence of casual inhomogeneities, since the inhomogeneity parameters are included in the correction Δ^{γ_m}, that in the first approximation is proportional to the total "volume" of inhomogeneity $\int_0^L \Delta A(x) dx$.

At relatively low overvoltage $[(U_{CB} - U_{cr}^N)/U_{cr}^N \ll 1]$, the parameter $\gamma_1 \approx 0$ and dependence τ_D on the inhomogeneity amplitude is determined by $\Delta \gamma_1$ (neglecting here the weak logarithmic function), and, therefore, τ_D^{-1} is linearly increasing with increase in amplitude and dimension of inhomogeneity. At high ΔA and/or adequate high overvoltage, the filament is formed on the background of total structure heating process practically without delay: $\tau_D \ll \tau_T$.

For an adequately long structure ($L \gg l_T$), the difference between values is rather low: $\gamma_m \approx \varphi_T'$. The case is quite typical when the values $I_C \gamma_1^N$, γ_2^N are approximately equal to zero simultaneously at given U_{CB}. In this case, according to (3.18), two different harmonic disturbances can rise simultaneously. The disturbance $m = 1$ corresponds to the filament in the structure edge, and $m = 2$, to the filament in the center. The final site of filament localization depends on the inhomogeneity parameters. At adequate overvoltage $\Delta\gamma_{1,2}^N \ll \varphi_T'$ according to (3.23),

$$\tau_{D1} - \tau_{D2} \cong \frac{\tau_T}{\phi_T'} \ln \frac{\Delta A_1}{\Delta A_2}. \tag{3.24}$$

If inhomogeneity is located near the center of the structure, then $\Delta A_2 \gg \Delta A_1$ and $\tau_{D2} < \tau_{D1}$, therefore filament is formed on inhomogeneity in the structure center. Similarly, the problem of filament localization can be solved for the structure with several inhomogeneities.

In conclusion, one of the important circumstances should be emphasized. Either measured delay time τ_D from the $U_{EB}(t)$ characteristic or the characteristic time for thermal transition process τ_T (in the case $\tau_D < \tau_T$) is practically coincident with a typical time of stabilization required for the stationary temperature distribution $T(x)$. However, the stabilization time of the current distribution τ_j can be significantly different from τ_T [37]. Apparently, this is the result of the much smaller active region dimension; where current redistribution is formed, in comparison with a large whole chip dimension. The last determines the value τ_T. The fact that the current at thermal filamentation is really redistributed faster than the temperature has been confirmed experimentally in [57].

3.1.6 Three-Dimensional Thermal Model and Design Optimization

A built-in structural inhomogeneity of the active region with heat generation area practically excludes application of the analytical methods toward estimation of thermal stability of real transistors. Numerical simulation on the basis of two-dimensional thermal models [49–51, 54] presents an opportunity to estimate the conditions and critical regimes for current filamentation. However, by neglecting the third dimension in 2D simulation cases, an error in the quantitative estimation practically brings the accuracy down to the "analytical" level. Another problem is that 2D simulation does not provide comparison for thermal filamentation stability in real transistor structures.

Three-dimensional stationary numerical thermal model is realized using (3.1)–(3.7). As demonstrated below, the model significantly facilitates the previous estimation and comparison of transistor designs from the point of view of the stability to thermal breakdown [58].

Two examples of transistors with a single chip of $W = 200$ µm thickness are considered below. The emitter region is assumed to be rectangular with the dimensions given in Fig. 3.13 (version 1). Assuming the given transistor structure as the baseline design, the following possible optimizations are discussed below. The voltage of hot spot formation U_{cr} can be assumed as an optimization criterion. As is well known, this voltage corresponds to appearance of a break or a jump in the dependence $U_{EB}(U_{CB})$ at fixed I_E. For calculation, the total resistance of emitter ballasting resistors is assumed equal to $R_E \approx 0.1$ Ω. This value does correspond to sheet resistance $r_E = R_E S_E = 10^{-4}$ Ω cm, where S_E is the area of emitter regions. For simplicity, the stabilization is assumed given by diffusion resistors in multiple emitter structure. The base resistance is neglected: $r_{Beff} = 0$.

Seven versions were selected (including the baseline version) for the emitter region topology (Fig. 3.13). The length, width, quantity, and mutual arrangement of emitter regions are varied. With this approach, the emitter area S_E was constant. This condition practically provides constant collector junction capacitance and in the range of operation frequencies. Minimal spacing between the emitter area and chip edge is not less than $2W/3$ and it corresponds to real transistor structures. Simulation results have demonstrated that increase in this parameter has practically no effect on the thermal stability.

The calculated U_{cr} values, $U_{EB}(U_{CB})$ characteristic, and $T_{max}(U_{CB})$ dependence for the device versions are presented in Fig. 3.14 for the same maintained constant level $I_E = 200$ mA.

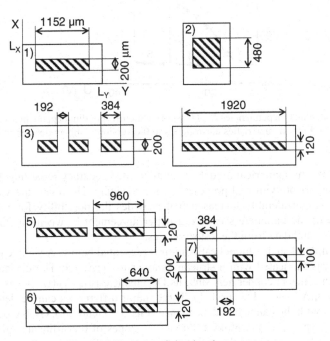

Fig. 3.13 Versions of simulated structures

It has been demonstrated that a significant increase in the device capabilities can be obtained by lengthening or splitting the emitter region under constant S_E. For example, U_{cr} of version 6 is 12 V higher than baseline version 1 (Fig. 3.14). The maximum rating of the dissipated power is 1.5 times higher. Narrowing of the emitter region provides a major effect (version 4, Fig. 3.14) with accompanied splitting in a few parts (versions 5, 6, 7). Increase of U_{cr} due to change in the emitter region topology is accompanied by a variation in the mode of filament excitation. In versions 4–7, the filament formation is abrupt and accompanied by a jump of peak temperature of more than 100 K. Thus, such an effect may result in real structure failure.

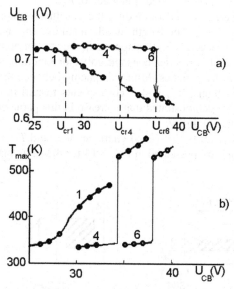

Fig. 3.14 Calculated dependencies $U_{EB}(U_{CB})$ (a) and corresponding dependencies $T_{max}(U_{CB})$ (b). The legend for characteristics corresponds to the version numbers defined in Fig. 3.13. $I_E = 200$ mA

Of course, any numerical model has quite limited accuracy regarding a simulation in comparison with real processes in the structure. However, the validity of the model is considerably increased if it provides the possibility for fitting and calibration of the parameters toward a better agreement between the simulation and experimental data from the test structures.

An example of such a model calibration is presented below. A sample of power transistor was fabricated with a structure similar to version 1. The device was attached to an efficient copper heat sink. Its DC characteristic $U_{EB}(U_{CB})$ was obtained at $I_E = 200$ mA (Fig. 3.15, curve 1). Three major differences can be detected by comparison with the calculated dependence (Fig. 3.15, curve 2): (i) 7-V difference between U_{cr} values; (ii) twofold difference in slopes of quasi-linear regions; (iii) jump mode filamentation in the real structure versus soft mode for the simulation.

What simplifications in the model (3.1)–(3.7) provide the reason for the deviations?

First, it was assumed that the temperature T_0 on the bottom of the chip is constant. In reality, several intermediate layers are usually incorporated between the chip and heat sink: beryllium ceramic substrate, metal flange, and various die attachment epoxy materials. Ignoring these facts results in lowering the RT value and reduction of the value of $|dU_{EB}/dU_{CB}|$.

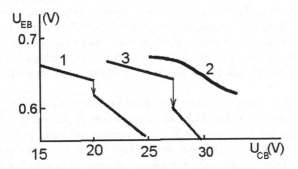

Fig. 3.15 Results of simulation parameter "fitting": experimental characteristic "1," initially calculated "2," and after calibration dependencies "3"

Second, the assumed condition of constant heat conductance does not correspond to reality, since the heat conductance κ of pure silicon varies with temperature as $T^{-1.3}$[38]. Decreased heat conductance can result in more abrupt mode of filament excitation.

Third, an adequate consideration of balancing emitter resistors presents a significant difficulty. Rather low r_E variation in (3.6) may result in significant U_{cr} variation.

Finally, it is practically impossible to consider various systematic and casual inhomogeneities. These structural defects reduce the limit capabilities of the structure as a rule.

Two modifications were introduced in the model in order to reduce the divergence with the experiment. The first consists in simulation of ceramics. For this the condition $T(z = W) = T_0$ is replaced by $T(z = W) = T_0 + U_{CB}I_C R_{TC}$, where $U_{CB}I_C$ is the dissipated power from the chip and R_{TC} is the thermal resistance of the heat sink. The second modification is done in order to take into account the dependence $\kappa(T)$: it is assumed that $\kappa(T) = \kappa_0(T/T_0)^s$, where $s < 0$. With the help of new variable $\Theta = T_0 + \int_{T_0}^{T} \kappa/\kappa_0 \, dT$ by standard methods this dependence can be taken into account. However, the exact s value lacks reliable data about $\kappa(T)$ for heavily doped silicon.

The model modification allows fitting the simulation data to the experimental results (Fig. 3.15). The remaining difference in the value of U_{cr} might be a result of inhomogeneities in the experimental structure design. This conclusion has been supported by measurement by scanning IR microscope where hot spot formation that is asymmetrically shifted to the edge has been observed.

3.1.7 Current Filamentation in Dynamic Regimes

RF signal operation is one of the major regimes for microwave bipolar transistors. Therefore, even from the application point of view, an understanding of the evolution of emitter current filamentation is very important. Below, a case study is discussed for the conditions of current through transistor and applied voltage variable in time by some periodic law. Similar analysis, however, is adequate for switching devices such as the case of power management applications for switching voltage regulators.

It is relevant to emphasize that the dynamic operation regime in general does not protect the transistor from thermal filamentation. Experimental study demonstrates that there are some critical levels of median dissipated power P_{Mcr} in all dynamic regimes (A, B, C) when hot spot formation in the transistor structure either decreases the transistor capability or results in catastrophic failure. The failure mode is similar to that observed in the DC regime: local metallization melting followed by short circuiting or, in opposite, formation of the open circuit conditions. Evolution of the failure is approximately the same as evolution in the DC regime when all the peculiarities related to the periodic signal shape are taken into account.

Since evolution of thermal processes in transistor structure has a typical time, it could be expected that the filamentation condition in the dynamic regime is frequency dependent. In fact, in the frequency range $f^{-1} \geq \tau_T$, where the relaxation of temperature oscillations is observed, a considerable increase of average (with respect to oscillation period) dissipated power for thermal instability is observed (Fig. 3.16).

It is rather logical to assume that the cause of this phenomenon is the decrease in T_{max}, down to some average temperature T_M realized under the condition of $f^{-1} \geq \tau_T$.

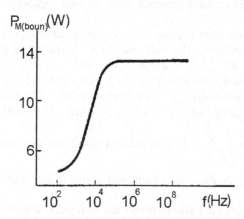

Fig. 3.16 Increase of critical dissipated power in the dynamic regime

In reality, this statement is true if the thermal filamentation of collector junction saturation current is analyzed. For this mechanism, the critical power is a trivial function of the junction temperature. Decrease in the junction temperature is a direct cause of the increased thermal stability.

However, according to experimental studies, such increase of P_{Mcr} is observed in transistors only at relatively low I_{CBO} at $T_M < 600$ K. In most practical cases, thermal instability is based on the emitter current filamentation process. For this mechanism, as known, critical power practically does not depend on the chip temperature. It has been demonstrated that relative increase in P_{Mcr} is strong only in the C-regime rather than the A-regime. In the A-regime at small amplitude of microwave component, P_{Mcr} practically coincides with P_{cr} for DC operation.

In order to clarify the reason for increase in median critical power dynamic regimes, a theoretical criterion of thermal stability is obtained in [59] for high-frequency regimes. The criterion enables one to use the average heating power value with respect to the oscillation period similarly to Eq. (2.7):

$$\tau_T \frac{\partial T_M}{\partial t} = l_T^2 \frac{\partial^2 T_M}{\partial x^2} + \frac{1}{\tau}\int_0^\tau R_T P(t)dt - (T_M - T_0),$$
(3.25)

where τ is the input signal oscillation period and $R(t) = j_C(t)U_{CB}(t)$ is the density of dynamic power.

The analysis [59] demonstrates that the thermal instability criterion for periodic signal with $w \gg 2\pi\tau_T$ can be presented in a form similar to (3.11):

$$\tilde{R}_T \frac{dP}{dT} > 1,$$
(3.26)

where the derivative $\dfrac{dP}{dT}$ is normalized for period τ.

$$\frac{dP}{dT} = \tau^{-1}\int_0^\tau \left[U_{CB}(t)\frac{\partial j_C}{\partial T}(t) \Big/ \left(1 + r_{eff}\frac{\partial j_C}{\partial V_E}(t)\right)\right]dt.$$
(3.27)

Taking into account both (3.2) and (3.6), criterion (3.26) can be rewritten in a more simplified form [60]:

$$\tau^{-1}\int_0^\tau \frac{aj_C(t)U_{CB}(t)}{1 + bj_C(t)}dt - 1 > 0.$$
(3.28)

For simplicity, the case of constant collector voltage is analyzed below as an example. The statement, that for given form $j_C(t)$ the thermal stability in the dynamic regime is in general higher than in the DC regime, is equivalent to a condition: $U_{CBcr}(j_C = j_C(t)) > U_{CBcr}(j_{cr})$. For this condition from (3.28) and (3.12) it follows that

$$\tau^{-1}\int_0^\tau \frac{aj_C(t)}{1+bj_C(t)}dt < \frac{aj_M}{1+bj_M},\qquad(3.29)$$

where $j_M = \dfrac{1}{\tau}\int_0^\tau j_C(t)dt$.

Assuming for simplicity $j_C(t)=j_0+j_a f(t)$, where j_0 is the constant component of current density, $f(t)$ is some periodic function, and j_a is the corresponding amplitude. Then, $j_C(t)=j_0+j_a k_S)$, where $k_S = \dfrac{1}{\tau}\int_0^\tau f(t)dt <1$ is the shape coefficient. For example, for classical meander (Fig. 3.17a): $j_0=0$; $k_S=1/2$; $j_M=j_a/2$.

In the A-regime at relatively low amplitudes of alternative component ($j_0 \gg j_a$), the left and right sides of (3.29) practically coincide, and the critical power is determined by j_0 value and is approximately equal to j_M.

In B- and C-regimes ($j_a \gg j_0$), many peculiarities are determined by values j_a and k_S. In fact, at small current amplitudes ($bj_a \ll 1$) the critical voltages in dynamic and DC regime are coincident and $U_{CBcr} \approx (aj_a k_S)^{-1}$. If the swing of oscillations is adequately high, then the left side of (3.29) is less than the right side. This case is easy to demonstrate in the example of the simplest meander pulse (Fig. 3.17a).

The left side of (3.29) is equal to

$$\tau^{-1}\int_0^{\tau/2}\frac{aj_a}{1+bj_a}dt = \frac{1}{2}\frac{aj_a}{1+bj_a}=\frac{aj_M}{1+2bj_M},$$

and thus always less than the right side. In limit($j_a\to\infty$), the critical power in the dynamic regime exceeds the corresponding DC power by $(k_S)^{-1}$ times.

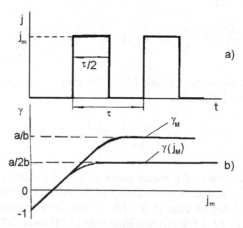

Fig. 3.17 Square pulse (a) and qualitative view of the dependence of rise increment on pulse amplitude (b)

A different interpretation can be given for the increase of thermal stability in the dynamic regime. The value of the left side of (3.28) is the rise increment γ_M of the amplitude δT_M of median temperature nonuniform disturbance $(\delta T_M \sim \exp\gamma_M t)$. Taking this into account, it can be stated (at $U_{CB} = $ const) the bigger $\gamma(j_M)$ over γ_M is realized, where $\gamma(j_M)$ is the rise increment of temperature fluctuation in the DC regime with $j_C = j_M$. Figure 3.17b essentially provides a demonstration of this fact for the mentioned meander. Thus, difference in thermal stability is observed at adequately high amplitude of alternative signal $j_a \gg j_0$, when significant nonlinearity of dependence $\gamma_M(j_a)$ becomes active. This situation is typical first of all for the C-regime.

However, the obtained results are not obsolete. In a general case of alternative signal, the level P_{Mcr} depends not only on $j_C(t)$, but also on $U_{CB}(t)$ and the phase ratio between them. At load mismatch in power amplifier, a significant variation of P_M is observed [61]. The critical level P_{Mcr} alsovaries. It is demonstrated [62] that at variation in $j_C(t)$ and $U_{CB}(t)$ in antiphase, the thermal instability can be observed at significantly lower P_{Mcr} in comparison with the DC regime. In other words, for complete analysis of the thermal stability in the dynamic regime the parameters of the external circuit must be taken in account.

Above, the conditions of emitter current filamentation for alternative signal were analyzed for frequency range $f \gg \tau^{-1}$ where thermal oscillation has been excluded. A single pulse regime with duration $t_P > \tau_T$ is discussed in Sections 3.3.2–3.3.5. The filamentation conditions in this regime are determined by criterion (3.12), the kinetics is determined by the ratio between the delay time τ_D, calculated in Section 3.3.5, and the pulse duration t_P.

A frequency range that does not allow one to neglect the thermal oscillations ($f \leq \tau^{-1}$) has not as yet received an accurate theoretical analysis. In this case, it is impossible to introduce the parameter of average temperature. The temperature and the input signal are varied within the same period τ. However, the amplitude of small nonuniform disturbances should be presented in the form [63]:

$$\delta \tilde{T}(t) = \delta T_0 y(t) \exp(\gamma_F t),$$ where $y(t) = y(t + \tau)$ is some periodic function and γ_F is the Floke exponent (for its fundamental value equation

$$dy/dt = y\left[A(t)/\tau_T - \gamma_F\right]$$ is true, where $A(t) = R_T \, dP/dT - \left[1 + (\pi l_T n)^2\right].$

Only in the case of $\mathrm{Re}\,\gamma_{Fn} < 0$ for all $n > 1$, the initial temperature distribution $T(t)$ can be considered as stable. Obviously, this problem is rather difficult to solve even for particular cases. However, numerical simulation [44] gives a rather clear and simple picture of thermal filamentation evolution in similar pulse-period regimes. For a sequence of power pulses with constant amplitude $P > P_{cr}$, duration t_P, and period t_0, the simulation demonstrated that conditions and filamentation mode depend on the ratio between the values t_P, t_0, and τ_T.

The case $t_0 \ll \tau_T$ is discussed above when t_P, t_0, and P_{Mcr} are determined by the pulse amplitude P_P and the signal shape.

The duty factor Q_{cr} corresponds to filamentation beginning, where $Q \equiv t_0/t_P$. At constant P_P, the critical value Q_{cr} corresponds to some value P_{Mcr} and does not depend on t_P, since $k_S \sim t_P/t_0$ remains constant. If t_P becomes comparable with τ_T, then Q_{cr} increases with t_P increase. This regularity is rather apparent since the longer t_P is, the stronger the current redistribution during the single pulse. Thus, in order to provide the same level of current redistribution, the duty factor must be lower (Fig. 3.18).

The maximum level of the current redistribution is achieved in the DC regime at $Q \rightarrow 1$. On the other hand, the minimum level is achieved at $Q > 2$ (Fig. 3.18). If t_P is comparable to the delay time of filament formation τ_D (curves "1" and "2," Fig. 3.18), then a significant redistribution is achieved practically immediately during the first pulse, and therefore no additional decrease is observed with increase in Q. Apparently, it is possible to select such t_P and Q $(t_P > \tau_D; Q \gg 2)$ parameters that will provide a decay of current and temperature filaments between the pulses $(Q-1)t_P$. In this case, the pulsed filaments are observed with corresponding frequency t_P^{-1}.

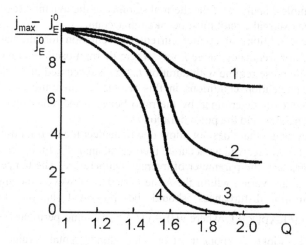

Fig. 3.18 Calculated dependencies $(j_{max}-j^0_E)/j^0_E$ on Q: "1," $t_P = 100$ μs; "2," 80 μs; "3," 50 μs; "4," 20 μs at constant $j^0_E = 200$ A/cm^2 and $U_{CB} = 40$ V

From the simulation results, it follows that the external characteristic $U_{EB}(t_P) = f(U_{CB})|_{I_E,Q,t_P}$ can be used for thermal stability boundary control and current redistribution detection in the pulsed-periodic regimes. This characteristic essentially presents itself as a quasi-dynamic analog of the corresponding DC characteristic $U_{EB}(U_{CB})|_{I_E}$.

The value U_{EB} is fixed at the end of the pulse at $t = t_P$ under the conditions of stabilized periodic regime. In this process, the variation in U_{CB} must be adequately smooth. Formation of pulsed or quasi-static filament is accompanied by increased values of modules $dU_{EB}(t_P)/dU_{CB}$ (Fig. 3.19). In this case, the shorter t_P is, the sharper the change of the derivative of this informative characteristic.

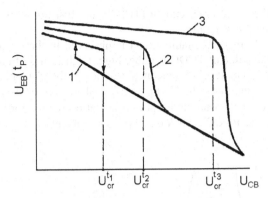

Fig. 3.19 Qualitative view of dependence of $U_{EB}(t_P)$ on U_{CB} for pulsed regime at fixed current pulse amplitude, given Q for various t_P: $t_P = t_1 \gg \tau_T (1)$; $t_P = t_2 < \tau_T (2)$; $t_P = t_3 \ll \tau_T (3)$

In conclusion, the peculiarities due to structural defects and inhomogeneities are discussed for the case of current filamentation in dynamic regimes. Cases of current filamentation both at $t_P \gg \tau_T$, $t_P \gg \tau_D$ and at $t_P \ll \tau_T$ are described by a criterion similar to (3.12). The conclusions about the role of inhomogeneity are the same too.

An interesting unusual behavior of the structure can be expected in the case of $t_P \approx \tau_T, \tau_D$. Under these conditions within the given t_P range, a stationary formation of two or multiple filaments can be observed. These filaments are localized on corresponding inhomogeneities with similar efficiency. The effect becomes possible only if $\tau_{D1}, \tau_{D2} < t_P < \tau_{D1-2}$, where τ_{D1}, τ_{D2} are the typical times of filament formation at the first and at the second inhomogeneity, and τ_{D1-2} is the typical time of pumping over effect between filaments.

3.2 Filamentation of Channel Current in MOSFETs

Undoubtedly, the thermal stability of MOSFETs is much higher than that of bipolar transistors under the same dissipated power conditions. In various types of power MOSFETs, particularly in planar MOSFETs, the hot spot formation mechanism is not realized at all at near room temperature of the heat sink. Thermal instability in these structures can be developed only due to the saturation current of the drain junction at temperature levels that are typical for thermal breakdown above 600–700 K [64–66]. This fact is provided by the negative temperature coefficient of drain current in typical MOSFETs due to the decreasing dependence of the electron mobility upon temperature realized in standard operation regimes.

However, for power MOSFETs with high transconductance values (over 1000 mS in MOSFETs with V- or U-shape recess), the thermal instability mechanism is actual.

Indeed, for these transistors the temperature dependence of the boundary voltage $U_{boun}(T)$ provides a positive feedback between the channel current density j_{CH} and the lattice temperature in a wide current range. In these structures with positive temperature coefficient of drain current (TCDC), the first hot spots were observed [66] in the DC operation regime

The most clearly thermal filamentation is observed in the range of relatively small I_C, where the value of TCDC is rather high. High values of breakdown voltage provide an adequate level of generated heat power for filamentation. In regard to output I–V characteristic $I_D(U_{DS})|_{V_{GS}}$, the S-shape region is observed (Fig. 3.20a). This region disappears under pulsed operation regime [66], when time to develop a thermal positive feedback on chip is insufficient.

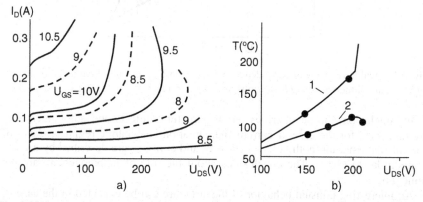

Fig. 3.20 (a) Characteristics of MOSFET at current filamentation: output I–V characteristic at temperature values 60°C (solid line) and 160°C (dashed line). (b) Dependence of hot spot (1) and median chip temperatures (2) on U_{DS}

The classical view of S-shape I–V characteristic (Fig. 3.20) points to some classical filamentation mechanism with the initial conditions at $dI_D/dU_{DS} \leq 0$. However, by direct measurement using IR microscope setup in the constant collector current regime with gate auto-biasing circuit (I_C = const), the hot spot formation (Fig. 3.20) is observed at positive differential conductivity ($dI_D/dU_{DS} > 0$). This indicates the presence of some hidden S-shape dependence in the case of thermal filamentation in MOSFETs. This mechanism is similar to the thermal filamentation of emitter current in bipolar transistors.

3.2.1 Criterion for Channel Current Filamentation

Criterion of current filamentation can be derived for a model similar to that discussed above. Considering a model of uniform transistor structure with the heat distribution given by (3.1) or by an equation similar to (2.7):

$$l_T^2 \frac{\partial^2 T}{\partial x^2} + R_T U_{DS} j_D(T) + (T - T_0) = 0, \qquad (3.30)$$

where T_0 is the supported heat sink temperature and j_D is the drain current density. j_D is the sum of the channel current density j_{CH} and saturation current density j_{DS0} of the "parasitic" bipolar transistor. The parasitic BJT is composed of the drain junction acting as a collector junction, the source acting as an emitter, and the substrate acting as a base:

$$j_D = j_{CH} + j_{DS0}.$$ (3.31)

For simplification in this section, not all the phenomena are taken into account. Avalanche multiplication at the drain junction, current injection from the source, and temperature dependence of thermogeneration currents are neglected for this approximation. Thus, the case is limited by the mechanism of "pure" channel current filamentation is assumed $j_{DS0} = 0$.

It is assumed that in the drain current range the transconductance characteristic has a classical view [38]:

$$j_{CH} = b(U_{GS} - U_{boun})^m,$$ (3.32)

where U_{GS} is the gate-source bias, U_{boun} is the boundary voltage, m is some empirical coefficient $m = 1-2$, and b is the temperature dependence parameter that is determined by the temperature dependence of current carriers:

$$b(T) = b_0 \left(\frac{T_0}{T} \right)^s,$$ (3.33)

where the coefficient s depends on the material and type of conductivity. For silicon, $s \approx 1-1.5$ [64], b_0 is constant [38], $T_0 = 300$ K.

The temperature dependence of boundary voltage can be written as

$$U_{boun}(T) = U_{boun}(T_0) - \beta_0(T - T_0),$$ (3.34)

where the coefficient β_0 depends on the gate oxide thickness and the doping level in the channel region [38].

The higher the substrate doping level is, the more significant the decrease of U_{boun} on increase in temperature as well as the increase of j_{CH}.

Thus, the channel current is given by

$$j_{CH} = b_0 \left(\frac{T_0}{T} \right)^s [U_{GS} - U_{boun}(T_0) + \beta_0(T - T_0)]^m.$$ (3.35)

At uniform channel current distribution and its fixed total value, the heating up of the structure by the electrical power on increase in U_{DS} is accompanied by the corresponding variation in U_{GS}. In this process, the slope of external DC characteristics $U_{GS}(U_{DS})$ is almost constant and can be determined from the condition

$$k_0 \equiv \left. \frac{\partial U_{GS}}{\partial U_{DS}} \right|_{j_{CH} = const} = \frac{\partial U_{GS}}{\partial T} \frac{\partial T}{\partial U_{DS}}.$$

Using (3.32)–(3.34) and taking into account $\dfrac{\partial T}{\partial U_{DS}} = R_T I_{CH}$,

$$k_0 = \left\{ \frac{s}{m} \left(\frac{j_{CH}}{b_0} \right)^{\frac{1}{m}} \left(\frac{T}{T_0} \right)^{\frac{s}{m}-1} \frac{1}{T_0} - \beta_0 \right\} R_T I_{CH}. \tag{3.36}$$

It is easy to see that k_0 can be both positive and negative. At some "thermostable" current I_{CH}^*, k_0 is close to zero practically in the whole discussed U_{DS} range. It should be noted that k_0 only slightly reacts on the structure heating. This parameter could be used as an indicator of thermal resistance value for a given structure. Combination of (3.35) with (3.30) for $T = \text{const}$ provides the output I–V characteristic of transistor. This characteristic exhibits an S-shape region at an adequate $\beta_0 > 0$. At the point $dI_{CH}/dU_{DS} = \infty$, the current and temperature instability takes place relatively variations in total current $\Delta I_{CH} > 0$.

In the current source regime, the instability due to nonuniform disturbances of current and temperature can begin at $dI_{CH}/dU_{DS} > 0$. A criterion of this instability at the total constant current condition ($\int j_{CH} dx = \text{const}$) is given in a form that is similar to (3.11) [67]:

$$\tilde{R}_T \frac{dj_{CH}}{dT} U_{DScr} - 1 \geq 0, \tag{3.37}$$

where $\dfrac{dj_{CH}}{dT} = \dfrac{s j_{CH}}{T} \left[\left(\dfrac{j_{CH}^*}{j_{CH}} \right)^{\frac{1}{m}} - 1 \right]$ and $\tilde{R}_T$ is determined from (3.11a).

For power MOSFETs with double diffusion (DMOS), the total channel current value I_{CH}^* may achieve the level of several amperes. Thus, the drain current range where (3.37) is true can be rather wide. Obviously, at $j_{CH} \geq j_{CH}^*$ the instability can arise under any conditions; this is observed in planar MOSFETs with I_{CH}^* of a few milliamperes.

It is not hard to see an analogy between the channel current and the emitter current filamentation processes. In both cases, formation of hot spots takes place in the regions of I–V characteristic with positive differential conductivity.

According to this analogy, the DC characteristics $U_{GS}(U_{DS})|_{I_{CH}}$ are expected to be rather informative. Indeed, the direct experiment [67] revealed (Fig. 3.21) that formation of hot spots in MOSFETs is accompanied by a break of $U_{GS}(U_{DS})$ characteristic. The formed filaments at $U_{DS} > U_{DScr}$ are capable of providing rather high local overheating (above 150°C), and a minor further regime deviation results in irreversible thermal breakdown (burnout).

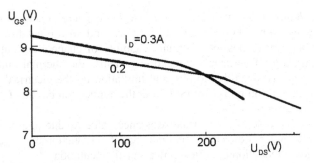

Fig. 3.21 Dependencies $U_{GS}(U_{DS})$ for MOSFET

It should be noted that the mechanism of channel current thermal filamentation can be simulated using 2D numerical model (3.15) together with the equation for current density (3.6) replaced by (3.35) [67]. The simulation results correspond to both the experiment and general theoretical conclusions. In particular, the informative capability of $U_{GS}(U_{DS})|_{I_{CH}}$ characteristic and transition characteristic $U_{GS}(t)|_{I_{CH}}$ is confirmed.

However, in comparison with similar characteristics of bipolar transistors, its detection ability to local overheating is not very high. This is the result of the fact that the positive feedback on temperature is not too strong (at $j_{CH} \to j_{CH}^{*}$ $dj_{CH}/dT \to 0$) in the case of channel current. Besides, as follows from (3.37), the rise increment of nonuniform temperature disturbances is only slightly dependent on the overvoltage $(U_{DS}-U_{cr})/U_{cr}$.

An important peculiarity of channel current filamentation is the shape of stability boundary $I_{CHcr}(U_{DScr})$ (Fig. 3.22). In fact, from (3.37) it follows that the voltage U_{DScr} has a minimum $U_{cr\,min}$ at $I_{CH} \to I_{CH}^{*}$ $I_{CH} \to 0$.

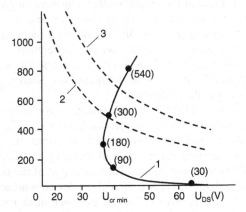

Fig. 3.22 Calculated boundaries of stationary stability: "1," for channel current filamentation (in parentheses T_{max} in °C is presented); "2," for thermal breakdown at maximum injection; "3," for thermal breakdown without injection

It can be demonstrated that $U_{cr\,min} \sim T_0/(R_T I_{CH}^*)$. Therefore, for a number of MOSFET types with relatively low I_{CH}^* values and low breakdown voltages U_{DSBR}, the boundary of possible thermal instability of channel current is shifted over the boundary of avalanche breakdown. Thus, the thermal filamentation mechanism is not realized due to physical limitation by the electrical filamentation. On the contrary, in power MOSFETs at the application circuits $U_{cr\,min}$ might be a few times lower than U_{DSBR}.

Apparently, increase of the thermal resistance value R_T due to a bad heat sink or emptiness in solder at die attachment may result in reduction of $U_{cr\,min}$ followed by corresponding reduction in critical power for filamentation.

Another basic defect that may decrease the thermal stability of MOSFETs is inhomogeneity of U_{boun} within the structure area and scattering in U_{boun} value between the chips in multiple die devices. If some defect region with low U_{boun} is presented, the initial current redistribution is observed. This may result in local overheating and hot spot formation at significantly lower values of dissipated power. Technological scattering of U_{boun} between chips in multichip power MOSFETs is one of the major reasons for chip overheating. The higher the transistor transconductance is, the lower the voltage scattering ΔU_{boun} for corresponding significant reduction in P_{cr}.

Obviously, to provide a lower level of limitation of transistor capability, the value U_{boun} must be provided within some given range either by yield and spec improvements or chip selection procedure.

3.3 Thermal Breakdown

A sharp increase of transistor structure temperature accompanied by an increase in total current and a decrease in the output voltage represents a typical scenario of thermal breakdown. Physical mechanism of this phenomenon is based on the thermal instability of thermogeneration current flowing through the reverse-biased p-n junctions or other regions with an intensive heat generation.

Thermal breakdown in its "pure" mode rather infrequently occurs in properly designed devices. Thermal emitter current filamentation (Section 3.1) usually precedes and dominates in bipolar transistors.

Both in bipolar and in field-effect microwave transistor structures, the thermal breakdown is accompanied by various injection processes. These processes enhance and amplify the breakdown to some extent. However, the basis of thermal breakdown is always the same: an increase of thermogeneration currents that is exponentially dependent on temperature. Therefore, one of the typical classifications of the thermal breakdown scenario for a given device is an actual dependence of the critical regime on temperature. Typical evidence of thermal breakdown is reduction of critical dissipated power with the heat sink temperature increase.

In several studies, thermal breakdown is called a "second(ary) breakdown" due to typical voltage decrease and current increase observed in earlier studies [31]. This effect was secondary with respect to the "primary" current increase due to, apparently, avalanche multiplication effect. The latest studies, however, demonstrated that the "secondary breakdown" can have not only a thermal mode but an electrical mode (current mode second breakdown) too. Furthermore, it cannot have a "secondary" feature at all. For example, in the case of pulsed regime due to multiplication of the displacement current, the "secondary breakdown" can occur without any primary breakdown features. Therefore, the term "secondary breakdown" is not considered useful for the focus of this book that involves more particular classification.

Thermal breakdown is a direct failure mechanism. During the thermal breakdown evolution, the temperature in local regions can reach the critical values of contact metallization melting or even the critical temperature for the semiconductor material itself. An uncontrollable current increase at thermal breakdown can result in breakage of the bonding wires due to melting them like a fuse by its own Joule heating power.

Suitable temperature level for thermal breakdown evolution in silicon devices is usually achieved at 300–400°C. Due to wider band gap, the same level for GaAs MESFETs is 500–550°C. For the same reason for SiC and GaN, this parameter is significantly higher, thus making them very attractive for specific applications with high ambient temperature such as automotive and nuclear energy applications.

The thermal breakdown corresponds to classical current instability mechanisms with typical S-shape $I–V$ characteristic.

According to the general understanding of thermal breakdown (Chapter 2), it can evolve either uniformly, for example in structures with strong thermal feedback, or can be accompanied by a current localization in distributed structures at an adequate differential resistance value $R_D < 0$. A quasi-uniform thermal breakdown is usually observed in low-power devices due to comparable linear dimensions of the device structure and the chip thickness. In contrast, in power transistor structures the current filamentation is typical for the thermal breakdown. In BJT, thermal breakdown usually develops locally in a hot spot region that is formed due to preliminary emitter current instability and redistribution. At the same time, destruction of the transistor is visually similar for both the "pure" thermal breakdown and the emitter current filamentation that preceded the final thermal breakdown (similar to for example Fig. 1.1a).

The delay time for thermal breakdown is mainly determined by the time of structure overheating up to the critical temperature level. This time depends on the chip dimensions and the level P > Pcr. It can be varied in a wide range (from microseconds up to seconds). The time of thermal breakdown evolution itself, i.e., the specific process of mutual current increase and voltage decrease, is usually a few orders of magnitude lower. Therefore, application of a protection circuit against thermal breakdown presents a challenge.

3.3.1 Thermal Breakdown in Bipolar Transistors

In the uniform transistor structure model, the temperature distribution is described by one-dimensional equation (2.7); the current density j_C and j_E through the collector and emitter junctions are given by the corresponding Ebers-Moll model [36,38]:

$$j_E = j_{SE}\left(\exp\frac{qV_E}{kT} - 1\right) + \alpha_2 M j_{SC},\tag{3.38}$$

$$j_C = \alpha_1 M j_{SE}\left(\exp\frac{qV_E}{kT} - 1\right) + M j_{SC},\tag{3.39}$$

where V_E is the voltage drop on emitter junction; j_{SE} and j_{SC} are the thermal current densities through the emitter and collector p-n junction: $j_{SE}, j_{SC} \sim \exp\left(\dfrac{E_G}{kT}\right)$; α_1 and α_2 are the direct and indirect current gains of transistor; M is the multiplication coefficient of carriers in collector junction.

In the simplest case the continuity equation for base volume element is given by

$$j_E - j_C - j_B = 0.\tag{3.40}$$

As known, the instability criterion for relatively uniform temperature and current disturbances is given by

$$R_T U \frac{\partial j_C}{\partial T} - 1 > 0,\tag{3.41}$$

where U is the voltage drop on collector p-n junction ($U \approx U_{CE}$ in collector-emitter circuit and $U \approx U_{CB}$ in collector-base circuit).

Instability criterion for relatively longwave disturbances $\delta T(x) \sim \cos(\pi x/L)$ is similar to (3.41):

$$\widetilde{R}_T U \frac{\partial j_C}{\partial T} - 1 > 0,\tag{3.42}$$

where $\widetilde{R}_T = R_T\left/\left(1 + \dfrac{\pi^2 l_T^2}{L^2}\right)\right.$.

It is important that only in the case of $j_C = j_C(T)$ (i.e., $\dfrac{\partial j_C}{\partial T} = \dfrac{d j_C}{dT}$), the instability criterion for uniform disturbances (3.41) corresponds to appearance of NDC region of the *I–V* characteristic: $\dfrac{dI}{dU} \leq 0$. In general, current filamentation is already possible when $\left|\dfrac{dI}{dU}\right|$ reaches the value to satisfy the criterion (3.42).

In the case of bipolar transistor structure, $j_C = j_C(V_E, T)$ and the condition $R_D = dI_C/dU < 0$ can be used as a criterion of thermal instability only at $V_E = \text{const}$, i.e., if the transistor operates in the regime of input voltage source. However, typical application condition for BJT is the current regime: $I_E = \text{const}$. In this case, j_C in principle depends on two parameters: T and V_E. These parameters are controlling the phenomenon of emitter current filamentation at $dI_C/dU > 0$ before thermal breakdown.

Combining (3.38) and (3.39) j_C is given by

$$j_C = \alpha_1 M j_E + j_{SC} M(1 - \alpha_1 \alpha_2 M). \tag{3.43}$$

In Section 3.1, j_{SC} was neglected in comparison with j_E for criterion (3.12) for the case of emitter current filamentation. Now, if the thermal current j_{SC} is significantly higher than j_E, the instability criterion (3.42) has a simple form:

$$\tilde{R}_T U_{CB} M(1 - \alpha_1 \alpha_2 M)\frac{\partial j_{SC}}{\partial T} - 1 > 0. \tag{3.44}$$

If the condition $j_{SC} \gg j_E$ is not true, criterion (3.44) becomes true only at $j_E = \text{const}$ and $\dfrac{\partial j_E}{\partial T} \ll \dfrac{\partial j_{SC}}{\partial T}$. Thus, emitter current distribution must remain unchanged with the chip temperature increase. Then criterion (3.44) becomes true before criterion (3.12). This is theoretically possible for example in a highly stabilized structure such as with high r_E.

However, in practical cases the emitter current filamentation always precedes thermal breakdown. Apparently, in this case the model of uniform structure is incorrect (due to break of the uniformity in initial temperature distribution), and, therefore, criterion (3.44) is not true. A rough estimation of the thermal breakdown regime can be obtained if the temperature in hot spot Tmax is known [the exact dependence $T_{max}(U_{CB})$]. Then the corresponding value T_{max} and the corrected value $\tilde{R}_T$ must be substituted into (3.44), since real structure dimension L is practically equal to the thermal filament dimension l_T^{fil}. If the hot spot is so narrow that $l_T^{fil} < W$, the correction of typical thermal dimension l_T is required.

The criterion of thermal breakdown in bipolar transistors depends on the test circuit. Criterion (3.44) apparently corresponds to CB circuits. For CE circuits (base current regime),

$$j_C = \frac{\alpha_1 M j_B}{1 - \alpha_1 M} + \frac{M(1 - \alpha_1 \alpha_2 M) j_{SC}}{1 - \alpha_1 M}. \tag{3.45}$$

Then the corresponding criterion of thermal breakdown is

$$\widetilde{R}_T U_{CE} M \left(1 - \alpha_1 \alpha_2 M\right)\left(1 - \alpha_1 M\right)^{-1} \frac{\partial j_{SC}}{\partial T} - 1 > 0 . \tag{3.46}$$

It is quite easy to see that the critical voltage for thermal breakdown in CB circuits is $\left(1 - \alpha_1 M\right)^{-1}$ times higher than the critical voltage in CE circuits.

As was mentioned above, thermal breakdown is a classical type of instability that is accompanied by an NDC region of S-shape I–V characteristic. Observation of such I–V characteristic under DC conditions presents difficulties and is practically impossible in power transistors due to the structure melting after the switching into the filament state. On the contrary, pulsed measurements might provide reversible measurements. However, in this case the shape of I–V characteristic depends on the pulse duration.

Quasi-static measurements of special structures with refractory metallization have provided some understanding about S-shape I–V characteristic at thermal breakdown (Fig. 3.23a,b). In CE circuit (Fig. 3.23a) just before thermal breakdown, a smooth current increase is observed. Such an increase was qualified as a "primary" breakdown in earlier studies. The cause of this increase is injection from emitter p-n junction. On approach of U_{CE} to U_α, the injection is increased, where U_α is the breakdown voltage in CE circuit for the condition $\alpha_1 M(U_\alpha) = 1$.

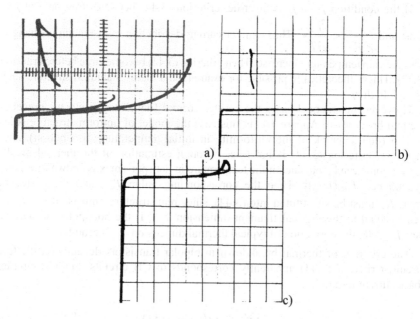

Fig. 3.23 Quasi-stationary I–V characteristics at thermal breakdown: CE circuit (a), CB circuit (b), and the "loop" at breakdown (c). (The current axis is vertical, the voltage is harizontal)

A heat-injection current component is added to the avalanche-injection component. The heat-injection component represents multiplication of thermally generated current. The component physically represents the result of $j_{SC}(T)$ increase that is amplified $(1 - \alpha_1) M^{-1}$ times. Thus, the critical voltage of thermal breakdown U_{CEcr} according to (3.46) is the result of cooperative action of both the thermal and the injection processes.

The injection process at thermal breakdown evolution in CB circuit is not as important as in CE circuit. Therefore, just before thermal breakdown a visual current increase might not be detected from the I–V characteristic. Since for the CB circuit the critical voltage U_{CBcr} is always higher than U_{CEcr}, the probability of emitter current filamentation exists before completion of the criterion for pure thermal breakdown (3.46). In this case, during hot spot formation a local thermal breakdown can occur followed by the switching into the filament region of the I–V characteristic with positive differential conductivity. This switching can be observed from the region without any "precursors"(Fig. 3.23b).

From measurements of I–V characteristic using a tester with 200-Hz scanning frequency, the typical sign of thermal breakdown beginning is appearance of a loop on the characteristics (Fig. 3.23c). Observed noncoincidence of the direct and reverse scanning path is the result of inertia in heating and cooling processes. It is possible since the typical times of these processes are comparable with the scanning period of the characteriograph. However, the loop is not always a direct consequence of thermal breakdown. Similar loops can be formed for example as a consequence of deep traps recharging at avalanche breakdown.

A detailed picture of thermal breakdown is quite complex due to dependence of M and α on T, j_C, U. In criteria (3.44) and (3.46), this fact is not taken into account. Besides, the "jumps" of current density (total and local) during thermal breakdown can cause development of other instability mechanisms, for example isothermal avalanche injection. That is why it is hard to find differences between the thermal and injection physical mechanisms of filament formation only from the S-shape I–V characteristics (Fig. 3.23).

Superposition of various factors can produce quite an unusual view of I–V characteristics near the critical point. This is demonstrated by experimental I–V characteristic (Fig. 3.24) for thermal breakdown in CE circuit. In region "AB" the current increase is the result of emitter injection.

Near point "B", the critical regime of thermal filamentation of emitter current is reached and in transistor structure the hot spot is formed. Inside the spot, the temperature and current density are quite high. Due to this effect, both the gain and the multiplication coefficient are decreased thus providing total collector current decrease. This results in formation of the N-shape region instead of the expected S-shape.

On measurement in given current regime (for stable measurement of NDR region of S-shape characteristic), the switching is observed from point "B" to point "C." In this case, the region of I–V characteristic is formed due to current increase j_{SC} just before thermal breakdown. However, often, the transition "B→C" ends up in the over-critical area. In this case, the transistor from point "B" turns on into the thermal breakdown mode followed by irreversible failure.

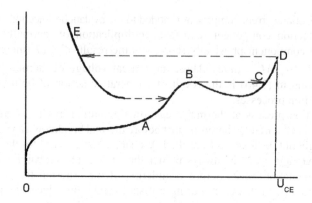

Fig. 3.24 N-shape and S-shape I–V characteristics

Due to complexity of experiments for thermal breakdown, numerical simulation is an important research tool. On the basis of the model (Section 3.1) for uniform structure in [44] DC I–V characteristic and kinetics of "pure" thermal breakdown were studied. Expression for j_C was assumed in the simple form

$$j_C = \alpha j_E + B \exp\left(-\frac{E_G}{kT}\right), \tag{3.47}$$

where j_E = const, and α, B are constants independent of temperature and current density.

In the simulated S-shape I–V characteristic, the "depth" of S-shape region provides evidence that isothermal mechanisms contribute to the thermal breakdown process.

According to simulation, the filament dimension at thermal breakdown is a few micrometers. Filament dimension for this "delta-shape" heat source slightly depends on chip geometry and can be less than a few micrometers. The maximum temperature in the filament exceeds the temperature of metallization melting.

Filamentation process by numerical simulation [44] under the condition of single power pulse can be subdivided into three stages. The first stage is a uniform heating at $0 < t < \tau_{cr}$ up to T_{cr} on criterion (3.44). The second is an increase of small nonuniform disturbances δT and δj_{CB0} at $\tau_{cr} < t < \tau_D$. The disturbances are mostly caused by design inhomogeneities and various structure defects. Increase of disturbances is observed almost without any observable changes in the external characteristics, i.e., the values T and j_{CB0} are practically unchanged in the second stage. Finally, in the latest stage ($\tau_D < t < \tau_{fil}$), a macroscopic redistribution of j_{CB0} and T is accompanied by U_{CB} reduction. This stage is finalized at $t > \tau_{fil}$ by a final stationary filament.

In order to obtain filamentation at thermal breakdown, the pulse duration t_P must exceed τ_D. The value depends both on the operation regime (U_{CB}, I_C) and on the presence of design and technological defects in the structure. The interval ($\tau_D - \tau_{cr}$)

is the most "sensitive" to these factors. It can degenerate down to zero at adequate overvoltage level ($U_{CB} - U_{cr}$) and defect amplitude.

In this case, the delay time is practically equal to the heating time up to T_{cr}. The interval ($\tau_{fil} - \tau_D$) of current and temperature redistribution can be less than 1 μs according to simulation results. This fact is confirmed experimentally and explains the speed of catastrophic changes at thermal breakdown. Obviously, the requirements for the protection circuit must be rather high.

In concluding this section, save operation area (SOA) of bipolar transistors must be discussed again. This is done so as to generalize the thermal instability mechanism in Sections 3.1 and 3.3 as a physical limitation factor.

Typical view of SOA is presented in Fig. 3.25 [68]. In addition to conventional limitations for current and voltage SOA, it now takes into account the limitation of maximum junction temperature (region 1). This part of the boundary presents itself as a part of line $R_T UI = T_{max} - T_0$ on double logarithmic scale.

Region 2 presents the boundary of emitter current thermal filamentation; for small currents it coincides with the boundary of thermal breakdown (dashed line, Fig. 3.25). The shaded region is the area of stable hot spot existence. At the right, this area is limited by the condition for local thermal breakdown (region 4).

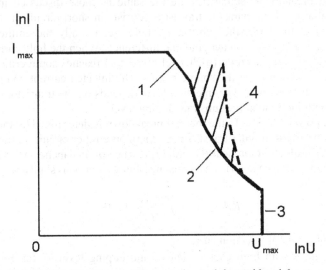

Fig. 3.25 SOA at thermal filamentation and thermal breakdown

3.3.2 Thermal Breakdown in MOSFETs

Experimental case studies and application of MOSFETs demonstrate significantly higher thermal stability of MOSFETs in comparison with power bipolar transistors of similar size and power. This fact is related to negative temperature coefficient of the drain current for a majority of MOSFET devices.

An exception to this rule is presented by V-recess MOSFET architecture (Section 3.2) that provides the positive TCDC in a wide drain current range. The thermal filamentation phenomenon is not as dangerous as in bipolar transistors due to significant dumping of I_D on approaching the thermostable value $I_D{}^*$. In the current range $I_D - I_D{}^*$ the thermal filamentation can take place only due to thermal breakdown.

Thermal breakdown as a failure mechanism can be observed in MOSFETs in the region of relatively high I_D values. The area of median and low I_D is limited usually by isothermal instability mechanisms (especially in n-channel transistors). These mechanisms will be discussed in the next chapter.

Experimentally measured boundary of thermal stability [65] exhibits the linear dependence on critical power that is typical for thermal mechanism.

In MOSFETs, the region of intensive heat generation is the subsurface space charge region of the drain junction. The channel current is collected in this region. Thermal breakdown begins in the structure when heating by the channel current reaches some critical value. Then further uncontrollable increase of saturation current j_{DS0} is provided through the drain-substrate junction according to $j_{DS0} \sim \exp(-E_G/kT)$.

Since any MOSFET structure includes so-called "parasitic" bipolar transistor, the thermal breakdown regularities are the same as those discussed in Section 3.3.1. Peculiarity of the parasitic transistor consists in short circuit of its emitter (the source) to the base (the substrate). Therefore, initially the emitter (drain) junction is closed by the contact potential difference. When the hole current component of j_{DS0} in the substrate (n-channel transistor) reaches some critical value, the electron injection from the source begins. The injected current contributes to the total structure heating. The injection level depends on the transistor design. It decreases with increase in the substrate doping level.

Thus, in power MOSFETs the thermal breakdown is determined by the thermal injection similarly to bipolar transistors in common-emitter circuit. In the absence of avalanche multiplication near the drain (presence of avalanche can excite electrical instability; see Chapter 4), the thermal instability criterion is similar to (3.44):

$$\tilde{R}_T U_{DS}(1-\alpha)\frac{\partial j_{DS0}}{\partial T} - 1 > 0, \qquad (3.48)$$

where α is the gain of injection current.

For structures with high channel length and doping level, α can be close to zero, i.e., all injected electrons will recombine in the substrate region. In this case, U_{DScr} is maximal similarly to bipolar transistor in CB circuit. The higher the gain of the parasitic transistor, the lower the U_{DScr} and the sooner thermal breakdown begins.

Thermal breakdown simulation [67] in the current mode demonstrates that during the instability process, the channel current j_{CH} can be reduced to zero. In this case, the whole transistor current can be provided by the pinched area of narrow filament formed by the saturation current of drain junction.

Both the critical value of dissipated power P_{DIS} and the corresponding value of critical temperature T_{cr} depend on the value j_{CH} and the substrate resistance r_{SUB}. Calculations demonstrate that j_{CH} increase results in increase in P_{BR} and T_{cr}, but increase of T_0 provides a linear reduction of these parameters. Increase of r_{SUB} is equivalent to increase of α. This effect shifts the stability boundary in the direction of highest U_{DS} (Fig. 3.22, curves 2,3).

Numerical simulation of thermal breakdown kinetics [67] has demonstrated that with slight increase of pulsed power P over P_{BR} there is a time interval $0 < t < \tau_D$ of practically uniform current density distribution. Subsequent macroscopic redistribution occurs in less than a microsecond. Current and temperature redistribution are accompanied by the same sharp and fast decrease of U_{DS} (Fig. 3.26). Finally, U_{DS} slightly increases and redistribution is formalized in accordance with the median temperature T_M reduction (curves 2, 3). During this process, the maximum temperature growth is observed (curve 1). Final stationary temperature and current filament are extremely narrow: its typical dimension cannot exceed a few micrometers independently of linear dimensions and structure thickness. The total current increase results in maximum temperature increase. During thermal breakdown process, the temperature exceeds values of structure melting at rather moderate average current densities (< 100 A/cm^2).

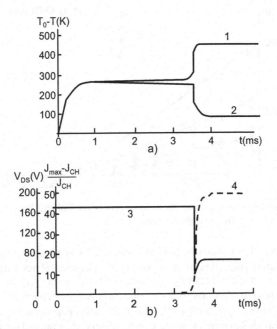

Fig. 3.26 Thermal breakdown kinetics of "parasitic" bipolar transistor: 1, $T_{max}-T_0$; 2, T_M-T_0; 3, $U_{DS}(t)$; 4, $(j_{max}-j_{CH})/j_{CH}$ ($j_{CH} = 100$ A/cm^2, $L = 240$ µm)

3.3.3 Thermal Breakdown in GaAs MESFETs

Gallium arsenide has a larger band gap than silicon. In power transistors with good thermal resistance, the thermal generation of electron–hole pairs is low. Therefore, as usual mostly failures of electrical nature are experimentally observed.

However, under certain conditions (particularly at high R_T) thermal breakdown can occur too [69–72] due to high overheating in on-state operation. Measurements of safe operation area demonstrate that the critical power of MESFETs in the DC regime is preserved in some current and voltage range. With ambient temperature increase, the critical power is linearly reduced. Linear dependence $P_{cr}(T_0)$ was observed in [72] for samples of various topology (Fig. 3.27). Extrapolation of these dependencies provides the same critical temperature $\approx 550°C$. Simple estimation demonstrates [69] that the semi-insulating substrate plays a major role at thermal breakdown in MESFETs. In fact, at 500–550°C the electron density due to heat generation in GaAs layer reaches $2 \times 10^{15} cm^{-3}$. This corresponds to surface concentration of $\sim 10^{12}$ cm^{-2} even in the case of rather thin 50-μm substrate. At the same time, the density of free electrons in thin active n-epi channel layer (~ 0.2 μm) has the same value $\sim 2 \times 10^{12}$ cm^{-2}.

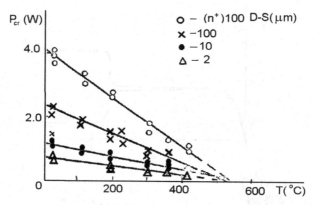

Fig. 3.27 Dependence of the thermal breakdown power on ambient heat sink temperature for GaAs MESFETs with different drain source spacing

Thus, the thermal breakdown in GaAs MESFETs is provided by the thermal breakdown of buffer and substrate regions. The experimental activation energy E_A can be obtained by the conductivity measurements $\sigma \sim \exp(-E_A/kT)$. It is equal to 0.75–0.85 eV. This corresponds to both ionization energy typical for compensated GaAs deep centers ($E_A = 0.75$ eV) and half of the band gap value ($E_A/2 = 0.7$ eV).

The thermal breakdown mechanism can be observed in pulsed regime [70] with a pulse duration of a few microseconds. This is confirmed by the dependence of power on ambient temperature (Fig. 3.28).

Typical I–V characteristics of such structures (Fig. 3.29a) exhibit a considerable current increase in prebreakdown area with $R_D \approx 0$. After the switching into

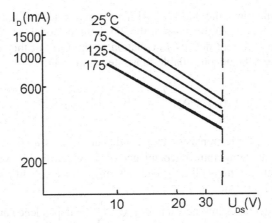

Fig. 3.28 SOA boundaries on double logarithmic scale for pulse regime at various ambient temperatures

high current mode, the failure is usually observed [73]. Self-heating in the structure results in appearance of N-shape region of *I–V* characteristic after primary current saturation at small $|U_{GS}|$ values (Fig. 3.29a). Decrease of I_D is caused by the falling dependence of electron drift velocity on the lattice temperature. With exponential increase of the additional substrate current with temperature. the polarity of $R_D = dU_{DS}/dI_D$ is changed and approaches zero.

Is the critical temperature of thermal breakdown in real GaAs MESFETs always close to 500°C? In [73] it was reported that the critical temperature is 350°C. The reason for such behavior was not given in [73]. A few assumptions can be expressed: (1) presence of deep levels with $E_A < 0.8$ eV and corresponding beginning of thermogeneration at low temperature; (2) melting of the drain and source eutectic at the sites of local overheating; (3) excitation of a new injection instability mechanism by thermal generation.

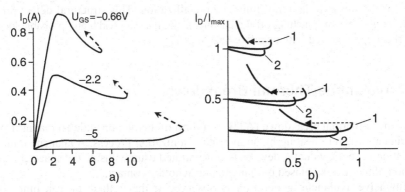

Fig. 3.29 Experimental (a) and simulated (b) *I–V* characteristics of GaAs MESFET at thermal breakdown [1, values v and κ constant; 2, with consideration of $v(T)$ and $\kappa(T)$]

Results of numerical simulations [73] mainly replicated the reality of thermal breakdown mechanism on the basis of the 2D thermal model [see (3.15), Section 3.1) and the drain current density j_D equation in the form of the sum of channel component j_{CH} and the uncontrollable substrate component:

$$j_C = j_{CH} + j_{BUF} = S_0 \left(\frac{T_0}{T} \right)^m \phi(U_{GS}) + j_{BUF}^0 \exp\left(-\frac{E_A}{kT} \right), \qquad (3.49)$$

where S_0 is the specific transistor transconductance, m is a constant coefficient that determines the temperature dependence of the electron drift velocity, $\phi(U_{GS})$ is a function that determines the dependence of j_{CH} on U_{GS}, and j_{BUF}^0 is a coefficient that depends on U_{DS}.

The final model takes into account $\kappa = \kappa_0 \left(\frac{T_0}{T} \right)^s$ dependence and possible inhomogeneities that were simulated by changing of j_{BUF}^0 in limits of segment ν.

The I–V characteristics (Fig. 3.29b) were calculated for typical numerical values of the parameters: $L = 960$ μm, $W = 100$ μm, $T_0 = 300$ K, $\kappa_0 = 0.45$, $m = 0.5$, $s = 1.25$, $\nu = 0.1L$ (edge inhomogeneity).

Simulation demonstrated (Fig. 3.29b) that taking into account the dependence $\kappa(T)$ reduced the level of critical power and the dependence $\nu(T)$ is the reason for N-shape formation.

The dependence $\kappa(T)$ results in significant increase of T_{max} in thermal filament in comparison with the case $\kappa = \kappa_0 = $ const. In this case, practically immediately after voltage switching T_{max} exceeds the temperature of GaAs-Au-Ge eutectic melting. The melting provides a direct reason for catastrophic failure [73].

There are several reasons for increase in real R_T in comparison with the calculated value. Among them is the inhomogeneity in current distribution. Significant inhomogeneity of current distribution in GaAs MESFETs apparently is caused both by design peculiarities and by technological defects. For example, one of the structural defects in large total gate width devices is local absence of the gate regions due to break or exfoliation of thin gate metallization. This structural defect provides a nonuniform pinching off the channel along the gate width and probable local overheating increased with negative bias increase.

3.4 Avalanche-Thermal Breakdown

Avalanche-thermal breakdown (ATB) is a form of thermal breakdown that utilizes a primary structure heating by an avalanche ionization process. Avalanche multiplication in semiconductor devices is concentrated usually inside the space charge region of the reverse-biased p-n junction or Schottky contact.

Intensive avalanche generation is observed if the voltage on p-n junction approaches close to some breakdown value U_{BR}. An important characteristic of avalanche breakdown is the multiplication coefficient M that is given by a simple empirical expression [38]:

$$M = \frac{1}{1 - (U/U_{BR})^n},$$ (3.50)

where n is some constant coefficient ($n > 2$).

The avalanche multiplication current j_A, through the reverse-biased p-n junction, is given by

$$j_A = Mj_S,$$ (3.51)

where j_S is the saturation current of p-n junction.

Since by definition $M = \infty$ at $U = U_{BR}$, the critical level j_A is always achieved at $U < U_{BR}$. With the structure self-heating increase, the component $j_S(T)$ in the total current is increased. In turn, the role of factor M is decreased since M usually decreases with the increase in temperature. In [74] the hypothesis was suggested that the thermal generation of carriers totally replaces the avalanche generation under conditions of developed ATB, especially in the case of current filamentation.

A simple instability boundary is given by (3.41):

$$R_T U j_A \frac{\partial}{\partial T} \left(\ln M + \ln j_S \right) - 1 = 0.$$ (3.52)

From the experimental results for ATB, as well as for the case of "pure" thermal breakdown, the constant critical temperature T_{cr} of instability existence is demonstrated, while the I–V characteristic exhibits the NDC region [74, 75]. From criterion (3.52), it does not follow with the same simplicity as for criterion (2.21), if heating is provided by an increase in ambient temperature.

Constant T_{cr} is adequately clear in the case when the heat source current only slightly depends on the temperature. This condition is realized for example for channel current of field-effect transistors. In the discussed case, j_A depends on T and U, therefore a fraction of "heat" component is impossible.

However, in reality at relatively low temperature the major impact on current increase is provided by the dependence $M(U)$. Sharp dependence $j_A(T)$ acts in a rather narrow temperature range $T \leq T_{cr}$ due to the exponential dependence $j_S \sim \exp(-E_G/kT)$. In this region of I–V characteristic, U is practically unchanged ($U \leq U_{BR}$). Therefore, in (3.52) the dependence on T dominates and results in the constant T_{cr}. This finding has been demonstrated in [76].

What is the significance of ATB as a failure mechanism? In transistor structures, it is realized rather rarely. This is the case because the avalanche breakdown of collector (drain) p-n junction usually results in avalanche injection and subsequent isothermal current instability. These effects occur before the level of critical temperature for ATB. However, under conditions where the output voltage is close to the breakdown voltage value, the higher ambient temperature is, the higher theATB probability. For GaAs MESFETs (and Si field-effect transistors controlled by p-n junction—JFETs), the ATB evolution is possible at Schottky gate breakdown. It is clear that for various diode structures (for example, IMPATT diodes), ATB is the major catastrophic failure and power limitation physical mechanism.

A classical experimental study of ATB was made in [74] employing diode structures (Fig. 3.30a). A very small active area of p-n junction in these structures produced a stable and reversible current filament in the region of S-shape I–V characteristics.

Using comparative analysis in different operation regimes of DC I–V characteristic of avalanche diode (Fig. 3.30b), the authors [74] fixed in photographs the stages of filament formation and evolution at ATB. The region "AB" corresponds to classical avalanche breakdown of reverse-biased p-n junction with the increasing resistance on increase in j_A [76]. The current increase approaching point "B" is accompanied by an increase of visible light emission that is typical for avalanche ionization in silicon. However, the observed light intensity increase was less than the current increase. This provided experimental evidence that part of the current is not provided by the avalanche generation.

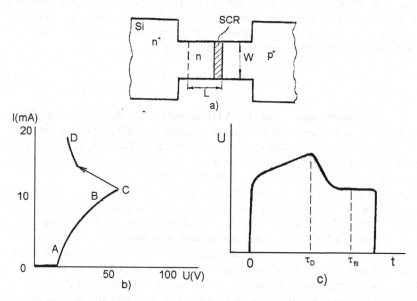

Fig. 3.30 Avalanche-thermal breakdown in p$^+$-n-n$^+$ diode: (a) structure (25 μm $< W <$ 75 μm; 2 μm $< L <$ 300 μm); (b) DC I–V characteristic; (c) dependence $U(t)$ in pulse

In the region "BC," the total level of avalanche light emission decreases sharply. Within the boundaries of p-n junction, the region of a dark spot with dumped avalanche luminescence is formed. Usually the spot is localized in the center of p-n junction with maximum temperature. At this site the filament (hot spot) is finally formed.

In the NDC region "CD," the subsequent dumping of avalanche electroluminescence and appearance of the red or infrared thermal light emission from the filament region was observed. Light emission spectrum measurements confirmed the thermal origin of ATB filament.

Measurement of $I–V$ characteristic of the diodes at various temperatures demonstrated that total reduction of the avalanche multiplication was at temperatures 300–330°C. The saturation current of p-n junction is practically totally provided by the carrier thermogeneration.

Considerable interest concerns the pulse measurements of ATB using a stroboscopic setup. The dependence of the ATB delay time on the pulse energy and median value of pulse current was studied. The delay time τ_D is defined by the time up to a falling region in the dependence $U(t)$ (Fig. 3.30c), where U is the voltage on the diode in the fixed current conditions. Increase of U in the interval $0 < t < \tau_D$ is the result of p-n junction heating and corresponding multiplication coefficient reduction (increase of U_{BR}).

In [74], the limits of delay time have been obtained for the transition from avalanche generation along the space charge region to thermal generation in hot spot. According to the measurements, the transition requires less than 100 ns.

For high overvoltage ($U \gg U_{BR}$), the τ_D is below 100 ns (up to sub-nanosecond range) and the obtained dependence is

$$P\sqrt{\tau_D} = \text{const},\qquad (3.53)$$

where P is the median power in pulse.

Constant $P\sqrt{\tau_D}$ is maintained in a wider range $\tau_D < \tau_T$, where τ_T is the structure thermal relaxation time. In the experiment, $\tau_T = 5$ µs and epitaxial film thickness of the diode did not exceed 3 µm.

Moving thermal filament inside high-resistance n-region from p-n junction (where it is initially formed) up to n^+-diffusion region of ohmic contact was observed. The effect occupied the major part of interval $\tau_D < t < \tau_T$. The velocity of movement is determined [74] by the rate of heating of the surrounding n-region. The occurrence of filament propagation in n-region was proved in step-by-step photographs. It was stated that spontaneous filament propagation takes place if T_{max} in hot spot exceeds an intrinsic temperature T_i of the semiconductor. Optical measurements show that T_{max} is usually within the limits of 700–1000°C. Heavily doped n^+-region stops the filament propagation since its T_i value exceeds 1000°C.

If T_{max} slightly exceeds T_i for n-region, then only partial filament penetration inside the structure is observed. If $T_{max} < T_i$, then a stable filament is formed near the p-n junction.

The results [74] have additional value since in real transistors with high p-n junction area in the DC regime ATB always results in structure damage and thus no opportunity for experimental being provided.

3.5 Peculiarities of Avalanche Thermal Instability in JFETs and MESFETs

In JFETs and Schottky gate GaAs MESFETs, the generation of electron–hole pairs of both thermal and avalanche nature results in appearance of the gate leakage current I_G. This leakage current creates an additional gate-source voltage drop on the external resistor in the gate circuit R_G; the current $I_G \Delta I_{GS} = I_G R_G$. Thus, at fixed gate voltage source Σ_G, the gate bias in p-n junction of Schottky contact can be decreased. With the U_{DS} increase, the gate bias decreases and the channel current increases I_{CH}. Under certain conditions, this produces an S-shape region in the output drain-source I–V characteristic. This phenomenon does not correspond to any intrinsic S-shape dependence [77]. Considering the case when the gate leakage is of thermal nature, the equation system is

$$\begin{cases} I_D = I_{CH} + I_G(T) \\ T = T_0 + I_D U_{DS} R_T, \\ \dfrac{dI_{CH}}{dI_G} = SR_G \end{cases} \tag{3.54}$$

where I_D is the drain current, R_T is the thermal resistance, and the last equation follows from transconductance definition $S = \dfrac{dI_{CH}}{dU_G}$. It is easy to see that the S-shape region ($\dfrac{dU}{dI} \leq 0$) in the output I–V characteristic appears at

$$(1 + SR_G)\dfrac{dI_G}{dT} U_{DS} R_T \geq 1. \tag{3.55}$$

Taking into account $I_G(T) \sim \exp\left(-\dfrac{E_G}{kT}\right)$, the condition becomes

$$(1 + SR)_G \, I_G(T)\left(\dfrac{E_G}{kT^2}\right) R_T U_{DS} \geq 1.$$

This condition practically corresponds to reaching in the device structure of some critical temperature that decreases with increased R_G.

Similar analysis can be made for the I_G increase due to avalanche ionization intensity increase. For GaAs MESFETs, the equation system is given by

$$\begin{cases} I_D = MI_{CH} \\ I_G = \gamma_P (M-1) I_{CH} \\ U_{GS} = \Sigma_G - I_G R_G \\ dI_{CH} \big/ dU_{GS} = S \end{cases}, \tag{3.56}$$

where M is the multiplication coefficient and γ_P is the portion of holes in SCR of the Schottky gate.

In GaAs MESFETs, U_{BR} in the general case is an increasing function of I_{CH}, since with I_{CH} increase the electric field strength in drain domain is increased too (Section 5.1). Considering this, it can be stated that $\left(dI_D \big/ dU_{DS} \right)^{-1} \leq 0$, i.e., S-shape region appears in output I–V characteristics at completion of the following condition:

$$M \geq \frac{1}{\gamma_P S R_G} + 1 - I_D \frac{d \ln M}{d \ln I_{CH}}. \tag{3.57}$$

At slight $M(I_{CH})$ dependence, the boundary of S-shape must be close to vertical:

$$U_{DS} = U_{cr} = \text{const}; M(U_{cr}) = \frac{1}{\gamma_P S R_G} + 1. \tag{3.58}$$

Since the dependence $M(I_{CH})$ cannot be neglected, the critical voltage U_{cr} will decrease with increased I_D. Increase of R_G shifts the boundary to the left.

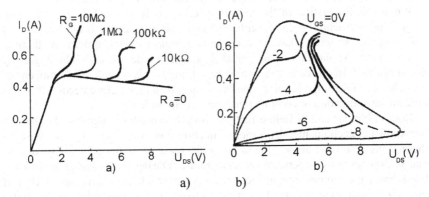

a)   b)

Fig. 3.31 I–V characteristics of GaAs MESFET: (a) at various resistance values in the gate circuit; (b) at various U_{GS} values ($R_G = 330$ kΩ)

In an experimental study [77] of GaAs MESFETs having a total gate width of 5.4 mm, it was stated that increase of R_G indeed results in reduction of U_{cr} of S-shape region formation (Fig. 3.31a). Relative effect of R_G on U_{cr} is significantly higher at small R_G and near the pinch-off condition. The transition of transistor in the open state can be realized by a jump similarly to the instabilities discussed in previous sections. However, it is clear that no filamentation process is observed. The shape of the stability boundary (dashed line in Fig. 3.31b) is evidence that the leakage current hole generation is of complex nature. At small current level, obviously, the avalanche generation mechanism dominates. At high currents, the decreased U_{cr} is provided by the thermal generation.

3.6 Summary

In this chapter, the major regularities and basic mechanisms for thermal and electrothermal filamentation are discussed.

Physical interpretation enables the creation of corresponding physical and mathematical tools for particular analysis. The analysis of different types of devices demonstrates that in most real cases, multiple forms of similar mechanisms rather than multiple unique mechanisms are involved. These multiple forms have design and technological peculiarities that are related to multiple design and technological peculiarities of real devices.

Thermal filamentation and thermal breakdown are most important in distributed power transistor structures and are accompanied in some part of the transistor structure by current localization, thermal current filamentation, or a "hot spot" region formation. Using equations and an instability boundary analysis for a simplified device model, an analytical mathematical description of the stability criterion for the critical regime of current filamentation can be obtained. This stability boundary acts as an ultimate physical limitation of the device depending on the operation regime.

While this fluctuation of the instability criterion is an ideal case, local inhomogeneities, structural defects, semiconductor material, and/or contact regions in real transistor structures have a considerable impact on the process of thermal filamentation. Application of analytical approach to a device with inhomogeneity is complicated. In this case, practically everything that is related to inhomogeneity becomes a problem that requires nonlinear analysis and therefore needs either significant simplification or computerized numerical modeling.

Both in bipolar and in field-effect microwave transistor structures, the thermal breakdown is accompanied by various injection processes. However, the basis of thermal breakdown is always the same: an increase of thermogeneration currents that is exponentially dependent on temperature. During the evolution of thermal breakdown, the temperature in local regions can reach critical values of contact metallization melting or even the critical temperature for the semiconductor material itself. An uncontrollable current increase during thermal breakdown can result in the breaking of the bonding wires due to their melting like a fuse by its own Joule heating power.

Chapter 4
Isothermal Current Instability in Silicon BJT and MOSFETs

4.1 Avalanche Injection in Elementary Semiconductor Structures

Typical requirements of high middle time before failure (MTBF) ~ 10^6 hours result in application of FET devices under conditions of rather low operation temperatures < 120°C due to the activation energy ~ 0.7–1.2 eV for basic degradation mechanisms.

In state-of-the-art FETs, the temperature of the active region under operation (calculated assuming uniform dissipated power distribution) is significantly lower than the critical temperature of thermal instability. Even a few orders of magnitude overload in the dissipated power cannot result in the critical thermal instability temperature of ~ 300°C for Si FETs or ~ 500°C for GaAs FETs. However, at some critical pulse or DC voltage level a catastrophic failure is observed. The reason for failure is an electrical conductivity modulation followed by current instability. The instability results in the current-controlled negative differential resistance (NDR) followed by uncontrollable current increase and filamentation. In a very short time, this high Joule power level is generated in the filament region and further results in the local burnout.

Since this type of current instability can be observed in a short pulsed regime (2–20 ns) at rather negligible level of Joule overheating, the basic physical mechanism of the electrical burnout is an isothermal instability phenomenon. In these conditions, the initial transistor temperature is just a passive parameter. As a rule, the typical time of isothermal instability evolution is a few orders shorter than the typical time of thermal instability. Thus, a simple identification of electrical or thermal nature of the instability can be provided by pulsed measurements.

As will be demonstrated in Chapters 4 and 5, the basis of electrical mechanisms of catastrophic failures is the breakdown and instability in elementary semiconductor structures (n-i-n, p-i-n, and so on). However, as usual in real transistors the current instability may not converge down to just the instability in the elementary diode structure. In most practical cases, the avalanche-injection breakdown in one part of the transistor may initiate instability in another part of the device.

For example, avalanche breakdown in the GaAs MESFET channel turns on the avalanche-injection instability in the n^+-i-n^+ structure formed by the substrate buffer region and the source and drain contact regions (Chapter 5).

Another example is the instability in the Si MOSFETs. In these devices, the substrate current impacts not only the critical drain current and voltage, but also the filamentation time and filament dimension (this chapter).

For all studied isothermal instability mechanisms, the impact ionization in high electric field precedes the catastrophic failure. This fact is very simple to confirm using failure analysis of samples that have failed catastrophically. In the devices with large contact width, the area of destruction always has a local nature; no uniform structure melting has ever been reported.

However, avalanche generation is not a direct cause of the catastrophic failure. Indeed, both the avalanche generation in reverse-biased p-n junctions and the current injection for example in forward-biased junctions in the short pulsed regime can be distributed rather uniformly, while the final failure event produces the localized picture.

Nevertheless, avalanche generation results in initialization of the injection mechanism followed by a positive feedback in the form of conductivity modulation of active regions, negative differential conductivity, current filamentation, generation of excessive local power that finally leads to local melting, irreversible change, and catastrophe. This is the dominant scenario of catastrophic failure in pulsed regimes including ESD operation (Chapter 7)

4.1.1 Reverse-Biased p-n Junction

Generation of electron–hole pairs due to impact ionization in semiconductor structures becomes evident at an electrical field level $E > 10^4$ V/cm. At some critical electric field level, this process has an avalanche nature. For analysis of the avalanche breakdown process, the impact ionization coefficients α_n for electrons and α_p for holes are used. The coefficient refers to the number of ions generated by the charged particle on the pathway unit. The coefficient is a strong function of the electric field. It is given by exponential dependencies [78, 79]:

$$\alpha_n, \alpha_p \approx \exp\left(-\frac{E}{E_o}\right).$$

The case of a one-dimensional asymmetric reverse-biased p^+-n junction may be written in the following way without taking into account the diffusion and recombination the balance equation:

$$-\frac{dj_n}{dx} = \frac{dj_p}{dx} = \alpha_n j_n + \alpha_p j_p, \tag{4.1}$$

where $j_n = qnv_n, j_p = qnv_p$ are the electron and hole current densities, v_n, v_p are the electron and hole drift saturation velocities that are independent of E in strong fields: $v_n \approx v_p = 10^7$ cm/s. The boundary conditions for the p $^+$n junction are given by

$$j_n(x = 0) = 0; \, j_p(x = W) = j_s,$$ (4.1a)

where W is the width of the space charge region (SCR) of the p^+-n junction and j_S is the saturation current density. The current discontinuity equation is added to (4.1):

$$j = j_n + j_p = \text{const}.$$ (4.2)

From the solution of (4.1) and (4.2), at $\alpha_n = \alpha_p = g$ the current density is

$$j = \frac{j_s}{1 - \int\limits_0^W g dx}.$$ (4.3)

The ratio of the total current density to the saturation current density j/j_S is called the multiplication coefficient M:

$$M = \frac{1}{1 - \int\limits_0^W g dx}.$$ (4.4)

For the conditions of high avalanche breakdown, $j \rightarrow \infty, M \rightarrow \infty, \int\limits_0^W g dx \rightarrow 1$.

Usually, as a criterion of avalanche breakdown, a certain given integral value $\delta \ll 1$ is accepted:

$$\int\limits_0^W g dx = \delta.$$ (4.5)

Since the total current density $j(E)$ is a sharp function of E, most of the avalanche current in the p^+-n junction is generated in a rather narrow layer $\Delta \ll W$. Therefore, the breakdown condition (4.5) can be approximated by

$$g(E_{BR})\Delta = \delta.$$ (4.6)

Thus defined electric field E_{BR} can be adopted as the avalanche breakdown value. This value changes slightly with Δ and δ variation.

For analytical estimation of the I–V characteristic in the avalanche breakdown mode, the following assumptions are used in addition to the above assumptions:

(1) Uniform distribution of the breakdown across the uniform p-n junction area with the lateral dimension of the p-n junction is significantly higher than its thickness and thus a one-dimensional problem is automatically justified.

(2) The drift velocity v does not depend on the electric field starting from the levels above $E > 10^4$ V/cm.

(3) $E(x)$ is a smooth function of coordinate with the simple function of ionization coefficient $g \approx E^m$.

Under these assumptions, the basic equation for I–V dependence is given by the Poisson equation for the one-dimensional case:

$$\frac{dE}{dx} = -\frac{\rho}{\varepsilon},$$

(4.7)

where ρ is the charge density, and ε is the dielectric constant of the semiconductor material.

In the case of developed avalanche breakdown ($j_S > j$), the space charge density in the n-region is given by

$$\rho = qN_D - \frac{j}{v},$$

(4.8)

where N_D is the donor concentration, j is the electron current density in the n-region, and the avalanche generation is concentrated in multiplication layer $\Delta \ll W$.

From [79] the voltage on the p-n structure U is

$$U = U_{br}\left(1 - \frac{j}{j_N}\right)^{-1},$$

(4.9)

where $U_{br} = \dfrac{\varepsilon E_{br}^2}{2qN_D}$; $j_N = qN_D v$ is the critical current density at total space charge density equal to zero.

At $j \to j_N$, the I–V dependence $I(U)$ of the p-n junction aspires to saturation. Under this condition, $\dfrac{dI}{dU} > 0$, i.e., the p-n junction has positive differential conductivity.

However, in some cases the avalanche breakdown of real p-n junctions is not uniform. In several early studies, regions of localized avalanche breakdown have been reported. These localized regions have been called microplasmas.

Microplasmas of various dimensions are observed from less than one up to hundreds of micrometers [80]. In most of the studies, microplasma mode of avalanche breakdown is related to the presence of various uncontrolled structural defects and nonuniformities in the structure, for example nonuniformity of ρ along the p-n junction that may result in E nonuniformity and provide local multiplication due to the sharp dependence $g(E)$.

There is a contrary understanding of microplasma physics as spontaneously formed multiple filaments in a nonlinear nonequilibrium system [81]. For this understanding, nonuniformities play a secondary role as a primary factor only. Microplasma physics is discussed in section 5.4.

4.1.2 Avalanche Breakdown in p-i-n Diode

At reverse bias, the main part of the potential difference U is supplied to the i-region of the p-i-n structure. At $U \approx U_{BR,}$ the electric field in the i-region is constant and impact ionization begins uniformly through the i-region. In the i-region, the generated electrons are accumulated near the n-region and the generated holes near the p-region. The result of this distribution is increase of the field near the junctions and decrease in the center of the structure due to space charge mutual neutralization. The p-i-n diode I–V characteristic can be determined by the following equations. The electric field in the i-region at $I = \delta I \neq 0$ is given by

$$E = E_{br} + \delta E(x, \delta I).$$ (4.10)

For initial part of the characteristic at $\delta E / E_{br} \ll 1$, the electric field changing δE is limited by the breakdown condition:

$$\int_0^W g(E)dx = \text{const}.$$ (4.11)

By substitution of (4.10) into (4.11), expansion $g(E)$ and limiting by the first three terms:

$$\delta U = \int_0^W \delta E\, dx = -\frac{1}{2}\frac{g'_{br}}{g''_{br}}\int_0^W (\delta E)^2 \, dx \,.$$ (4.12)

Since $g'_{br} \equiv \left(\dfrac{dg}{dE}\right)_{E_{br}}$, $g''_{br} \equiv \left(\dfrac{d^2g}{dE^2}\right)_{E_{br}}$ are positive, the increment of the voltage drop δU is negative. This simple fact means that at least internal differential resistance of p-i-n diodes at avalanche breakdown is negative.

The breakdown of p-i-n structures is similar to the reverse-biased Schottky and is discussed on the basis of numerical simulation results in Section 5.3.

4.1.3 Avalanche Injection in n⁺-n-n⁺ Diode

Structure of the n^+-n-n^+ diode (Fig. 4.1a) is helpful for modeling of avalanche injection in n^+-p-n-n^+ bipolar transistors for active regime operation (positive n^+-contact is collector analogue). The same n^+-i-n^+ structure determines GaAs MESFET's instability as well as the n^+-p-n^+ structure in Si MOSFETs.

For the n^+-n-n^+ diode at a fixed current I, Ohm's law is true at small current levels and the resistance of the n-region R_0 is equal to

$$R_0 = \frac{L}{q\mu_n N_D A},$$

(4.13)

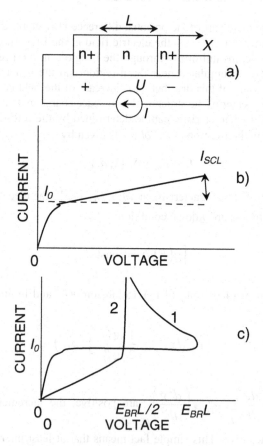

Fig. 4.1 Structure of the n^+-n-n^+ diode (a) and typical I–V characteristic (b) for the ultimate cases of SCN type "1" and SCL type "2"

where N_D is the donor concentration and A is the total diode area ($\sqrt{A} >> L$). With further current increase, the electric field $E = U/L$ increases up to the level $E > 10^3 \, \text{V/cm}$ where the drift velocity is saturated to the level $v \approx 10^7 \, \text{cm/s}$.

The density of free electrons in the n-region is equal to $n = j/v$. For dielectric relaxation time in the n-region less than electron drift time, $n = N_D$ is true and the current density is saturated at level $j_0 = qvN_D$.

In the opposite case, $n > N_D$ and the total space charge is negative $\rho = qN_D - j/v$, the I–V dependence of the n^+-n-n^+ diode at $j > j_0$ is not saturated and has a slope equal to $dI/dU = 2\varepsilon \, v \, A/L^2$.

The experimental I–V characteristic of the n^+-n-n^+ diode is presented in Fig. 4.1b. For sufficiently high U, the diode current I is equal to the sum $I = j_0 A + I_{SCL}$, where I_{SCL} is the space charge limited current component (SCL). Ratio between I and $I_{0\,SCL}$ depends on U, N_D, and L.

For theoretical analysis, an understanding of two boundary cases is rather helpful: SCN diode and SCL diode (Fig. 4.1).

In SCN diode $\rho = 0$, the electric field is distributed uniformly: $E=U/L \approx \text{const}$, and the I–V characteristic has saturation region at $I = I_0$.

In SCL diode $\rho < 0$, the electric field is increased linearly in the anode contact direction: $E_{MAX} = E(0) = 2U/L$. Near the cathode contact, $E(L) \approx 0$ and the electron injection in the n-region is rather limited.

Since both the doping concentration and the length of the diode play major roles, a relative boundary between SCN and SCL diodes for Si devices is $N_D L = 6 \cdot 10^{11} \, \text{cm}^{-2}$ [82]. At $N_D L << 6 \cdot 10^{11} \, \text{cm}^{-2}$ the diode is SCL; at $N_D L >> 6 \cdot 10^{11} \, \text{cm}^{-2}$ the diode is SCN. However, this statement is true based on the assumption that the diffusion length of major carriers is larger than the device length L, $L_D \sim L$.

At a small diffusion length or a long n-region, for example due to high recombination rate, the SCL case is dominant. This situation is also realized in GaAs n^+-i-n^+ diodes (Chapter 5).

Using the avalanche breakdown criterion (4.6) in SCN diodes, the avalanche multiplication begins at $U \approx E_{BR}L$. Electrons are accumulated near the anode electrode; the field is increased in this region and shifted in the multiplication region near the n^+-n boundary. The holes drift to the cathode thus being injected into the n-region. This phenomenon incorporates the notion of "avalanche injection."

This uncompensated hole space charge causes injection of electrons from the n-n^+ junction. At $I >> I_0$, a multiplication region is formed at the anode region, the maximum field slightly exceeds E_{BR} due to a sharp dependence $g(E)$, and the negative space charge $\rho = N_D - n + p$ in the multiplication region provided

electric field decrease up to $E(L) \approx 0$. This SCL diode regime is realized at
$$U = \frac{E_{BR}L}{2}.$$

With the current increase above $I > I_0$, the SCN diode voltage is decreased twofold (Fig. 4.1c), and S-shape I –V characteristic is observed in the critical regime $I = qN_D vA; U \approx E_{BR}L$. Obviously for SCL diode, avalanche multiplication begins at once at $U = E_{BR}L/2$ and a negative differential conductivity region in I–V characteristic is not seen (Fig. 4.1c).

In the n^+-n-n^+ structure $M = j/j_n$ in the avalanche injection condition [5], therefore in the condition of developed avalanche injection (at $j \to \infty$ $j_n \to j_p$) and $M \to 2$. This fact provides the difference between the avalanche breakdown process in reverse-biased p-n junctions, when $M \to \infty$, and the avalanche-injection conductivity modulation that occurs at rather low $M \to 2$.

From the experimental study [82] of samples with various N_D and L, it follows that the I–V characteristic of real n^+-n-n^+ diodes has an intermediate shape between the boundary cases of SCN and SCL diodes.

It should be emphasized that the above conclusions are true for cases where the diffusion length $L_D > L$. For a diode structure with $L_D < L$, the S-shape I–V characteristic is realized independently of the type and doping level. For example, in n^+-i-n^+ structures the deep S-shape I–V characteristic is observed both in experiments and in numerical simulation studies (Chapter 5).

4.1.4 Kinetics of Avalanche Injection in p-n-n⁺ Structures

The reverse-biased p^+-n-n^+ structure is a part of any power bipolar n^+-p-n-n^+ transistor, where a lightly doped n-collector region is introduced to provide the desired level of collector-emitter breakdown voltage.

The behavior of the p-n-n^+ structure under the avalanche breakdown conditions is similar to the above-discussed (Section 4.1.2) p-i-n diode. The negative differential conductivity is observed at the DC regime above some critical current level $I_{CR} > 0$. The current I_{CR} [84] depends on the donor concentration N_D in the n-region. The lower estimation for the I_{CR} value is $I_0 = AqN_D v$.

Dynamics of avalanche injection in p-n-n^+ structures was studied in [83] using numerical simulation. The equations used in [84] are presented below in order to demonstrate both the approach and the complexity of theoretical analysis for the instability problem. For a 1D p-n-n^+ structure in circuit with voltage source U_S and the load resistor R_L, the electric field concentrated in the n-region is assumed; the n-region length is 10 μm and the donor concentration $N_D = 10^{15}$ cm^{-3}. The latter is a typical value for several power epitaxial bipolar transistor architectures. The field distribution is given by the Poisson equation:

$$\frac{dE}{dx} = -\frac{q}{\varepsilon}(p - n + N_D).$$ (4.14)

Nonstationary electron and hole concentration n, p are given by the continuity equations:

$$\frac{\partial n}{\partial t} = -\frac{\partial}{\partial x}\left(v_n n - D_n \frac{\partial n}{\partial x}\right) + \alpha_n n|v_n| + \alpha_p p|v_p|,$$

$$\frac{\partial p}{\partial t} = \frac{\partial}{\partial x}\left(v_p p + D_p \frac{\partial p}{\partial x}\right) + \alpha_n n|v_n| + \alpha_p p|v_p|,$$ (4.15)

where D_n, D_p are the diffusion coefficients and the ionization coefficients α_n, α_p are given by the expression $\alpha = A \exp(-B/E)$, where A, B are some empirical constants.

Three ranges of dependencies are introduced for electron and hole velocities:

$$v \approx E \qquad E < E_C$$
$$v \approx \sqrt{E} \qquad E_C < E < E_S$$ (4.16)
$$v = v_S = \text{const} \quad E > E_S.$$

Total current in the external circuit is given by

$$I = \frac{S}{W}\int_0^W q\left(\left|v_n - D_n\frac{\partial}{\partial x}\right|n + \left|v_p - D_p\frac{\partial}{\partial x}\right|p\right)dx,$$ (4.17)

where S is the device area and W is the n-region length.

In the initial time moment, the electron density n is assumed equal to $n_0 + n_{IN}$, where n_0 corresponds to the intrinsic current and n_{IN} corresponds to the initial injection current (analogue of the emitter current). External circuit equation is given by

$$U = U_S - IR_L,$$ (4.18)

where $U = \int_0^W E dx$ is the voltage supplied to the sample.

Equations (4.14)–(4.18) have been solved numerically with 1-ps time step.

In Fig. 4.2, the calculated dynamics of $E(x)$ is presented for the load characteristics (4.18) when there is no point of intersection with stationary S-shape I–V characteristic in uncontrollable "secondary breakdown" in the structure. At the initial moment, the field has a linear distribution in the n-region with amplitude decrease from the p-region toward the n^+-region according to the donor concentration N_D and given injection value n_{IN}. At $t = 0.14$ ns, the linear field distribution is disturbed and broken toward increase of the amplitude in both edges of the n-region due to avalanche generation occurring in both p^+-n and n-n^+ junctions. This kind of avalanche generation is called a "double avalanche injection." The intensity of the double avalanche injection is increased sharply and the field in the center of the n-region is decreased. This results in decreased total U voltage.

Sharp voltage decrease and current increase and transition of $E(x)$ to the oscillation shape (Fig. 4.2) are observed after ~1 ns from the pulse start. The transient process duration does not exceed 1 ns. In [84], the numerical solution became unstable thus providing the structure transition into the oscillation regime where electron and hole transport through the n-region have wave properties. In Chapter 5, the stable numerical solution for similar GaAs structures is presented using 2D numerical simulation.

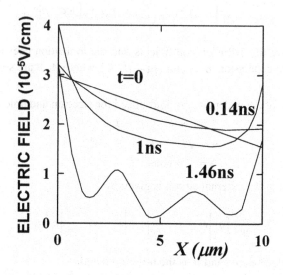

Fig. 4.2 Dynamics of isothermal instability in p-n-n$^+$ structure

In real p-n-n$^+$ structures, simultaneously with uniform isothermal instability current filamentation is observed in the plane that perpendicular to the field direction (Chapter 5).

4.2 Isothermal Instability in Bipolar Transistors

Isothermal instability (or the current mode secondary breakdown) was observed in the original power epitaxial n$^+$-p-n-n$^+$ structures. The first studies demonstrated that the thickness L and the doping level N_D in the n-region have a considerable impact on the thermal instability condition. In particular, it was determined in [85] that in the structures with a thick epitaxial layer and in the transistor structures without an n-layer, S-shape I–V characteristics are not formed in a rather wide current range. At the same time, in other BJT structures the isothermal current instability has been observed at relatively small collector current followed by catastrophic failure.

Different types of the breakdown S-shape $I–V$ characteristics were measured using high-speed experimental technique. The particular type depends on the structure, operation regime, transistor type and varies from device to device.

However, this variety is not the result of different isothermal instability mechanisms. According to understanding of the avalanche-injection mechanism, described above, it produces the instability. However, the interaction itself between the impact ionization and injection processes might have rather different scenarios. This conclusion will be demonstrated below using several simple models of structures chosen according to the experimental results.

4.2.1 Isothermal Instability in Diode Operation Regimes

For simplification of the following discussion, the one-dimensional n^+-p-n-n^+ diode structure will be described under the avalanche breakdown regime. This structure represents the open base NPN BJT in so-called BVCEO conditions.

The case of open base breakdown (BVCEO) $I_B = 0$

The operation regime of transistors with the floating base is roughly similar to the n^+-n-n^+ diode described in the previous chapter. However, this transistor operation has several significant peculiarities.

Two versions of the transistor structure with different doping level in the n-collector region present a particular interest: A-structure with $N_D L \approx 6 \cdot 10^{11}$ cm^{-2} $(L = 10$ μm$)$ and C-structure with $N_D L = 6 \cdot 10^{12}$ cm^{-2} » $N_D L = 6 \cdot 10^{11}$ cm^{-2} $(L = 9$ μm$)$.

C-type n^+-n-n^+ diode corresponds to the SCN type. A-type diode is intermediate between the SCL type and the SCN type.

Shape of $I–V$ characteristics of the transistor structures and the electric field distribution in the n-region are presented in Fig. 4.3 in comparison with $I–V$ characteristics of the corresponding n^+-n-n^+ diode structures.

In the operation regime with zero base current $I_B = 0$, the collector current is equal to the emitter current and is proportional to $(1 - \alpha M)^{-1}$. The increase in current begins under conditions $U_{CE} \rightarrow U_\alpha$, $(\alpha M \rightarrow 1)$, i.e., when the electric field approaches a certain maximum value $E_{MAX} \approx E_{BR}$ on the p-n junction (Fig. 4.3b). Since the value E_{MAX} is limited by the condition $\alpha M = 1$ below E_{BR} the initial current increase is observed at the collector-emitter voltage equal to $U_{CE} \approx \int_0^L E dx$ increase.

In structure "A" at $j = j_0 = q N_D v$, the electron space charge results in leveling the electric field in the n-collector region. Due to lower n-collector doping level

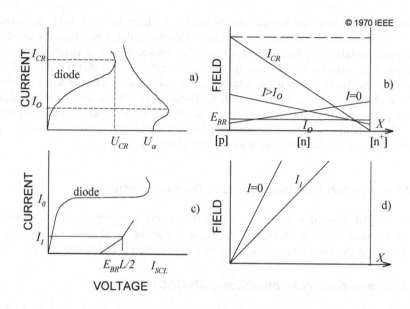

Fig. 4.3 Calculated I–V characteristics and electric field distributions for the structures with $N_D L = 6 \times 10^{11}$ cm^{-2} (a,b) and $N_D L = 6 \times 10^{12}$ cm^{-2} (c,d) at different current levels [82]

at $j > j_0$ the avalanche multiplication region is formed on the n-n$^+$ junction followed by the voltage decrease. The I–V characteristics of A-type transistor structure have negative slope and aspire to the I–V characteristics of corresponding n$^+$-n-n$^+$ diode structure. At significant current density $I > I_{CR}$, both the transistor and the diode have similar characteristics.

For the more heavily doped structure "C," the I_0 value is so high that breakdown on the p-n junction is initiated under the condition where the space charge region (SCR) is confined within the n-collector region (Fig. 4.3d). In this case, I–V characteristics have positive slope which increased at $U \to E_{BR}L/2$, when SCR attains the n-n$^+$ junction. Therefore, for C-type structure with higher n-collector doping level, the differential resistance remains positive $\frac{dI}{dU} > 0$ up to the current level I_0 that exceeds the corresponding level in A-type structure by approximately one order of magnitude.

The case of grounded base breakdown operation $U_{EB} = 0$ (BVCES)

In the case of shorted emitter and base terminals, the initial location of avalanche breakdown is at the collector p-n junction. In this case, the avalanche breakdown is not accompanied by injection from the closed emitter junction due to the contact potential difference. Therefore, the collector current is provided by the multiplication of the saturation current I_{C0} $I_C \sim M I_{C0}$. The breakdown voltage is $U_{CEBR} > U_\alpha$ and practically equal to the collector-base breakdown voltage U_{CBBR}.

The avalanche generated holes flow out from the base in the negative base current regime. This hole current produces additional voltage drop across the p-base that results in increase of the base potential at the emitter junction on corresponding value $r_B I_C$, where r_B is the corresponding internal base resistance. The r_B can be extracted for the simulation or measurement data. Thus, when the voltage level $r_B I_C$ is equal or above the potential of the emitter junction opening (~ 0.6 V), the injection of the electrons from the emitter begins followed by positive feedback mechanism. Thus, this voltage drop controls both the critical current level and the critical voltage of instability.

In experimental conditions, the beginning of the instability practically coincides with the condition of the emitter junction opening. Therefore, for estimation of the critical regime, the following equation may be used [85]:

$$I_{CR}(R_B + r_B) = 0.6, \qquad (4.19)$$

where R_B is the external resistance in the base circuit. With R_B increase, the critical current decreases [85] and breakdown voltage value U_{CEBR} practically is unchanged. The latter results in decrease of any thermal instability mechanism probability.

Condition (4.19) also covers the case with the base grounded through the external resistor of value R_B. In the device terminology, this case corresponds to so-called BVCER breakdown voltage.

The open emitter case $I_E = 0$

In the case of floating emitter circuit, the injection of electrons becomes impossible. However, from an experiment [85] it follows that in a number of power transistors with open emitter, the instability is observed in $I_E = 0$ regime too. In this case, however, the critical current level is approximately ten times higher. The reason for this effect is an increase of the emitter potential with increase of the residual current in the base up to some critical value that corresponds to a level of surface breakdown voltage in the region between the emitter and grounded base. Due to this specific breakdown scenario, the electrical contact between the emitter and base is restoring and providing a possibility of injection from the emitter junction. The critical voltage remains the same, but the critical current in reality corresponds to increase of the emitter potential up to the value U_{EBBR} (that for the case discussed in [85] was ~ 6 V). Therefore, critical current for the positive feedback and corresponding instability is $I_{CR} = \dfrac{0.6 + U_{EBBR}}{R_B + r_B}$. Apparently, this condition can be satisfied by implementation of proper device architecture.

4.2.2 Avalanche Injection in Common-Emitter Circuit

From numerous experimental studies of isothermal instability, it follows that formation of the negative differential conductivity in the pulsed regime results in current filamentation in power transistors and ESD devices. The pulse duration t_P is adequate for filamentation process (for example, $t_P \geq 1$ μs). Therefore, for estimation of the critical conditions, it is sufficient to determine the regimes with $\dfrac{dI}{dU} \leq 0$.

The positive base current regime $I_B > 0$

In common-emitter biasing circuit, the base current is constant and positive: $I_B = $ const, $I_B > 0$. Neglecting the saturation current contribution versus high base current, the collector current is equal to

$$I_C = \frac{\alpha M I_B}{1 - \alpha M}. \tag{4.20}$$

Since the multiplication coefficient is a function of the collector base voltage and the collector current $M = M(U_{CE}, I_C)$, (4.20) determines the device I–V characteristic. By differentiating the right side of (4.20) on U_{CE} at $\alpha = $ const, $T = $ const, and $\partial M / \partial U_{CE} > 0$, the simple criterion of instability is

$$\frac{I_C}{M(1 - \alpha M)} \frac{\partial M}{\partial I_C} > 1. \tag{4.21}$$

Unfortunately, (4.21) and the more universal criteria suggested in [87] are obtained involving both the current gain $\alpha(I_C)$ and the multiplication coefficient $M(T)$ dependencies. Thus, the criteria is hard to use in practical cases since it requires the exact dependencies $\alpha(I_C)$ and $M(I_C, U_{CE})$.

For some particular cases (see Section 4.2.4), it is possible to complete the calculation of the instability boundary. Meanwhile, the result from this calculation is no more informative than the qualitative estimation [82] presented below.

At low base currents when $j_C < j_0 = qN_D v$, the transistor behavior is simply similar to the case of $I_B = 0$ presented in Section 4.2.1. In the vicinity of U_α, the collector current begins increasing according to (4.20) up to the critical value I_0 when the negative differential conductivity is formed followed by the current instability.

At corresponding base current level when the collector current $j_C > j_0$, the maximum of electric field is formed on the subcollector n-n$^+$ junction and the conditions are formed for avalanche injection and corresponding instability immediately with the avalanche generation, i.e., at the electric field $E_{MAX} = E_{BR}$. This effect is observed at $U_{CEBR} < U_\alpha$ [see Section 3.3.1, where U_α is the breakdown voltage in CE bias configuration derived from the condition $\alpha_1 M(U_\alpha) = 1$].

It is quite clear that the I_C increase will decrease the value of $U_{CE\,BR}$. For $I_C > 2I_0$, the thickness of space charge region, W, still will be lower than the thickness of n-epilayer, L. Assuming a triangular electric field distribution, at the critical point the integral value of the collector-emitter voltage is $U_{CECR} \approx \dfrac{E_{BR}W_{CR}}{2}$, where $W_{CR} \approx \dfrac{E_{BR}\varepsilon}{|j/v - qN_D|}$. Therefore, for sufficiently high current $I_C \gg I_0$, the isothermal stability boundary must be close to hyperbolic function:

$$U_{CR}I_{CR} \approx \frac{\varepsilon E_{BR}^2 v}{2}. \tag{4.22}$$

The case of $I_B < 0$ regime

The negative base current regime is not so exotic. Examples of its realization are at overloads in collector circuit the pinch-off condition or in large-signal operation at the load mismatching and of course ESD and EOS events. The importance in terms of reliability is defined by the fact that for a number of power transistors the level of critical current at $I_B < 0$ is sharply decreased to less than 1 mA.

What is the main feature of the negative base current operation? The hole current flow from the base produces instability of the emitter current: the higher j_B is, the higher the value of base potential near the emitter junction, therefore the injection level of electrons from the emitter is higher too. Flowing along the emitter junction (Fig. 4.4) in the base contact direction, the $I_B(y)$ increased emitter current density $j_E(y)$ in the center of the emitter strip and dumped it on the edge of the strip. This effect is a mirror to the Fletcher effect. At fixed emitter current I_E (in common base configuration), the negative base currents might result in the filamentation of emitter current in the "OX" direction to the base current flow I_B in the "OY" direction (Section 4.2.3).

Under negative base current conditions in the common-emitter circuit, the following expression can be used for the base current I_B:

$$I_B = -\frac{I_{C0}}{\alpha} + I_C \frac{(1 - \alpha M)}{\alpha M}. \tag{4.23}$$

For sufficiently high potential on the emitter junction, the current is determined by the avalanche multiplication in the collector region: $M \gg 1$ and $I_B \approx -I_C$. This is similar to the diode circuit for the case $I_E = 0$ discussed above.

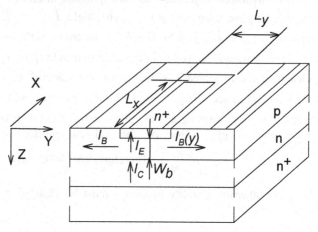

Fig. 4.4 Simplified structure of power bipolar junction transistor with implanted emitter in case of negative base current operation in common-emitter circuit

For the pinch-off regime, the injection efficiency of emitter junction is very low, $\alpha \ll 1$. Therefore, regarding the essential current generation the conditions $M \gg 1$ and $U_{CE} > U_\alpha$ are necessary. Appearance of holes in the base enhances injection and results in increased emitter current thus immediately increasing α to its maximum value. In this case, high M values are not required anymore in order to support already increased current level. At given current source in the collector circuit, the multiplication coefficient M is decreasing due to U_{CE} reduction. Thus, the S-shape region is formed. This negative differential resistance region was well studied. It can be realized at rather small current levels under $\frac{d\alpha}{dI_E} > 0$. At the same time, in the structures with high values of the saturation currents I_{SE} and I_{SC} the S-shape may not form at all.

Obviously, from the reliability point of view, this S-shape is not "dangerous," because current redistribution is realized only at rather small current levels and automatically terminated by saturation of the gain α. While the filamentation at small current levels does not produce failure, it should be considered that any NDC region on the I–V characteristics may be accompanied by the spontaneous switching of the transistor into a new state according to the load characteristic. This may be critical for operation thus creating a typical example of the latch up scenario.

The I_c increase at $U_{CE} > U_\alpha$ is determined by ratio (4.23). According to the avalanche injection concept, the critical current level is the level when j_c (at least locally) increases up to j_0 value. At negative base currents ($\alpha M \geq 1$), the emitter current redistribution is possible, therefore critical level of the current density may be realized at lower I_c levels in comparison with $I_B = 0$. Typical instability boundary for the transistors with this effect is presented in Fig. 4.5a [82].

To demonstrate the transistor behavior at negative base current $I_B < 0$, it is useful to analyze the common-emitter circuit with the resistor R_{BE} between the base and emitter at a given collector current I_C [82]. This circuit can be used both to measure the pulsed stability boundary $U_{CECR}(I_C)$ at different R values (from $R_{BE} = 0$ to $R_{BE} = \infty$) and to simulate various base current levels.

A high, but finite, resistance results in decrease of the gain coefficient in comparison with the open base circuit case $R_{BE} = \infty$ ($I_B = 0$). Some portion of holes flows out of the base. This corresponds to the condition $\alpha M \geq 1$ and $U_{CE} \geq U_\alpha$ (4.23).

A small negative base current $I_B < 0$ results in a double effect. On the one hand, negative base current $I_B < 0$ reduces the emitter injection in comparison with $I_B = 0$; on the other hand, it enhances the local current density in the central part of the emitter too. In the region $U_{CE} > U_\alpha$ (Fig. 4.5a), the redistribution j_C is dominant: while the peak current density j_{max} is increasing, the total critical current is decreasing. With R_{BE} decrease, the total injection level is reduced. For the conditions of emitter junction opening, the current increase of I_C ($I_C \approx -I_B$) is necessary:

$$I_{CCR}(R_{BE} + r_B) \approx 0.6.$$

For the transistor with $N_D L >> 6 \cdot 10^{11}$ cm^{-2} the falling region of I_C at $U_{CE} > U_\alpha$ may not even be present [82]. Indeed, in those structures the main part of the collector current I_C is provided by avalanche generation. In this condition, the injection efficiency of the emitter is rather low, so the emitter impact in

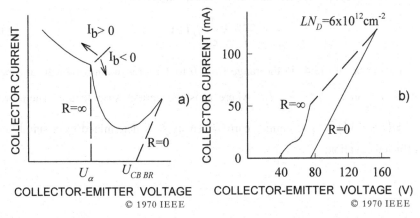

Fig. 4.5 Stability boundary in the negative base current regime ($I_B < 0$) for transistors with the collector region parameters $N_D L \ll 6 \times 10^{11}$ cm^{-2} (a) and $N_D L \gg 6 \times 10^{11}$ cm^{-2} (b) [82]

the field redistribution in the n-collector region is negligible. Apparently, in this case the negative base current flow does not cause any significant redistribution of emitter current. Therefore, the stability boundary has a shape similar to that in Fig. 4.5.b. Practically, at $R_{BE} = 0$ the I–V characteristic is equivalent to the base-collector diode structure.

Evaluation of the critical regime for the avalanche-injection isothermal instability under the $I_B < 0$ condition is not as simple as the positive base current case $I_B > 0$. However, analytical estimation for this effect is in rather good agreement with the experimental data [86].

In [86], the measurement setup provided the measurements of instability evolution reversibly due to implemented high-speed protection circuit. The experimental structure of power epitaxial transistor ($L = 40$–80 μm, $N_D = 2 \times 10^{14}$ cm^{-3}) has been switching in common-emitter circuit conditions. The current source provided the switching from a forward base current I_{B1} to a reverse base current I_{B2}. Using the current I_{B1} the collector current I_C is supported (in the range 6–12 A). The collector current is stored during a time after the switching of base current. The inductance in the collector circuit (800 μH) has supported the increase of collector voltage up to 1000 V during a time less than 1 μs, but the collector current I_C remained constant within a few percent accuracy. When U_{CE} was increased to a critical value, the instability was formed: U_{CE} decreased rapidly. In this case, the protection circuit (controlled by dU/dt value) provided turn off of the power supply from the transistor. The time of the transistor transition into a low-voltage state was approximately 10 ns. Time of the protection circuit response was below 350 ns. This was sufficient to avoid any degradation in the device.

In [86], the measured value of the critical voltage U_{CECR} was compared with the calculated value. Since the agreement between the values was less than 15%, the equation system used [86] is reproduced below.

The analytical solution is presented for a two-dimensional problem in the plane ZY (Fig. 4.4). Zero coordinate point is selected in the middle of the emitter region. The base current per emitter unit length is determined by

$$\frac{di_B}{dy} = \frac{1}{G} j_E(y). \tag{4.24}$$

Integration of (4.24) in the range from 0 to $1/2 l_Y$ results in a typical ratio between I_B and I_C: $I_B = \frac{1}{G} I_E$, where l_Y is the emitter strip length. Therefore, $G \approx (1 - \alpha M)^{-1}$. The emitter current density j_E is determined by a simplified diffusion equation:

$$j_E(y) = \frac{qD_n}{W_B(y,z)} n(y) \left[1 + \frac{n(y)}{N_B + n(y)}\right],$$ (4.25)

where N_B is the doping concentration in the base and W_B is the total width of both the metallurgical junction and the induced base. Presence of Z coordinate in $W_B(y,z)$ points to the dependence of W_B on the field distribution $E(z)$ in the collector region.

The density of injected elections in the base $n_E(y)$ is determined by

$$n_E(y) = \frac{n_i^2}{N_B} \exp \frac{qV(y)}{kT}, (n_E < N_B),$$ (4.26)

for the high injection level:

$$n_E(y) = n_i \exp \frac{qV(y)}{2kT}, (n_E > N_B),$$ (4.26a)

where $V(y)$ is the electrical potential at the emitter p-n junction that could be found from the equation

$$\frac{dV(y)}{dy} = -\frac{i_B(y)}{W_B(y,z)\sigma(y)}.$$ (4.27)

The conductivity $\sigma(y)$ includes an unmodulated component σ_0 and a modulated component σ_1. The last is determined from the equation [86]

$$\sigma_1(y) = \frac{2}{3} q \left(\frac{\mu_n}{2} + \frac{3}{2}\mu_p\right) n(y),$$ (4.28)

where μ_n, μ_p are the electron and hole motilities.

The total emitter current is

$$I_E = 2l_X \int_0^{y_0} j_E(y)dy,$$ (4.29)

where y_0 is the distance from the middle of the center of the emitter to a point where the emitter electron injection ability is lost, or in other words the electrical potential in the point y_0 is $V(y_0) = \frac{2kT}{q}$.

Finally, the following boundary conditions are added to (4.24)–(4.29): $y = 0$, $dV/dy = 0$, and $V(y) = V(0)$.

The solution of (4.24)–(4.29) is the distributions $j_E(y)$ and $V(y)$. It is significant that $V(y)$ does not exceed V_{EBBR} at any point. The distribution $j_E(y)$ provides $j_{max} = j_E(0)$ for the Poisson equation for the n⁻ region:

$$\frac{dE}{dz} = \frac{q}{E}\left[N - \frac{j_{MAX}}{qv}\right].$$
(4.30)

Final solution of (4.30) provides the critical voltage V_{CECR}.

4.2.3 Avalanche Injection in Common-Base Circuit

The conditions of the isothermal instability in the common-base bias circuit (CB) and common-emitter bias circuit (CE) slightly different from each other at $U_{CB} < U_\alpha$. For a current density above the critical level qN_Dv, the instability begins immediately with the avalanche injection at U_{CB} of avalanche ionization: $U_{CB} \approx \varepsilon E_{BR}^2 \Big/ 2\left(\dfrac{j_C}{v} - qN_D\right)$. At $j_C > 2j_0$, the transistor turns on in the negative base current regime. For CE circuits, this effect is accompanied by loss in collector control by the base current. For CB circuits, the voltage range $U_\alpha < U_{CB} < U_{CBCR}$ is still operational.

As was mentioned, the negative base currents may result in a redistribution of the emitter current. In the CE circuits, this redistribution involves a part of injected electrons from the emitter followed by a redistribution of the total collector current that has an avalanche component. In the CB circuits at $U_{CB} < U_{CBCR}$, the collector current is practically equal to the emitter current ($j_C = \alpha M j_E$). Therefore, a degree of current redistribution might be significantly higher. At $I_E =$ const, it can result in a scenario where the isothermal filamentation of emitter current is not translated to the external circuit as negative differential resistance, but essentially takes place under positive differential conductivity $I(U_{CB})$.

This filamentation mechanism was first observed in [85]. At $U_{CB} > U_\alpha$, the hole space charge increases the forward bias level of the emitter junction. A non-uniformity of this process results in the redistribution of emitter current as follows: the higher the hole density is, the higher both the emitter forward bias and the emitter current density. When flowing through the avalanche multiplication region, the electron current density increases the avalanche hole density. If in these conditions the current density j_{Cmax} in the middle of the filament becomes equal to the critical value j_0, then the triggering on is observed according to the classical avalanche injection mechanism.

Although there is no direct confirmation, from a theoretical point of view, the isothermal instability in CB might exhibit two-stage evolution. The evolution of avalanche injection in the collector n-layer is not a necessary condition of the transistor failure. Emitter current filamentation at $\alpha M > 1$ by itself cannot cause the catastrophic consequences because the total dissipated power $P_C = U_{CB} I_C$ is practically unchanged. The current redistribution results in a sharp increase in maximum specific power $j_{CMAX} U_{CB}$.

A distinct feature of the isothermal instability in CB at real operation is the low level of excessive currents before voltage redistribution (at least one order of magnitude lower than in the case of CE bias circuit). Critical excessive current ΔI_{CR} means the difference between the values I_{CR} and I_C before the NDC formation (Fig. 4.6a). In CE, the value ΔI_{CR} may be achieved at ~ 0.1 mA, but in CB ΔI_{CR} in $U_\alpha < U_{CB} < U_{CBCR}$ range, its value is an order of magnitude higher.

The experimental shape of the stability boundary (Fig. 4.6b) corresponds to a concept of primary emitter current filamentation. Decrease of I_{CR} at $U_{CB} > U_\alpha$ is

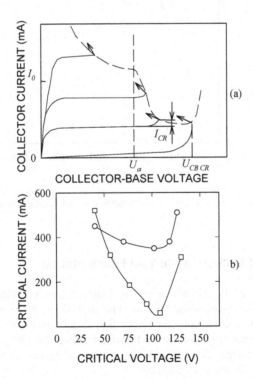

Fig. 4.6 Isothermal instability in common-base circuit: *I–V* characteristic (a) and experimental stability boundary for two samples (b)

apparently the same as the decrease of similar value in CE circuits. It should be considered that the value is varied for various devices (for a given transistor type) and is the result of the device parameter scattering and the presence of local structural defects (Fig. 4.6b). This problem will be discussed in the following section for the filamentation criterion.

The experimental results of isothermal instability in CB circuits are not complete. The oscillogram fixed the sharp decrease of U_{CB} and the increase of I_C (Fig. 4.7) gives only indirect confirmation of the double avalanche injection. A degree of the current filamentation is estimated only indirectly. In [83], it is suggested that the area of filament is one tenth of the total emitter area. An experimental confirmation of the emitter current filamentation in negative base current mode in CB is not as yet complete.

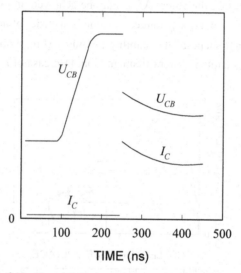

Fig. 4.7 Oscillogram of $U_{CB}(t)$ and $I_C(t)$ dependence at isothermal instability in common-base configuration

4.2.4 Criteria of Isothermal Current Filamentation

A simple analysis of the critical regime for filamentation at avalanche breakdown of collector junction is presented below. The analysis is based on application of approximation for linear theory for fluctuation instability to the equation system for processes in p-base and n-collector transistor regions [87, 88].

The collector current density including impact ionization in the space charge region of collector junction is given by

$$\tau_\Lambda \frac{\partial j_C}{\partial t} = \alpha j_E - j_C \{1 - \psi[E(j_C, U_{CB}, z)]\}, \tag{4.31}$$

where $\tau_\Lambda = (W_n + W_p)/2v$; W_n, W_p are the width of n and p parts of the space charge region of the collector junction; $\psi(E) = \int_0^{L_C} g(E)dz$ is the integral multiplication; $E(j_C, U_{CB}, z)$ is the field distribution in the collector junction; $L_C = (W_n + W_p)$; $g(E) = CE^m$ are the electron and hole avalanche ionization coefficients.

Equation (4.31) can be obtained after normalization on the thickness L_C of the continuity equations for electrons and holes inside the space charge region [79]. The integral multiplication ψ in the quasi-static approximation can be determined from the Poisson equation [79]:

$$\frac{dE}{dz} = \frac{q}{\varepsilon}[N_D(z) - N_A(z)] + \frac{j_C}{qv}\left[2\int_z^{W_n} g(E)dz - 1\right], \tag{4.32}$$

where $q(N_D - N_A)$ is the charge density created by donor and acceptor ions.

For simplification, it is assumed that the pinching of emitter current in a filament takes place only in the OX direction (Fig. 4.4) and the current density j_C, j_E, j_B and voltage U_E on emitter junction do not depend on the y-coordinate. In this case, all effects of current redistribution along the emitter strip could be neglected considering that in the case of highly doped base this effect is secondary in comparison with the current filamentation in the x-direction along l_X.

The distribution $U_E(x)$ is provided by the current continuity equation for unit volume $W_B dxdy$ of the transistor base [87]:

$$\tau_U \frac{\partial U_E}{\partial t} = L_U^2 \frac{\partial^2 U_E}{\partial x^2} - (j_E - j_C - j_B)\frac{kT_0}{qj_S(T_0)}, \tag{4.33}$$

where $\tau_U = C_E kT_0/qj_S$ is the characteristic time of U_E variation; $L_U = [\sigma W_B kT_0/qj_S]^{1/2}$ is the characteristic length of U_E variation; W_B, σ are the thickness and the conductivity of base; C_E is the specific capacity of emitter junction; T_0 is the ambient, heat sink, or package temperature.

Normalized to emitter junction area, the base current density j_B is expressed through U_E by the following ratio:

$$j_B = [U_{EB} - U_E(x)]/r_B, \tag{4.34}$$

where U_{EB} is the potential difference between emitter and base electrodes; $r_B = ab/2\sigma W_B$ [87]. Ratio (4.34) is true if an electrical flow in base along the X-axis could be neglected. Emitter current density is given by

$$j_E = j_{SE}\left[\exp\left(qU_E/kT\right)-1\right]$$ (4.35)

and the total current value I_E is given by

$$I_E = b\int_0^{l_X} j_E(x)dx.$$ (4.36)

The following equation system may be obtained by linearization of (4.31), (4.33)–(4.36) relatively nonuniform disturbances δU_E, $\delta j_C \sim \cos k_X \exp(\gamma t)$ ($k_X = \pi n/l_X$, $n = 1,2,...$). The disturbances provide neither any change in the total current through the external circuit nor the voltage on the structure ($\delta I_C = \delta I_E = \delta I_B = \delta U_{EB} = \delta U_{CB} = Q$) under the increase increment γ:

$$\gamma \tau_U \delta U_E = -L_U^2 k_X^2 \delta U_E - A'_{U_E}\delta U_E - A'_{j_C}\delta j_C$$ (4.37)

where $A'_{U_E} \equiv \partial A/\partial U_E$ and $A'_{j_C} \equiv \partial A/\partial j_C$ are the corresponding partial derivatives; $A = (j_E - j_C - \dfrac{U_{EB} - U_E}{r_{BEFF}})$; $N = j_C\{1-\psi\}-\alpha j_E$ for CB circuit and $N = j_C\{1-\alpha\psi\}-\alpha j_B$ for CE circuit.

From (4.37), it follows that in CB circuit the instability ($\gamma > 0$) is formed under one of the following conditions:

$$1-\alpha j_E \frac{\partial M}{\partial j_C} < 0$$ (4.38)

or

$$\tilde{r}_{BEFF}\frac{q}{kT}j_E\left[\alpha M(U_{CB})-1\right]-1 > 0,$$ (4.39)

where $\tilde{r}_{BEFF} = \left(W_B\sigma k_X^2 + \dfrac{2\sigma W_B}{ab}\right)^{-1}$ is the effective base resistance for the case of nonuniform disturbance; $M = \{1-\psi\}^{-1}$.

In CE circuit at $I_B = $ const, the instability criterion is (4.21):

$$1-j_C(1-\alpha M)^{-1}\frac{\partial M}{\partial j_C} < 0.$$ (4.40)

Conditions (4.38) and (4.40) are identical to the conditions of negative differential resistance $R_G < 0$ region on I–V characteristics. To determine the boundary of $R_G < 0$, it is necessary to find the explicit functions $\psi(j_C, U_{CB})$ and $M(j_C, U_{CB})$.

Let us consider the example of asymmetric p-n$^-$-n$^+$ junction for the CE bias circuit. In this case, the space charge region is formed in the n$^-$-layer, i.e., $W_P = 0$, $W_N = W$. Thus, in the CE circuit $(M-1) \ll 1$ in Poisson equation (4.32), the hole density can be neglected. Then the equation can be simplified to the view [89]

$$\frac{\partial E}{\partial z} = \frac{q}{\varepsilon} N(z) - \frac{j_C}{\varepsilon v}. \tag{4.41}$$

Solution of this equation at the boundary conditions for $E(z)$ results in the dependence $M(j_C, U_{CE})$:

$$M(j_C, U_{CE}) \cong \left(\frac{j_C}{j_0} - 1\right)^{m/2} \left(\frac{U_{CE}}{U_\alpha}\right)^{m/2}. \tag{4.42}$$

Substitution of (4.42) in (4.40) provides the equation for stability boundary:

$$\frac{m}{2} M(1 - \alpha M)^{-1} \frac{j_C}{j_0} \left(\frac{j_C}{j_0} - 1\right)^{m/2 - 1} \left(\frac{U_{CE}}{U_\alpha}\right)^{m/2} = 1. \tag{4.43}$$

Thus, at $j_C \gg j_0$ a hyperbolic dependence $j_{CR} U_{CR} = $ const is similar to the one obtained in Section 4.2.2 from (4.22) using the simplest qualitative arguments.

From the application, it is known that in the CB circuit the condition (4.34) is true at smaller U_{CB} than (4.38), i.e., the current filamentation is observed at positive differential resistance [90].

Taking into account that in the case of the CB circuit the collector current $I_C \gg \alpha M I_E$, it is easy to demonstrate that criterion (4.39) essentially requires exceeding in the local base potential above the thermal potential $j_B \tilde{r}_{BEFF} > \frac{kT}{q}$ rather than contact potential difference 0.6 V. Condition (4.39) is rather weak and is not automatically implemented immediately with the appearance of negative base current.

Apparently, U_{CB} increase decreases the critical current j_{CR}. Exceeding of the value $j_0 = q N_D v$ in the filament region results in realization of the avalanche injection mechanism in the collector n-layer, NDC formation followed by filamentation of the collector current.

At low level of emitter current density j_E, redistribution of the emitter current may be insufficient for $j_{Emax} > j_0$. In this case, an additional current is required for

S-shape $I–V$ characteristic formation. This current is provided by the corresponding increase of avalanche multiplication near the collector junction. This effect results in increase of I_{CR} at $U_{CB} \rightarrow U_{CBCR}$ (Fig. 4.6).

Are there any parameters of $I–V$ characteristics under constant I_E and U_{CB} that may provide some information about the isothermal filamentation of emitter current event in CB circuits?

The necessary condition for this mechanism is existence of negative base currents. However, in the case of strong inhomogeneities the condition $\alpha M \gg 1$ may be true locally and the current filamentation may begin under $I_B > 0$. Thus, the change of sign of total base current can not be always used as an effective filamentation criterion.

Using an analogy with the emitter current thermal filamentation, it should be considered that an adequate sign of current redistribution is the change in monotonic pulsed characteristic $U_{EB}(t)\,|\,I_E$.

In fact, the avalanche holes complete a function of emitter junction forward biasing. Under the condition of constant I_E, the external potential difference U_{EB} must be decreased. In the experimental oscillogram [86] at the time moment preceding the sharp NDC, the violation of monotonic distribution for $U_{EB}(t)$ is observed. This may be treated as an indirect consequence of I_E redistribution.

4.2.5 Some ways of Isothermal Instability Suppression

In production, a limitation of statistical deviation from SOA is provided to choose proper absolute maximum and maximum ratings for the device including the technology limitations. Nevertheless, several ways of dealing with the instability on the device and circuit level could be derived from the avalanche injection understanding provided above.

One of the straightforward ways of suppressing and even eliminating the avalanche-injection current instability is suggested in [82, 85] to be increase of the value of product LN_D in the device structure. However, in most practical cases this approach results in worse frequency properties and decrease of the breakdown voltage.

A compromise solution could be implemented by a circuit solution that consists in connection of additional protection elements in parallel to the collector or modification and optimization of transistor design.

Another way is additional protection diodes. In this case, the overstress protection approach is similar to ESD protection approach. For example, a protection diode structure can be formed by p^+ diffusion regions in the n^--layer. This punch-through diode structure essentially presents itself as an open base pnp bipolar device. The base spacing is adjusted to the desired turn-on voltage to obtain the shutdown before the avalanche breakdown at the surface or in the bulk region of active structure. The total diode area is chosen adequately to high power dissipation.

One of the modifications of transistor n^+-p-n^--n^+ structure is formation of an additional n-epilayer between the n^- and n^+ regions. The doping level and thickness of this layer are optimized for reduced electric field E value on the n-n^+

boundary. The device suppresses the instability evolution when the electron density exceeds the donor concentration in the n^- layer, thus presenting itself as a compromise version for LN_D product increase.

To reduce the possibility of avalanche-injection filamentation of emitter current, the transistor is designed as follows. The transistor has the middle part of the emitter narrower and lighter doped in comparison with the periphery [91]. The doping level in the middle of the emitter is approximately equal to the doping level of the p-base. Therefore, the injection coefficient in this part of the emitter is negligible. The gain coefficient in the middle of the emitter is also significantly lower due to larger base width. These factors decrease the avalanche injection efficiency.

It is important to emphasize that the stability of transistors to the isothermal current filamentation might not be the same for devices of the same type. A number of local structural defects can significantly reduce the critical regime (see Section 4.5).

4.3 Isothermal Current Instability in MOSFET

Simultaneously with the first studies of secondary breakdown in bipolar transistors, the same phenomenon was observed in MOSFET too. In [92], an S-shape breakdown $I–V$ characteristic (Fig. 4.8) and current filamentation were observed. Although the mechanism of the observed phenomena was not explained in [92], today, after multiple studies of this problem, the isothermal instability nature of observed effects is perfectly clear.

The first measurements of S-shape $I–V$ characteristic of MOSFET exhibit major peculiarities that are typical for isothermal current filamentation in these devices. Namely, two S-shape regions (Fig. 4.8) were observed. The first S-shape region ($\Delta U_1 \approx 8$ V) corresponds to a relatively small current redistribution according to the filament region of $\sim 1/4W$, where W is the gate width ($W = 500$ µm). The switching on the second region with the holding residual voltage $U_{RES} \approx 13$ V (at $U_{CR} \approx 33$ V) has corresponded to appearance of an infrared lighting spot of approximately gate length dimension $L = 10$ µm. It has been identified as a very narrow current filament. Filamentation at the drain current I_D levels (Fig. 4.8) was reversible (apparently due to low-power transistor type). It was reported that the filament can change its localization in the structure.

The interest regarding current instability in MOSFET was reborn with the appearance of power device family [93–95]. At the same time, so-called grounded gate snapback NMOS became one of the key ESD protection devices for local clamps realized in integrated CMOS and BiCMOS processes.

The basic idea of conductivity modulation and current filamentation effects is presented in [94], namely, NDC region formation is due to "parasitic" bipolar transistor built-in structure. As usual for this parallel structure the source can be considered as an emitter, the substrate as a base, and the drain diffusion region as a collector.

As was mentioned above, under normal operation conditions of power N-MOSFETs, the source is usually connected to the p-substrate and the "parasitic" NPN transistor is normally disengaged due to the contact potential difference

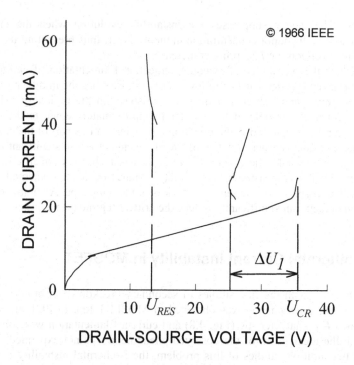

Fig. 4.8 Experimental I–V characteristic of MOSFET at secondary breakdown [92]

created at the emitter junction that prevents injection. Therefore, the parasitic bipolar n^+-p-n^+ (or p^+-n-p^+) device operation is realized in parallel to MOSFET structure under the condition with $U_{EB} = 0$.

However, this equilibrium situation changes at a certain avalanche breakdown level. Similarly to the grounded base NPN, discussed above, the beginning of avalanche ionization at the drain junction provides nonequilibrium hole carriers that penetrate into the substrate and provide the decrease of potential barrier at the source-substrate junction. At some adequate critical amount of these carriers, the source junction can be opened and the injection of the minor carriers into the substrate begins. The injection results in enhanced avalanche multiplication near the drain and appearance of the positive feedback. Realization of this mechanism provides the S-shape region in I–V characteristic $I_D(U_{DS})$.

In particular, this effect provides the physical limitation for layout design rules for P-well connection in the case of power and high-voltage NLDMOS arrays due to relatively high substrate current and distributed PWELL resistance.

Due to the importance of this class of devices. this upfront primitive phenomenological scheme deserves more details. Thus, the examples of power MOSFET structures and the critical isothermal instability regimes are discussed below.

4.3.1 Field Distribution in MOSFET Structures and Critical Operation Regimes

Intensive impact ionization is one of the necessary conditions for isothermal instability in MOSFETs. Therefore, the value of the critical voltage U_{CR}, when the derivative dI_D/dU_{DS} has change in polarity, and the breakdown voltage value U_{DSBR} are usually close to each other. However, an analytical estimation of U_{DSBR} value is not so simple due in principle to the two-dimensional distribution of the electric field in MOSFET structure in comparison with the one-dimensional approach in bipolar transistors. Therefore, simple analytical equations for U_{BR} estimation for bipolar transistors are not directly applicable for MOSFETs. One of the real ways of U_{DSBR} estimation is numerical simulation of ionization integral on the basis of Poisson and current balance equations with surface boundary conditions [96].

As known, a relatively low breakdown voltage value U_{DSBR} in low-power transistors is the result of formation of a sharp electric field distribution with maximum near the Si-SiO$_2$ interface under the drain side of the gate region. A decrease of absolute maximum of electric field and leveling of field distribution by various design methods presents a conventional way of U_{DSBR} increase toward MOSFET robustness to the isothermal filamentation. This is also related to the hot carrier reliability improvement.

There are several ways to decrease the electric field in the source-drain spacing. They are mainly based either on implementation of the graded junction, drift region, or reduced surface field (RESURF) techniques. State of the art implementation of the last technique is implementation of the multi-RESURF or so-called super junction approach.

For low-voltage devices, the major method is implementation of a lightly doped n$^-$-drain region and additional P-diffusion, so-called "halo". These two implants allow forming a graded junction with proper characteristics.

Several architectures are utilized for high-voltage MOSFETs. The major one is design of n$^-$-drain region or field facing of the distributed source (Fig. 4.9a) or drain (Fig. 4.9b) electrodes, or design of the Ndrift region in drain extended MOS (DeMOS) and lateral double diffusion MOS (LDMOS).

Presence of the lightly doped n$^-$-drain allows shifting the induced channel region away from the n$^+$-drain region and thus "stretching" the electric field distribution decreasing its peak value in particular. The field facing allows additional decrease in the field along the n$^-$-region. Apparently, the ratio between L_α and L_n values (Fig. 4.9a) or L_β and L_n requires optimization due to significant impact on the on-state resistance of the MOSFET device. The latter is critical for switching and power capabilities. For example, strong increase of the source electrode above the matched channel region (small L_α) results in breakdown voltage decrease. With L_α increase, the U_{DSBR} may be increased up to the value of bulk avalanche breakdown of drain n$^+$-p junction [97]. The same dependence is observed with L_n parameter increase.

Another way to increase U_{DSBR} is seen, for example, in the high-voltage MOSFETs ($U_{DSBR} > 1000$ V) [91].

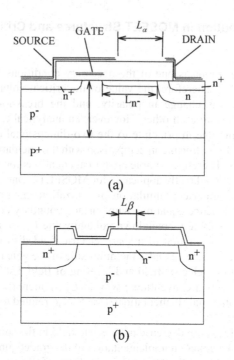

Fig. 4.9 MOSFET structure with source (a) and drain (b) field electrodes

However, U_{DSBR} increase by field reduction along the n^--region for MOSFETs with horizontal channel leads to decreased MOSFET stability to voltage overload. This means that a negligible increase in the drain-source voltage U_{DS} over the U_{DSBR} level may immediately result in current instability due to rather low critical drain current I_{DCR}. This effect is based on the fact that the operation of p- n^--n^+ structure with quasi-uniform electric field distribution in the n^--region is similar to p-i-n diode under avalanche breakdown conditions.

In [95], the critical drain current I_{DCR} dependence is determined on the p^+-substrate and L_α parameters (Fig. 4.9a). The tenfold decrease in p^+-substrate resistance decreased the critical current level only twofold (Fig. 4.10), i.e., doping level in p^+-substrate only slightly impacts the critical regime. From Fig. 4.10, at small L_α the I_{DCR} decrease is considerable and optimum value L_α is 10 μm. Together with other optimized parameters (ρ_{p+} = 0.01–0.02 Ωcm, ρ_p = 10 Ωcm, d_p = 20 μm), this provides for MOSFET structure with high stability to the secondary breakdown current [95]. However, major figures of merit for normal device operation for such a large device might be off target for the desired customer spec values.

An important parameter for the critical regime estimation (U_{DSCR} and I_{DCR}) is the substrate current level I_R. To some extent, this current is an equivalent of base current for parasitic NPN formed by drain bulk and source regions, respectively. In the case of n-channel MOSFETs, this is the current of avalanche holes. It creates

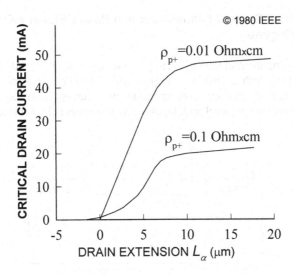

Fig. 4.10 Experimental dependence of the critical drain current on drain extension for two different substrate resistance values $I_{DCR}(L_\alpha)$ [95]

the forward potential difference of $I_R R_S$ value on the source-drain junction, where R_S is the substrate resistance.

At zero external substrate-source potential difference, but at $I_R R_S \approx 0.6$ V contact potential difference, the source junction becomes opened and electron injection from the source begins. The electron injection does not result in an immediate isothermal instability and S-shape I–V characteristic. However, for estimation in applications the criterion for critical regime may be assumed to be $I_R R_S = 0.6$ V. In the simplest case, the I_R value is given by

$$I_R = I_{CH}(M - 1), \qquad (4.44)$$

where I_{CH} is the channel current, and M is the multiplication coefficient in the drain space charge region, where I_{CH} flow occurs. More precisely, the I_R value is given by the ionization integral I_{ION} [79]:

$$I_{ION} = 1 - 1/M = \int \alpha_n \exp\left(\int(\alpha_p - \alpha_n)dx'\right)dx, \qquad (4.45)$$

where integration is along the current lines.

In MOSFETs. both coefficients α_n and α_p must be taken into account. For example, $\alpha_n \ll \alpha_p$ results in increased U_{DCR} in p-channel MOSFETs [96]. Moreover, the reason for isothermal stability in these devices is difference between the electron and hole mobilities. High electron mobility decreases the total resistance R_S of n-substrate and increases I_{CR}. Exact critical regime estimation can be accomplished using 2D numerical simulation (Section 4.4). However, in [98, 99] a simple analytical approach is suggested for estimation of the isothermal instability condition.

4.3.2 Isothermal Current Filamentation in Power Planar MOSFETs in DC Regime

Several early experimental studies were performed on the power MOSFET structures (Fig. 4.11a) with a builtin n⁻-region and field-plate drain electrode. Two types of investigated structures correspond to the commercial MOSFETs "1" and "2" with different doping level and depth of the n⁻-region (Fig. 4.11b,c).

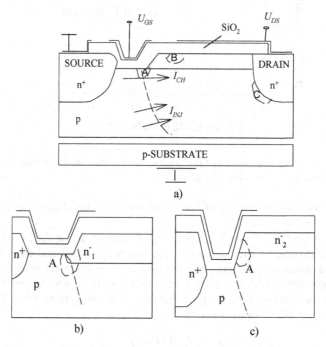

Fig. 4.11 (a) Experimental MOSFET structure with marked field maximum localization; (b) structure "1" ($n^-_1 = 5 \times 10^{16}$ cm⁻³, $p = 10^{15}$ cm⁻³); (c) structure "2" ($n^-_2 = 7 \times 10^{15}$ cm⁻³, $p = 10^{15}$ cm⁻³)

Isothermal instability in the structures results in sharp decrease of the drain-source voltage U_{DS} and failure due to local melting of the drain metallization. Dimension of the melted region is ~10 μm. In an optical microscope, the narrow metal short region is observed from the n⁺-drain along the n⁻-region in the gate direction. The short time of the instability evolution (< 1μs) and the small dimension of the melting region provide evidence of current filamentation of isothermal nature. However, a direct reliable observation of DC S-shape I–V characteristics was not possible at the time. In the pinch-off condition, the current limitation is less than 1 μA. It was revealed that the transistor failure at DC current limitation of ~ 50–150 mA/mm results in a transient process in the external circuit after the switching into the high-conductivity state. This process depends on the external circuit parasitic components.

The failure is caused by the current jumps with amplitude that exceeds the limited DC level. Appearance of the current overloads is the result of fast switching and presence of the circuit capacitance C between the drain and source terminals. Usually, C is the output capacitance of the current source, measurement network, and contact fixture or printed circuit board capacitance. The failure took place when the overload duration exceeded a typical heating time of the structure with the filament region above some critical temperature.

For the particular devices under DC conditions, the reduction of either the magnitude or the duration of overload provided elimination of the failure mode during the transient process. This was realized by decrease of the parasitic C capacitance or by including an additional limiting resistor R_{LIM} at the closest possible location in the drain terminal (Fig. 4.12b).

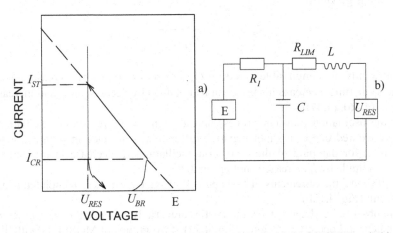

Fig. 4.12 S-shape I–V characteristic (a) and equivalent circuit for MOSFET switching

The transient process after switching can be described analytically as follows. For simplification, the MOSFET structure is taken into account only as an active element with zero differential resistance (Fig. 4.12a) and with immediate switching abilities. In the equivalent circuit (Fig. 4.12b), E and R_{LO} are the E.M.F. of the current source and the drain load resistance, and L is the total inductance of R_{LIM} and the circuit wires. The characteristic equation of the RLC circuit (Fig. 4.12b) has the roots $a_{1,2} = -a \pm \sqrt{a^2 - w_0^2}$,

where $a = \dfrac{R_{LIM}}{2L} + \dfrac{1}{2R_{LO}C}$, $w_0^2 = \dfrac{1}{LC}\left(1 + \dfrac{R_{LIM}}{R_{LO}}\right)$.

Usually, in application $R_{LO} \gg \sqrt{L/C}$. For this case at $a < w_0; \left(R_{LIM} < 2\sqrt{L/C}\right)$, the transient process has the oscillations with decrement a and frequency $\sqrt{w_0^2 - a^2}$. The maximum current is given by $I_{MAX} = I_{ST} + \Delta I_{MAX}$, where I_{ST} is the settled current value (Fig. 4.12a)

$$I_{ST} = I_{CR} + \frac{U_{CR} - U_{ST}}{R_{LO} - R_{LIM}}.$$

The maximum current amplitude at $R_{LIM} << \sqrt{L/C}$ is given by

$$\Delta I_{MAX} \approx \sqrt{L/C}\left(U_{CR} - U_{RES}\right). \tag{4.46}$$

At $a > w_0 ; \left(R_{LIM} > 2\sqrt{L/C}\right.$, the transient process is aperiodic and the current jump is given by

$$I_{MAX} \approx \frac{U_{CR} - U_{RES}}{R_{LIM}}. \tag{4.47}$$

Apparently, it is impossible to reduce I_{max} down to zero just by increasing R_{LIM}. At the same time, decrease in ΔI_{MAX} can be obtained by decrease in C and increase in R_{LIM} [(4.46), (4.47)].

In the oscillation transient process and at high R_L the relaxation oscillations may be excited due to a slight negative differential resistance in filament state. Therefore, for damping of the relaxation oscillations, a decrease in the current jump amplitude by L increase is not optimal.

In [98], DC measurements of S-shape $I-V$ characteristic were completed using the circuit (Fig. 4.12b).

The observed S-shape $I-V$ characteristics are strictly dependent on the structure parameters and operation regimes (Fig. 4.13). Two groups of MOSFETs with different regularities have been revealed.

In the first group

The avalanche breakdown is accompanied by a drift of the breakdown voltage U_{BR}. A drift up to 10 V of U_{BR} increase is observed. Typical for low-power MOS-FETs, the dependence $U_{BR}(U_{GS})$ is observed: $U_{BR} \cong U_{BR}\left(U_{GS} = 0\right) - |U_{GS}|$. At some critical current I_{CR1}, the first switching is observed with the residual voltage U_{RES1}. Value $\left(U_{BR} - U_{RES1}\right) = 30\text{–}40$ V. I_{CR1} in the pinch-off regime is less than 10 μA. With $I_{DS}\left(U_{GS} > 0\right)$ increase, the U_{CR1} value is decreased down to U_{RES1}. In this case, from a certain value $U_{GS} > 0$ the S-shape region has disappeared (Fig. 4.13a). In the case of floating substrate contact, the $I-V$ characteristic has not changed. On the contrary, the floating source configuration results in U_{BR} increase of 20–30 V and disappearance of the first break (dashed line in Fig. 4.13a).

In the second group

Here, the drift of U_{BR} is not observed, and the gate bias U_{GS} does not have an influence on U_{BR} at $U_{GS} < 0$. At some critical current I_{CR2} the switching of U_{DS} on the second vertical region of S-shape $I–V$ characteristic with $U_{RES2} = 15$–25 V (Fig. 4.13b). The value I_{CR2} in the pinch-off condition is small, but $I_{CR2} > 0.3$–0.7 mA. The switching in the second vertical region is observed in all current ranges (Fig. 4.13b). In the floating source configuration, the breakdown voltage U_{BR} itself was unchanged, but the current level for reversible snapback switching was significantly increased.

The regularities of the first group are typical for the first type of structure (Fig. 4.11b), and those of the second group are typical for the second type (Fig. 4.11c).

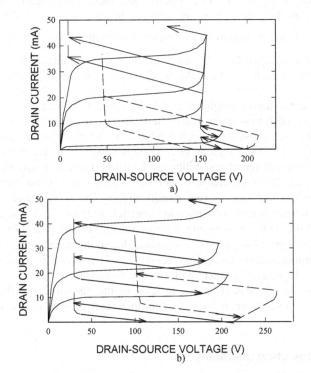

Fig. 4.13 Typical $I–V$ characteristic for first (a) and second group (b) (dashed line corresponds to floating source configuration)

To clarify the switching mechanism, it is necessary to analyze the peculiarities of the drain avalanche breakdown in the power MOSFET. According to simulation results [97], three multiplication regions can be formed in the device— "A,"

"B," and "C" (Fig. 4.11a). Region "A" is induced by the gate potential in the edge field of the p-n⁺ junction, region "B" is the result of the field effect of the distributed drain electrode, and region "C" is due to curvature of the cylindrical drain diffusion p-n⁺ junction. Domination of each region during avalanche breakdown depends on both the MOSFET structure parameters and the operation regime.

The drift of dependence $U_{BR}(U_{GS})$ and U_{BR} provides the conclusion that avalanche generation begins near the gate, i.e., in region "A." In fact, in the first type of structures this is the result of relatively high doping level of the n⁻-region and a high curvature of the p-n⁻ junction near the drain (Fig. 4.11b). The drift of U_{BR} can be explained by hot carrier degradation mechanism, for example the capture of hot holes on surface levels near the Si/SiO₂ interface followed by a corresponding change in the gate potential. The typical time of the drift observed in the study was ~ 0.1 sec, and the typical time of U_{BR} restoration was a few minutes. Application of the positive gate potential resulted in speeding up of the U_{BR} relaxation process.

Independence of $U_{BR}(U_{GS})$ proves that the avalanche multiplication region is farther away from the gate (region "B" or "C"). In fact, in the second type (Fig. 4.11c) the field in the n⁻-region was distributed more uniformly $\left(n_2^- < n_1^-\right)$ and the p-n⁻ junction at the surface was smoothed by mesa structure. Absence of the U_{BR} drift confirms that avalanche multiplication occurs in the "C" region at a certain depth.

Summarizing the above empirical conclusions, the switching on the first S-shape region is related to primary avalanche multiplication in region "A," but the switching on the second S-shape region is related to multiplication in region "C."

Presence of two S-shape regions of I–V characteristic provides an idea about involvement of two different instability mechanisms. Indeed, it will be demonstrated below that the first mechanism is based on the source-drain current gain dependence on the injection current density $\alpha(j_{IN})$. The second mechanism is based on the dependence of multiplication coefficient M on j_{IN}. The analysis of peculiarities of these mechanisms and the experimental results [82, 100] provided evidence that the first S-shape region is formed due to the dependence $\alpha(j_{IN})$ and is realized at multiplication located in region "A." The second is formed due to the dependence $M(j_{IN})$ at multiplication located in region "C."

This analysis is reproduced below. In spite of the particular specific device structures, the same analysis is valid for multiple cases of both power and ESD devices.

Criterion of injection instability

In the case of avalanche multiplication in region "A" (Fig. 4.11a) the flow path of holes into the substrate is located in the vicinity of the source-substrate junction. Therefore, the positive space charge of holes rather effectively increases the substrate potential near the drain and causes positive feedback in terms of a high injection current. This electron current results in intensive recombination of almost all the injected holes. The majority of electrons reach the space charge region of the drain junction. For the channel current I_{CH}, the source injection current I_{IN}, and their corresponding parts that flow through the multiplication region I_{CH}^* and I_{IN}^*, the balance equation is given by

$$I_{D0} + (M-1)I_{CH}^* + \alpha(M-1)I_{IN}^* = (1-\alpha)I_{IN}, \tag{4.48}$$

where I_{D0} is the saturation current of the drain junction.

At a small spacing between the source and multiplication region location, $I_{CH}^* \approx I_{CH}$, $I_{IN}^* \approx I_{IN}$. Thus, the multiplication of I_{D0} can be neglected.

On the basis of (4.48), for I_{IN} the following expression can be obtained:

$$I_{IN} = [(M-1)I_{CH} + I_{D0}]/(1-\alpha M). \tag{4.49}$$

The electron flow through region "A" slightly influences the field amplitude in this region. Since the width of the depletion region in the p-layer l_p for this region is less than the length of the depletion region L_n, then with the I_{IN} increase the field and multiplication coefficient are slightly decreased. Neglecting this dependence,

$$M = 1/\left[1 - (U_{DS}/U_{DS\,BR})^n\right], (n \approx 3 \div 5) \tag{4.50}$$

The dependence $\alpha(I_{IN})$ can be approximated by the function

$$\alpha\begin{cases} \alpha_{min} & \text{for } I_{IN} < I_{IN\,max} \\ \alpha_{max} & \text{for } I_{IN} > I_{IN\,max} \end{cases} \tag{4.51}$$

$$\frac{d\alpha}{dI_{IN}} = \text{const} > 0 \quad \text{for } I_{IN\,min} \le I_{IN} \le I_{IN\,max}$$

This dependence can be explained by the recombination decrease at small currents in the space charge region of source junction with I_{IN}.

Using (4.49) and $I_C = I_{IN} + I_{CH}$, the instability condition $R_g \equiv dU_{DS}/dI_{IN} \le 0$ is given by

$$I_{D0} + (M-1)I_{CH}^* + (\alpha M - \alpha)I_{IN}^* = (1-\alpha)I_{IN}. \tag{4.52}$$

At given I_{CH}, (4.49)–(4.52) determine the critical point ($V_{CR\alpha}$ and $I_{CR\alpha}$) when R_g becomes equal to zero. The stability boundary can be analyzed as follows. At $I_{CH} = 0$ (pinch-off condition $V_{GS} \sim 0$), the initial I_{IN} increase is the result of M increase with V_{DS} increase according to (4.49). The condition for instability is formed if at I_{INmin} the criterion (4.52) is true. This criterion can be transformed as

$$\frac{d\alpha}{dI_{IN}} I_{IN\min} \left(I_{IN\min} - I_{C0} \right) \geq \alpha I_{C0}. \tag{4.53}$$

If $I_{IN\min} \geq I_{D0}$ (4.53) is true, the conditions for instability occur immediately at $I_{IN\min} \left(\dfrac{d\alpha}{dI_{IN}} I_{IN} \sim \alpha \right)$. At $I_{D0} \rightarrow 0$, the critical voltage $V_{CR\alpha}$ approaches its maximum value: $V_{BR}(1-\alpha_{\min})^{1/n}$. With I_{D0} increase, $V_{CR\alpha}$ is decreased.

If $I_{IN\min} < I_{D0}$, then $\alpha = \alpha_{\max}$ before carrier multiplication and (4.63) cannot be satisfied. In this case, V_{DS} increase is limited, $I-V$ characteristic has a vertical region that corresponds to $M = a_{\max}^{-1}$, and thus no S-shape region is observed. The case of $I_{CH} > 0$ is the same as discussed above. In this case, the term $(M - 1)I_{CH}$ simply plays the role of I_{D0}.

The experimental regularities of the first group at first switching (Fig. 4.13.a) are in good agreement with the above discussion. A very small critical current I_{CR1} in the pinch-off condition can be explained by a small critical $I_{IN\min}$. The dependence of V_{CR1} on I_{CH} corresponds to the dependence $V_{CR\alpha}(I_{CH})$. The increase of the die temperature results in V_{CR1} decrease due to saturation current increase. The value $V_{RES1} \approx V_{BR} (1 - a_{\max})^{-1}$ can be sufficiently high. At floating source ($\alpha = 0$), the S-shape region is not observed due to impossibility of source injection. This results in V_{BR} increase up to a value approaching $M = \infty$ level.

The discussed conductivity modulation mechanisms by themselves do not cause the spatial current instability and filamentation [100]. In the measured 500-μm total gate width MESFETs, the current redistribution was slight and hadno impact on reliability parameters [92]. However, at drain current limitation with $R_{LIM} = 0$, the instability results in current jump that turns on the devices on the second S-shape region of the $I-V$ characteristic.

Analysis for the case when avalanche multiplication is started in the region far from the source-substrate junction (region C, Fig. 4.11a) is presented below. In this case, the holes flow to the substrate at some high distance from the source-substrate junction, therefore the hole space charge slightly influences the substrate potential near the source. This potential is determined by the current flow of holes through the substrate. The injection current is given by

$$I_{INJ} = I_{D0} \left[\exp\left(\frac{RI_R q}{kT} \right) - 1 \right], \tag{4.54}$$

where I_{D0} is the reverse current of the source junction, R is the effective substrate resistance, and I_R is the hole current flowing through the substrate.

Taking into account partial recombination of the holes,

$$\begin{aligned} I_R = \alpha \left(M - 1 \right) I_{IN}^* + \left(\alpha - 1 \right) I_{IN} + \left(M - 1 \right) I_{CH}^* , \\ + M I_{D0}^* + I_{D0}^* \left(1 - k \right) \end{aligned} \tag{4.55}$$

where $I_{IN}^* = \beta \, I_{IN}, I_{CH}^* = \gamma \, I_{CH}, I_{D0}^* = k I_{D0} \, (k, \beta, \gamma < 1)$.

The total hole current from the avalanche region to the substrate is given by

$$I_T = \alpha(M-1)I^*_{IN} + (M-1)I^*_{CH} + MI^*_{D0} + I^*_{D0}(1-k).$$ (4.56)

The major part of the electron injection current flows through the wide deple-tion region of p-substrate and penetrates into the n^+-drain region. If the increment of total hole current ΔI_T causes the increment of injection current resulting in ΔI^*_{IN} $> \Delta I_T$, then with the I_{IN} increase in the drain space charge region. the density of electron negative charge $\Delta j^*_{IN}/v$ is increased. According to the Poisson equation, this increases the field and M factor in the region near the n^+-drain (i.e., at the boundary p-n^+ or n^--n^+), therefore $dM/dI_{IN} > 0$.

From (4.54)–(4.56), it follows that $R_g = dU_{DS}/dI_D \le 0$ is true [$I_{DS} = I_{CH} + I_T +$ $\alpha(I_{IN} - I^*_{IN})$] at $dI^*_{IN}/dI_T \ge 1$ and $dI_R/dI_{IN} \ge 1$. This gives critical currents:

$$I_{RCR} = \frac{kT}{qR} \ln\left[\frac{kT}{qR(\beta+\alpha-1)I_{D0}}\right],$$

$$I_{INCR} = \frac{kT}{qR(\beta+\alpha-1)} - I_{D0}.$$

From (4.55), it follows that the second term is true at $M > 1 + (2 - \alpha)/2\beta$. By substitution of I_{RCR} and I_{INCR} into (4.55) and (4.50), the second equation for the stability boundary is given by

$$U_{CRM} = U_{BR}\sqrt{\frac{I_{INCR}(1-\alpha) + I_{RCR} - kI_{D0}}{I_{INCR}(1-\alpha+\alpha\beta) + I_{RCR} - \gamma I_{CH} + (k-1)I_{D0}}}.$$ (4.57)

Since in pinch-off conditions $I_{CH} = 0$, the spreading is high $0 < \beta \ll 1$ and U_{CRM} $\approx U_{BR}$. If $\gamma \ne 0$ (part of the channel current flows through the multiplication re-gion), with the channel current increase the U_{CRM} is decreased according to (4.57) and $I_{CRM} = I_{CH} + I_{TCR} + \alpha(1 - \beta)I_{IN}$. If $\gamma = 0$, then U_{CRM} does not depend on I_{CH}. At high I_{CH} at $dI_{IN}/dI_T \ge 1$ and $M < 1 + (2 - \alpha)/\alpha\beta$ true of $dI_R/dI_{IN} \ge 1$ might require an injection current additional to I_{INCR}. However, due to high $I_{CH}(dM/dI_{INT})$, this current is negligible and the stability boundary is determined by (4.57) with good accuracy.

The discussed instability mechanism explains the experimentally observed transition into the second S-shape region with small U_{RES2}.

For the studied structure at $R = 1500\ \Omega$, the calculated critical current $I_{TCR} \sim 0.1$ mA corresponds to the observed experimentally critical current I_{CR2} in the pinch-off condition. I_{CR2} does not depend on U_{GS} at $U_{GS} < 0$.

U_{CR2} corresponds to the U_{BR} value for a cylindrical p-n^+ junction: $U_{BR} < 250$ V at $p = 10^{15}$ cm^{-3} at a curvature radius ~ 6 μm.

Observed stability boundary (Fig. 4.14) in this case is relatively low dependent on I_{CH} corresponding to $I^*_{CH} \ll I_{CH}$, i.e., the major part of the channel current flows far away from the multiplication region. Apparently, this is evidence of the dominant multiplication in region "C." In the case of multiplication in region "B" (Fig. 4.11a), U_{CR2} must decrease sharply with I_{CH} increase.

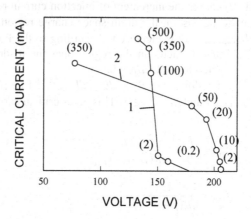

Fig. 4.14 Experimental boundary of injection instability for the first (1) and second (2) device types. The saturation current value is shown in parentheses

The experimental increase of the excessive current $(I_{CR} - I_{CH})$ with I_{CH} increase (Fig. 4.13b) is the result of injection current spreading $(\beta < 1)$ and $dI_R/dI_{IN} \geq 1$ holds.

After the second switching, the failure is observed due to local aluminum electromigration transport from the drain contact across the Si/SiO2 interface toward the gate. This is evidence of narrow current filament formation. Aluminum melting temperature is achieved in the filament region followed by melted metal transport due to the electrostatic force effect.

For the discussed electrical mechanism, a narrow filament is typical. A minimal filament dimension l_{min} can be estimated according to [101]

$$l_{min} = \sqrt[3]{\frac{|y_1| I_D}{g_\infty^2 vb\varepsilon}},\qquad(4.58)$$

where $g = g_\infty \exp(-b/E)$ is the ionization coefficient, the parameters $g_\infty = 7 \times 10^5$ cm^{-1}, $b = 1.2 \times 10^6$ V/cm, $y_1 = (-b/E_0)$, and $v = 6 \times 10^6$ cm/s.

From (4.58), for $I_D = 150$ mA/mm it follows that $l_{min} \approx 1$ μm and $U_{RES} \approx 20$ V. This practically corresponds to the measured values. For a filament model as a plane heat source of l_{min} radius on semiconductor surface and with κ thermoconductivity parameter, the temperature in the center is given by

$$T_{max} - T_0 = \frac{U_{RES2}}{\pi\kappa} l_{min}^{-1} I_D.$$

Therefore, with I_D increase T_{max} increases and at $I_D \approx 0.1$ mA $T_{max} \approx 600°C$, which corresponds to experimental data.

At floating source (dashed line in Fig. 4.13b), U_{BR} is practically unchanged ($\alpha = 0$). Increase of the avalanche current is limited by high substrate space charge region ~ 7 kΩ. The following behavior is similar to the case discussed in Section 4.2.1 for $I_E = 0$. The instability is realized only at the beginning of surface source-substrate breakdown. Since U_{RES} in this case is sufficiently high, $U_{RES} \approx U_{BR}/2$ (Fig. 4.13b).

A direct confirmation of the avalanche injection in MOSFETs was obtained via electroluminescence study of the structure surface before and after switching [102]. The spatial and spectral distribution of light emission intensity were measured with a linear resolution ~ 2 μm using a mechanical scanner setup with sensitive photomultiplier and filters.

It has been demonstrated [102] that before the switching, the light emission is distributed uniformly in the structure surface and the wide light emission spectrum is typical under avalanche breakdown conditions.

After switching into the first region, the light intensity is still uniform, but the portion of longwave light intensity is increased in the light emission spectrum (Fig. 4.15).

After switching into the second region, the light intensity distribution is localized in a region of ~ 10 μm; as a rule, this is the highest primary light intensity region, and the intensity of the longwave component in the light emission spectrum is increased even more. Thus, according to these measurements for given device structure, the filament dimension is ~ 10 μm.

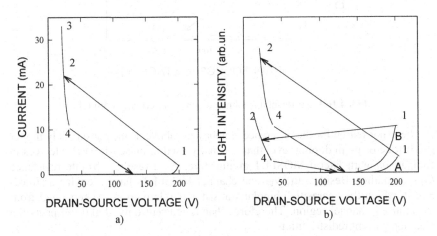

Fig. 4.15 Electroluminescence at MOSFET breakdown: *I–V* characteristic (a) and light intensity (b) in infrared (A) and visual range (B)

In DC conditions at high I_{CH}, the elevated lattice temperature cannot be neglected anymore and thermal heating plays a significant role in instability appearance due to dominant thermal carrier generation in the space charge region.

4.3.3 Isothermal Instability in Vertical MOSFETs

Physical principles of isothermal instability in vertical and trench gate double diffusion MOSFETs (DMOS) are similar to lateral MOSFETs, lateral DMOS (LDMOS), and drain extended (DeMOS) devices. However, in several experimental studies, higher stability to overload and isothermal instability of vertical MOSFETs is reported. Critical current value is an order of magnitude higher in the pinch-off condition in comparison with lateral MOSFET devices. Moreover, the switching into the current filament state does not always result in device failure. In some cases, the filament region of I–V characteristic as usual may be observed reversibly without any special protection measures (Fig. 4.16).

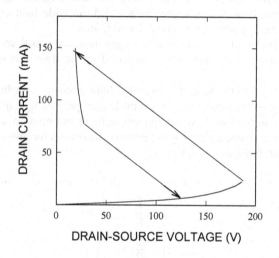

Fig. 4.16 Experimental I–V characteristic of vertical MOSFET

The major reason for the difference, emphasized above, is the considerably higher p-substrate or p-epi doping level that is usually used in these MOSFET devices in comparison with lateral MOSFET architecture. As a result, substrate resistance R_{SUB} is smaller. Besides, the vertical channel MOSFETs have a device architecture in which the "parasitic" bipolar transistor is relatively more separated from the induced channel region. Therefore, channel current only slightly impacts the injection current redistribution.

In this section, analysis for the critical regime is presented using a different approach than that in Section 4.3.2. This approach achieves rather good agreement with the experimental data [103].

In [103], a typical power V-MOSFET device (Fig. 4.17a) with equivalent circuit that includes a parasitic bipolar structure (Fig. 4.17b) is evaluated.

As first approximation, the critical regime (Fig. 4.17c) is reached when the potential $V_{SUB} = I_{SUB}R_{SUB} = 0.6$ V of the source-substrate junction is induced in the channel region [103].

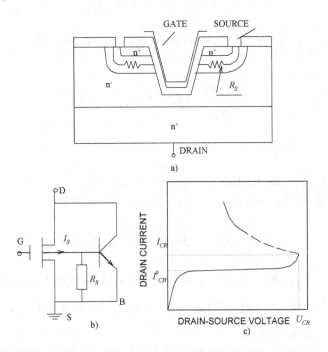

Fig. 4.17 (a) Structure, (b) equivalent circuit, and (c) S-shape I–V characteristic of V-MOSFET [103]

At a fixed V_{GD} value, the total channel current is given by

$$I_{CH} = I_{CH}^0 + \gamma_S V_{SUB} = I_{CH}^0 + \gamma_S R_{SUB} I_{SUB},$$

where I_{CH}^0 is the saturation current at small drain voltage (Fig. 4.17c) and γ_S is an analogue of the internal gate transconductance:

$$\gamma_S = \frac{\Delta I_{CH}}{\Delta V_{SUB}}. \tag{4.59}$$

Increase in channel current due to substrate potential increase may be calculated as

$$I_{CH}^0 = \frac{W}{2L} \mu C_I \left(U_{GS} - U_{CR}\right)^2, \tag{4.60}$$

where W, L are the gate width and length, respectively, and C_I is the MOS capacitance.

In short-channel MOSFETs, for high I_{CH} this ratio has another form [103]:

$$I_{CH}^0 = \frac{W}{L} C_I v (U_{GS} - U_{CR}),$$ (4.61)

where v is the saturation drift velocity under inverse channel conditions.

The U_{CR} is given by

$$U_{CR} \sim \frac{\sqrt{2\varepsilon q N_A}}{C_I} \left[\sqrt{2|\psi_B| - V_{CR}} - \sqrt{2|\psi_B|} \right],$$ (4.62)

where $\psi_B = \frac{kT}{q} \ln \frac{N_A}{n_I}$.

For high I_{CH}^0, the median γ_S only slightly depends on V_{SUB} and I_{CH}^0 and at fixed W and L is dependent only on the substrate doping level [103]:

$$\gamma_S \sim \left(\sqrt{2|\psi_B|} - \sqrt{2|\psi_B| - 0.6} \right) \frac{W}{L} v \sqrt{2\varepsilon q N_A}.$$ (4.63)

At $U_{DS} < U_{DS\,BR}$ at the beginning of avalanche multiplication I_{SUB} is given by

$$I_{SUB} = I_D - I_{CH} = (M-1)I_{CH} = (M-1)(I_{CH}^0 + \gamma_S R_{SUB} I_{SUB}),$$ (4.64)

where M is given by (4.50).

Critical regime for instability coincides with opening of the parasitic bipolar transistor at $I_{SUB} R_{SUB} \approx 0.6$ V. Then the critical regime I_{CR}, U_{CR} (Fig. 4.17c) is determined by

$$I_{CR} = I_{CH}^0 + \frac{0.6}{R_{SUB}} + 0.6\gamma_S;$$ (4.65)

$$U_{CR} = U_{CRBR} \left[1 + R_{SUB} \left(\gamma_S + \frac{I_{CH}^0}{0.6} \right) \right]^{-\frac{1}{n}}.$$ (4.66)

In this case, U_{DSBR} is the drain p-n$^+$-junction breakdown voltage. From solution of (4.65) and (4.66) the stability boundary is given by

$$U_{CR} = U_{CRBR} (0.6 / R_{SUB} I_{CR})^{-\frac{1}{n}}.$$ (4.67)

The instability begins at the moment of source junction opening $I_{CR}R_{SUB} = 0.6$ V. In this case, I_D increase up to I_{CR} is the result of channel current increase due to forward substrate biasing by the flowing hole current.

Experimental confirmation of this theoretical analysis is reported in [103] for the case of specially designed MOSFET test structures. Small contact width ($W = 400$ μm) of the structures provided reversible measurements of DC S-shape I–V characteristics. Other parameters of the test structure have been selected to be the same as typical power vertical MOSFET parameters.

The measured I_{CR} and U_{CR} values of the test MOSFET are compared with calculated data using analytical expressions (4.65) and (4.66). The R_{SUB} and γ_S values are determined empirically using

$$I_{CR}(R_{SUB} + R) = 0.6, \tag{4.68}$$

$$\gamma_S = \frac{I_{CR} - I_{CH}^0}{0.6}, \tag{4.69}$$

where R is the external resistance in the substrate circuit. $R_{SUB} \approx 0.85$ kΩ is determined by linear extrapolation of $I_{CR}^{-1}(R)$. The dependence $\gamma_S(I_{CH}^0)$ is presented in Fig. 4.18. With current increase, γ_S approaches a constant value according to (4.69). R_{SUB} and γ_S have been substituted in (4.65) and (4.66) by the empirical values to provide a good agreement of the measured and calculated U_{CR} values.

An idea that the critical condition is achieved due to the substrate biasing effect rather than due to the injection from the source for small I_{CH} levels was supported in [104] too.

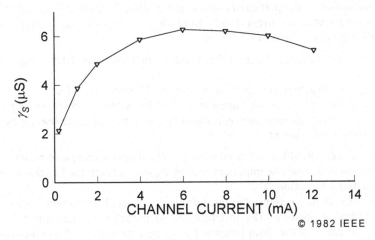

© 1982 IEEE

Fig. 4.18 Empirical dependence of the internal gate transconductance on channel current [103]

This approach [104] to determine for the shape of stability boundary (for horizontal channel MOSFET) at small drain current levels when the condition $M - 1 \cong \left(1 - \dfrac{1}{M}\right) \sim U_{DS}^4$ is true and therefore the critical point $\dfrac{dU_{DS}}{dI_D} = \dfrac{d(M-1)}{dI_D} = 0$.

In an inverted channel regime with the drain current dependence $I_D \sim (M-1)^2$, the stability boundary is given by

$$U_{CR} \sim N_A^{-3/4} R_{SUB}^{-1/4} I_D^{-1/8} . \tag{4.70}$$

For the light inversion regime,

$$U_{CR} \sim N_A^{-3/4} R_{SUB}^{-1/4} I_D^{-1/4} . \tag{4.71}$$

The experimental measurements [104] for current range $I_D = 10^{-4}$–1 A confirmed the presence of two stability boundary regions that correspond to dependencies $U_{CR}(I_{CR})$.

The nature of preliminary critical current, probably, is related to an increase of the surface potential. However, at the instability ($I > I_{CR}$) due to avalanche-injection mechanism is dominant. The negative "base" current flow may considerably enhance an inhomogeneity of the source junction and may further result in filamentation of injection current. Realization of such instability mechanism is treated in "bipolar" concept [105] and suggested ratios (4.39) and (4.40) may be used as criteria. However, this analogy requires experimental validation.

Indirect information on the filament nature may be provided by measurements on four-terminal MOSFET with additional contact with the substrate.

Identification of the particular mechanism is difficult because there are at least three ways for critical current density to reach $j_0 = qNv$ that usually significantly exceeds experimentally observed levels:

(1) j_0 may be exceeded during the switching process under NDC conditions at $I > I_{CR}$;
(2) j_0 may be achieved locally due to current filamentation effect;
(3) over critical j_0 may be caused by an instability that is accompanying channel current increase and redistribution due to the substrate biasing by avalanche hole current.

Since power MOSFETs are used mainly in the dynamic and pulse regimes, it is important to underline one important fact taking into account the kinetics of injection instability evolution.

From the above sections, the uncontrollable avalanche injection begins after source junction opening. The instability is supported by enhancement in avalanche generation at the drain junction by electron current that flows through the base of the "parasitic" bipolar transistor. Obviously, electron diffusion through the base (substrate) is a major factor that limits the speed of instability evolution. Minimal diffusion time is proportional to L^2/D_n (where L is the channel length, D_n is the electron diffusion coefficient in p-substrate). It determines the frequency dependence of the gain coefficient in the "parasitic" bipolar transistor. Operation frequency of MOSFETs is limited by the electron transit time through the device channel. In the saturation regime, this time is L/v_S, where $v_S \approx 10^7$ cm/s. Since the diffusion velocity of electrons D_n/L is at least one order of magnitude slower than the drift velocity v_S, there is an operation frequency range $\left(\dfrac{D_n}{2\pi L^2} < f < \dfrac{v_S}{2\pi L} \right)$ where the "parasitic" bipolar transistor is not operable. Thus, MOSFETs may avoid the avalanche-injection current filamentation at operation in the high-frequency range.

4.4 Numerical Simulation of Electrical Instability and Current Filamentation in NMOS Devices

Avalanche-injection instability has been observed in various MOSFETs and bipolar transistors [108–114], GaAs MESFETs [37, 38], and MODFETs [39]. This is one of the major physical limitations of the safe operating regime under short pulse conditions.

One of the most common practical examples is NMOS operation under ESD conditions or under electrical overstress. In this case, NMOS can be considered both as a device that may fail due to pulse overload and as a protection device. One of the most popular ESD protection solutions for CMOS integrated circuits is the grounded gate NMOS where an additional distributed drain ballasting region is added to the standard NMOS structure [105–108]. ESD applications will be discussed in Chapter 7. The purpose of this section is to demonstrate the in-depth processes in integrated NMOS devices in the case of electrical conductivity modulation using two-dimensional numerical simulation followed including the phenomenon of avalanche-injection current filamentation.

Numerical model

In the original study [114], the models implemented in noncommercial simulator were used. Today, several commercial numerical simulators are widely available. They can easily repeat similar numerical experiments.

The complete two-dimensional model and numerical simulation approach are described in Chapter 5, Section 5.2.4 [114] for the case of quasi-hydrodynamic equations required for compound semiconductor devices.

For silicon NMOSFET devices, the drift-diffusion approximation is valid. Thus, the model includes only balance equations for carrier density and total current. Nonflux boundary conditions have been used at the lateral sides. The Dirichlet boundary condition has been used at the contacts for electrostatic potential and carrier density. The absolutely stable, completely conservative finite-difference scheme and calculation algorithm for numerical solver is presented in [117].

The following silicon material parameters are used. The band gap energy E_G = 1.125 for T_0 = 300 K [78]. The ionization coefficients are given in [118]. For recombination, the Shockley-Read-Hall [78] and the Auger models are used in the form:

$$R(n,p) = (np - n_i^2)\left[\frac{1}{\tau_n(p + n_i) + \tau_p(n + n_i)} + C_n n + C_p p\right],$$

where n_i is the intrinsic carrier concentration,

$$\tau_n = \frac{\tau_{n0}}{(1 + N_I/N_0)^\delta}, \tau_p = \frac{\tau_{p0}}{(1 + N_I/N_0)^\delta}, \qquad N_I = N_D + N_A,$$

$$\tau_{n0} = \tau_{p0} = 10^{-5}\,\text{s}, \quad N_0 = 10^{15}\,\text{cm}^{-3}, \delta = 0.5, \quad C_n = C_p = 10^{-31}\,\text{cm}^6\text{s},$$

and N_D, N_A are the donor and acceptor concentration. The mobility is expressed by

$$\mu_n(E) = \frac{\mu_{0n}[\text{V/cm}]}{\left[1 + \left(\frac{E}{8000}\right)^2\right]^{1/2}}, \mu_p(E) = \frac{\mu_{0p}[\text{V/cm}]}{1 + \frac{E}{19500}},$$

where

$$\mu_{0n} = 65 + \frac{1265}{1 + \left(\frac{N_T}{0.85 \cdot 10^{17}}\right)^{0.72}}\left[\frac{\text{cm}^2}{\text{V} \cdot \text{s}}\right],$$

$$\mu_{0p} = 47.7 + \frac{477}{1 + \left(\frac{N_T}{0.63 \cdot 10^{17}}\right)^{0.76}}\left[\frac{\text{cm}^2}{\text{V} \cdot \text{s}}\right], E \text{ is the electric field.}$$

NMOS structures

The original simulation was done for the simplified long-channel NMOS structures (Fig. 4.19.a) to represent a planar MOSFET structure with grounded p^+-substrate. The grounded p^+-contact at the bottom provides rather low channel to substrate resistance.

The doping level in the n^+-drain (source), p-region, and p^+-region is $N_D = 10^{18}$ cm^{-3}, $N_A = 10^{16}$ cm^{-3}, and $N_A = 2 \times 10^{18}$ cm^{-3}, respectively. The gate oxide thickness is 0.12 μm. The n^+-region length $L_D = L_S = 0.41$ μm. The n^+-p-n^+ structure for filamentation simulation is presented below.

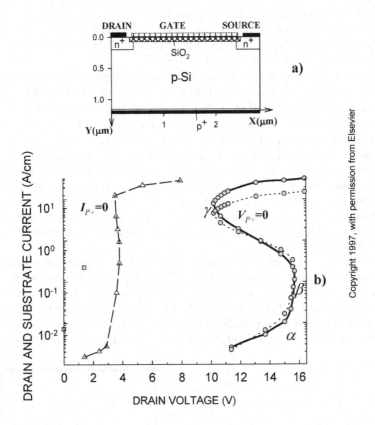

Fig. 4.19 Simulated cross sections for MOSFET structures with p$^+$-substrate (a) and calculated I_D–V_{DS} characteristics (b) for the biasing circuit with grounded gate and substrate in the NMOS circuit (solid line), consequent substrate current (dashed lines) and for the case of floating substrate contact ($I_{P+} = 0$) [121]

Transient I–V characteristic and spatial distributions were calculated in [121] at a given drain current value up to the time when a stationary spatial distribution of carriers was formed.

Simulation of NDR in the NMOS structures

The calculated I_D–V_{DS} characteristics of the NMOS structures ($V_{GS} = 0$) (Fig. 4.19a) are presented in Fig. 4.19b. The characteristics are typical for MOSFETs at zero gate bias. The characteristics immediately demonstrate the influence of the substrate resistance on the p-region conductivity modulation. The substrate current (I_{P+}) through the p$^+$-contact is presented by dashed lines (Fig. 4.19b). With substrate resistance decrease, the substrate current is increased and the critical voltage of NDR formation is increased too.

One can divide the I_D–V_{DS} characteristics into the following parts: region of drain junction saturation current I_{D0}, drain avalanche breakdown region, NDR region, and high-conductivity region with positive differential resistance. The regions can be understood from the analysis of the electric field, and the electron and hole density distribution in the structure (Fig. 4.20).

The distribution along the X coordinate (at the depth $Y = 0.14$ μm near the SiO$_2$/Si boundary) is presented in Fig. 4.20a–c for three states of the grounded substrate NMOS structure (Fig. 4.19a): avalanche breakdown (α), NDR (β), and high conductivity (γ). With a current increase, the field near the drain is increased up to the avalanche impact ionization state (α). Some of the generated holes penetrate to the source and enhance injection of electrons. The space charge of injected electrons compensates the space charge of holes in the drain-source spacing and increases the electric field in the avalanche ionization region near the drain n$^+$-p junction. This results in formation of a quasi-neutral region near the source n$^+$-p junction (Fig. 4.20b,c). At some critical level of the drain current, the voltage drop decrease due to the space charge neutralization exceeds the voltage drop increase in the drain avalanche region.

With increase in current, the neutralization is enhanced, thus further decreasing the voltage drop across the p-region (Fig. 4.20a) and provides the current-controlled NDR of the structure. The field and carrier depth profiles for the high-conductivity state (β) presented in Fig. 4.20d–f. In the high-conductivity state (γ), the voltage drop across the n$^+$-regions provides the positive differential resistance. Finally, the maximum current density is limited below the current saturation in the n$^+$-region.

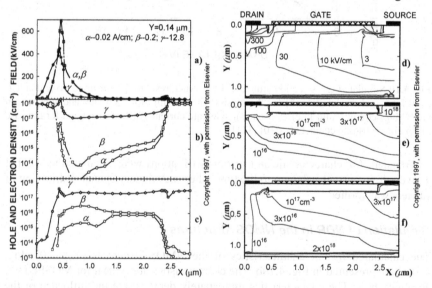

Fig. 4.20 The field (a), electron (b), and hole (c) density distribution in the NMOS structure (Fig. 4.19a) for: drain avalanche breakdown state (α) and (β) and NDR high-conductivity state (γ), marked by points "α," "β," "γ" in Fig. 4.19c. (d–f) The field and carrier depth profiles for high-conductivity state γ[121]

In the NMOS structure with local p^+-region (Fig. 4.19b), the avalanche-injection current instability due to p-region conductivity modulation, NDR region, and current saturation in the n^+-regions are similar to the structure (Fig. 4.19a).

The calculated I_D–V_{DS} characteristic of the snapback NMOS (Fig. 4.19a) structure with floating p^+-contact ($I_{P+} = 0$) is also presented in Fig. 4.19c. This characteristic has a low critical voltage and a small NDR region. However, the conductivity modulation and NDR are observed at a small multiplication coefficient due to high saturation current I_{D0} of the drain n^+-p junction. The conductivity modulation is observed at a low electric field. In this configuration, the snapback NMOS behavior is equivalent to a bipolar transistor with floating base discussed in the previous chapter.

Simulation of current filamentation

The second step in numerical simulation [114] was to obtain the numerical solution for isothermal avalanche-injection current filamentation. In semiconductor structures of an adequate high contact width W, spatially uniform current distribution along the contacts in current-controlled NDR state is in general unstable [119]. Thus, in [114] it was assumed that the current filamentation might be obtained by numerical simulation as well for the case of filament dimension less than W.

Prior to study [114], the problem of 2D simulation of isothermal current filamentation had been solved for the case of avalanche-injection instability in the GaAs n^+-i-n^+ structures, and the double avalanche-injection instability in the Schottky M-i-n^+ and p-i-n^+ structures are presented in [114, 120] and [121, 122], respectively (Chapter 5). In these studies, the parasitic device has been adequately represented by the two-terminal 2D distributed diode structure.

Unlike in GaAs structures, the challenge of the case of NMOS parasitic BJT simulation is the necessity to take into account the third terminal.

In the case of NMOS, according to the presented simulation results (Figs. 4.19 and 4.20) the substrate current determined the conductivity modulation mechanism in the drain-source spacing and NDR. Escape of holes through the p-substrate decreases the electron injection from the source up to the drain avalanche breakdown condition. Therefore, a simple 2D n^+-p-n^+ structure [114] is not suitable for 2D simulation of filamentation in parasitic NPN BJT of NMOS and MOSFET structures.

A pioneering approach has been suggested in [114] to reduce the 3D problem down to 2D numerical simulation capability by a composition of 2D triode n^+-p-$n^+[p^+]$ structure. This quasi-3D structure is suggested as a mathematical approximation (Fig. 4.21a). It presents itself as a periodic structure with multiple n^+- and p^+-cells of the source and substrate contact regions.

It is important that the dimension of n^+- and p^+-cells of the source and substrate contacts is less than the expected filament dimension. Similarly, the structure can be understood as a distributed cross section of very-low-gain multifinger

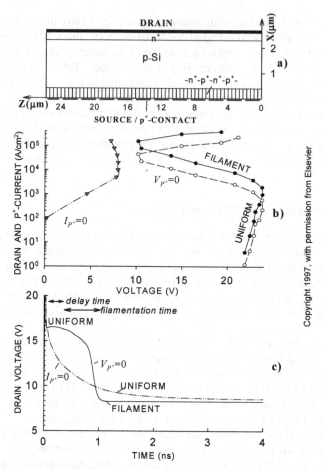

Fig. 4.21 Multicell n^+-p-n^+[p^+] structure (a), calculated I_D–V_{DS} characteristics for the grounded and floating p^+-cell configuration (b), and the drain voltage versus time after a stepwise current increase from $I_D = 0$ up to $I_D = 20$ kA/cm^2 (c) [121]

NPN BJT where emitter and base regions are reduced below the reasonable process limits and butted, while the P-base region is rather large.

The p^+-diffusion region provides partial escape of hole current from the bulk p-region thus dumping electron injection from the multiple source n^+-regions. To decrease the influence of cells on the electric field distribution along the drain, their dimension is decreased down to 0.8–0.1 μm.

This n^+-p-n^+[p^+] structure has been numerically simulated for three major configurations: grounded p^+-cells (fixed zero potentials of p^+ and source n^+-cells), floating p^+-cells (fixed zero total current I_{P+} through the p^+-cell contact), and at a given current I_{P+}. Essentially, these conditions correspond to the grounded base, open base, and positive base current through the parasitic NPN device.

In the n^+-p-$n^+[p^+]$ structure with the drain contact width $W \leq 0.5$ μm (the cell width 0.1 μm), current distribution along contacts is uniform in the NDR state. The I–V characteristic, field, and carrier density distribution along the X coordinate qualitatively correspond to the grounded substrate MOS structure with the same p^+-contact region efficiency (I_{P+}/I_D).

In the n^+-p-$n^+[p^+]$ structure of $W = 26$ μm drain contact width, the filamentation is observed at the drain current density that corresponds to the NDR region. The calculated I–V characteristic of the structure is presented in Fig. 4.21b. The shape of the NDR region of the I–V characteristic differs from that for the NMOS structure (Fig. 4.21b). The NDR branch in the n^+-p-$n^+[p^+]$ structure corresponds to the filament state.

The calculated transient voltage dependence for the fixed drain current 20 kA/cm^2 is presented in Fig. 4.21c. At grounded p^+-cells for a current step exceeding the critical current of NDR, a stable filament state is formed. After accomplishing the fast (~ 0.1 ns) transient process, a quasi-stable state with uniform current distribution ("uniform") along the drain contact is observed during the delay time interval of 0.4 ns. Then a new nonuniform state ("filament") is formed during a time interval of 0.5 ns (Fig. 4.21c).

An interesting observation from the methodology [114] is an automatically given possibility to utilize the fluctuation of the numerical solution errors to enable transient solutions for the current filamentation. This numerical "noise" due to finite accuracy of the numerical solution provided conditions for filament excitation with no additional fluctuation source included in the equation model [114]. Thus, the spatial instability of the uniform state is caused by numerical fluctuations of the finite-difference scheme itself.

The stages of front creation and rudimentary filament stagnation [123] are observed in the filamentation process. To enable the numerical solution for the current filament in the simulations small numerical errors are increased spontaneously; in real devices the current fluctuations (especially at the pulsed stress) and local inhomogeneity (doping profile, contacts, surface traps, etc.) determined the real critical current for instability, which is less than that calculated for the ideal structure.

For floating p^+-cell configuration, the stable state is formed during the first 3 ns. This state is uniform. However, for the 100-μm-contact-width n^+-p-n^+ structures, a filamentation is observed. For 20 kA/cm^2 current step, the filamentation time is 5 ns and the filament dimension ~ 40 μm. The 2D distribution for the floating p^+-cell structure is close to the structure with a condensed source n^+-region (without p^+-cells).

The distribution of the field, electron, and hole densities is presented in Fig. 4.22 for the filament state of the grounded p^+-cell structure. Across the filament, the field and carrier density distribution corresponds to the distribution in the high-conductivity state of the NMOS structure (Fig. 4.20).

After a current increase, the filament is broadened up to the structure width. The dependence of the filament amplitude and width on the drain current is presented in Fig. 4.23a. Dependence of the delay and filamentation time and filament amplitude on the p^+-region current to drain current ratio is presented in Fig. 4.23b.

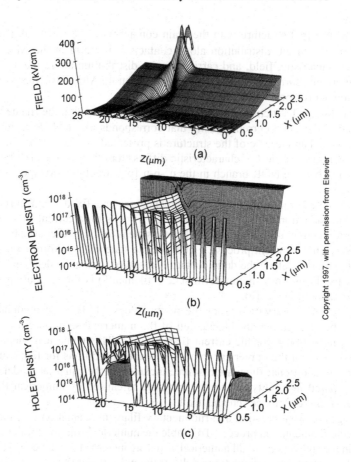

Fig. 4.22 Calculated distribution of the electric field (a), electron (b), and hole (c) density in 2D n⁺-p-n⁺[p⁺] structure for current density $I = 20$ kA/cm² [121]

The filament amplitude is limited below the current saturation in the drain region. A decrease of the donor concentration in the drain n⁺-region results in a decrease of the filament amplitude.

According to the presented results, the filamentation time, filament width, and critical drain current depend drastically on the substrate hole current level. The physical mechanism of p-region conductivity modulation of the n⁺-p-n⁺ structure with p⁺-contact corresponds to a phenomenological mechanism [113]: the positive feedback between the hole generation level in the drain region of avalanche multiplication and electron injection from the source n⁺-contact resulting in neutralization of space charge in the p-region.

In planar MOSFETs in a low thermal overheating condition, output I_D–V_{DS} characteristics are limited by the avalanche-injection NDR resulting in the snapback characteristic. To demonstrate this, the output I_D–V_{DS} characteristic of the MOSFET structure (Fig. 4.19a) at various gate biases V_{GS} is presented in Fig. 4.24. From the I_D–V_{DS} dependence and the carrier density and field distribution analysis, it follows that the limitation of V_{DS} is due to the same avalanche injection mechanism. The avalanche-injection filamentation in the NDR state must be observed at various gate biases V_{GS}. The same conclusion can be obtained for double-diffusion high voltage switching MOSFETs. For such device geometry, the filamentation in multicell n$^+$-p-n$^+$[p$^+$] structure is a good approximation.

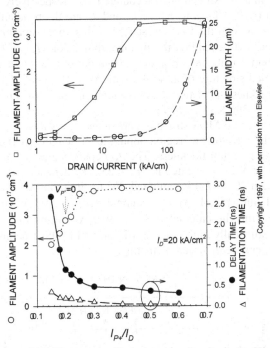

Fig. 4.23 Dependence of the filament amplitude (electron density in the $X = 1.64$ µm plane) and width on the drain current (a) and dependence of the filament amplitude and delay and filamentation time on the given I_{P+} current to drain current ratio (b) [121]

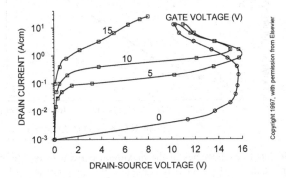

Fig. 4.24 Calculated output I_D-V_{DS} characteristics of the MOSFET for different gate voltages [121]

The maximum temperature in the filament state T_{max} can be roughly estimated for the filament as a plane heat source of diameter w at the semiconductor surface. Since filament and MOS structure dimensions are much less than the substrate (chip) thickness, the maximum temperature is given by $T_{max} - T_0 = 2P/(\pi w \lambda)$, where $T_0 = 300$ K, P, and $\lambda = 1.5$ W/cmK are the ambient temperature, the dissipated power, and the silicon thermoconductivity, respectively. The dissipated power is given by $P = V_R J_M w$, where V_R and J_M are the residual voltage and the linear current density in the filament state. For this approximation, the maximum temperature does not depend on the filament dimension: $T_{max} - T_0 = 2V_R J_M/(\pi \lambda)$. The linear current density can be obtained from the simulation results (Fig. 4.19c). For the simplified structure (Fig. 4.19a) with $N_D = 10^{18}$ cm^{-3}, $L_D = L_S = 0.4$ μm, and p-region thickness $d = 1$ μm, the maximum current is low ($J_M \approx 50$ A/cm) and $T_{max} - T_0$ is only ~ 200 K ($V_R \sim 10$V).

However, using real MOS structure parameters, the values J_M and $T_{max} - T_0$ are multiplied by three factors: n$^+$-region doping level, length (area) of the n$^+$-contact regions, and the p-region thickness. The J_M dependencies for the structure (Fig. 4.19a) at $V_{DS} = 15$ V are presented in Fig. 4.25. The dependence of J_M on the contact region length ($L_D = L_S$) is saturated due to limitation by the small p-region thickness ($d = 1$ μm) (Fig. 4.25). For example, at $N_D = 2 \times 10^{18}$ cm^{-3}, $L_D = L_S = 5$ μm, and $d = 4$ μm, the calculated J_M is 1950 A/cm and the estimated $T_{max} \approx$ 8000 °C is unrealistically high. That is why in real MOS structures T_{max} exceeds the critical temperature of structure melting.

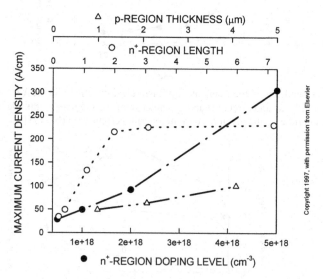

Fig. 4.25 Calculated dependencies of the maximum current J_M on the length and doping level of the n$^+$-region and p-region thickness of the structure (Fig. 4.19a) in the high-conductivity state at $V_{DS} = 15$ V

Consideration of the time dependence of the heating (thermal filamentation time usually $\sim$ 1 μs [124]) confirms that avalanche-injection filamentation must precede formation of thermal instability conditions. Therefore, the catastrophic failure due to the drain-source local burnout of NMOS and MOSFETs begins with local overheating inside the avalanche-injection filament. Meanwhile, the typical temperature of metallization melting is $T_{BURN} \sim 600°C$. Therefore, before melting the intraband carrier generation can enhance current density in the filament.

In this condition, the avalanche-injection filament acts as a local heat source for following thermal burnout. Therefore, the electrical filamentation must determine the NMOS protection structure capability at ESD. Exceeding of the critical drain current density 0.05–0.5 A/cm in the NMOS structures provides NDR due to avalanche-injection conductivity modulation of the source-drain spacing. At the NDR, the drain current distribution along the contacts is unstable and a 1- to 10-μm filament is formed during 0.5–3 ns. The maximum current density in the filament is limited below saturation in the n^+-regions. With substrate resistance decrease or substrate (p^+-region) current increase, the filamentation time and filament width are decreased.

4.5 Summary

This chapter has discussed isothermal instability and current filamentation in silicon devices based on experimental data and analytical and numerical simulation models. The important regularity was demonstrated: the basis of electrical mechanisms of catastrophic failures is the breakdown, isothermal conductivity modulation, and spatial instability in elementary semiconductor diode and triode structures (n-i-n, p-i-n, n-p-n). In some real transistors, the electrical burnout scenario may not be simply converged down to single instability in the elementary structure. For example, avalanche-injection breakdown in one part of the device may initiate instability in another part of the device. However, in each case the mechanism of isothermal SOA limitation can be "decoupled" on different stages of avalanche-injection processes in the elementary parasitic structures.

One of the important regularities is that the impact ionization in high electric fields precedes the catastrophic failure in the case of isothermal instability mechanisms. Another important regularity is that in the devices with sufficient contact width, the area of destruction always has a local nature. No cases of uniform structure melting have ever been reported.

At the same time, the avalanche generation is not a direct cause of the catastrophic failure. Both the avalanche generation in the reverse-biased junctions and the current injection can be distributed uniformly, especially in the case of short pulse regimes. Thus, in each case of physical limitation of SOA due to electrical catastrophic events, a positive feedback can be identified. This positive feedback results in conductivity modulation and requires both the initial avalanche generation and the current injection as necessary conditions. At some critical electrical regime, this positive feedback can produce local negative differential conductivity.

However, even this is not the final stage of the catastrophic failure scenario. For example, operation in the conductivity modulation mode is widely used in pulsed condition to create self-triggering in snapback ESD protection devices and clamps. In this case, special measures are taken to provide a uniform current distribution in the device. Thus, the following regularity is a possibility of current filamentation that results in generation of excessive local power in the device structure and finally leads to thermal breakdown, local melting of the structure material or creating cracks in dielectric due to electrothermal stress and other irreversible mechanical changes in structure region that can identified as catastrophe. In this scenario, the originally isothermal avalanche-injection current filament acts as a local heat source for following rapid thermal burnout and this process can be terminated under short pulse conditions. Thus, after all, the local overheating of the device or accelerated electromigration in the backend metallization is the final cause of the irreversible changes in the device structure.

The above dominant scenario of catastrophic failure under electrical overload in pulsed regime presents the physical limitation under ESD operation too (Chapter 7). However, this scenario is realized after modulation of the distributed contact diffusion regions. This general scenario is universal for different device architectures. The following chapter demonstrates that the same regularities can be derived in compound semiconductor devices, although with some important peculiarities.

Chapter 5
Isothermal Instability in Compound Semiconductor Devices

Catastrophic failure events in GaAs MESFETs and MODFETs (HEMTs) are usually preceded by a set of rather complex phenomena: formation of stationary or traveling domain of electric field, avalanche and tunneling breakdown, and formation of strong nonequilibrium states and filaments. A typical microwave compound semiconductor FET has submicrometer dimensions of the active region, multifinger array structure, complex profiles of the multiple-layer double recess structure, complex doping profiles, surface depletion of the Schottky gate region, and buffer layer. Heterojunction devices add variable compound semiconductor material. The specific of the carrier transport results in nonlinear and nonlocal dependence of the mobility and velocity on the electric field. Therefore, an accurate solution for the electric field and current distribution in these devices is more complex than in silicon. In particular, it requires a numerical simulation of a hydrodynamic equation model.

Two basic modes of avalanche breakdown in the common source circuit can be determined as a limitation of safe operation area: the drain-source breakdown and the gate breakdown. The first is realized when the high-electric-field region is localized near the drain n^+-n junction. In this case, the drain-source voltage increase results in an increase of the avalanche current component of the drain current. The gate breakdown is observed as a current increase in the gate circuit.

In most practical cases of the device design, the drain-source breakdown dominates over the gate breakdown in the on-state operation regimes. On the contrary, the gate breakdown dominates over the drain-source breakdown in the pinch-off condition. Similarly to silicon devices, the breakdown in MESFETs, MODFETs, or HEMTs is not of a "pure" avalanche nature. At some critical drain breakdown current, a positive feedback is observed due to injection from the source junction. In this case, current instability results in a sharp redistribution of the breakdown current along the contacts into the current filament state. Depending on the load parameters, the operation regime, and input power, a scenario with overheating inside the current filament and local burnout is realized. Thus, the instability can determine the drain-source voltage limitation and failure mode rather than the breakdown current level.

5.1 Avalanche Breakdown in MESFETs

At MESFET operation the electric field distribution corresponds to the states with static domain, traveling domain, and without domain (or JFET regime) [125, 126].

In properly designed commercial MESFETs, the traveling Gunn domain should not be observed within standard operation regimes mainly due to the submicrometer dimension and non uniform distribution of the electric field introduced by Schottky gate. However in some experimental long channel and active region with thicker n-GaAs film, the traveling domain can be realized at the drain current of I_{DSS}. Appearance of the traveling domain results in unstable operation. Criterion of traveling domain formation differs from classical criteria of Gunn oscillations [127].

The conditions of domain formation are suggested in [125] on the basis of numerical simulation. In Gunn diode theory, two electric field values are introduced. The critical electric field E_t corresponds to maximum value of the drift velocity v and the electric field E_r to the level far from the domain region. At increase of average electric field over the critical level $E > E_t$, field redistribution occurs and results in simultaneous field increase in the domain and decrease outside of the domain down to the level $E_r \cong E_t/2$. In MESFETs, the initial electric field distribution is not uniform: the field has a local maximum E_{MAX} near the drain side gate edge. Increase of U_{DS} results in increase of the electric field in the whole gate-drain spacing.

If gate E_{MAX} becomes equal to E_t at $E < E_r$ in the other part of the channel, then a stable static domain state is realized. If the field far from the gate exceeds E_r at $E_{MAX} < E_t$, the Gunn oscillations due to traveling domain are realized.

The regions of the MESFET states [126] can be presented by the directly measured values. An empirical ratio for the traveling and static domain states is given by

$$4\left(\varphi_K - U_{GS} - 2U_r\right) = U_p, \tag{5.1}$$

where φ_K is the contact potential difference of Schottky gate, U_P is the pinch-off voltage, and U_r is the saturation voltage, $U_r = E_r L = E_t L/2$, where L is the gate length.

If the channel is pinched off at $E_{MAX} < E_t$, then the domain is not formed. Corresponding boundary is given by

$$\varphi_K - U_{GS} - 2U_r = U_p. \tag{5.2}$$

Therefore, the region of static domain formation is bounded by (5.1) and (5.2). The possible states of MESFET are presented in Fig. 5.1 [126].

Drift-diffusion model for MESFETs

The drift-diffusion model (DDM) assumes a local dependence of the carrier drift velocity on the electric field $v(E)$. For example, for electrons in GaAs $v(E)$ is expressed by [125, 128]

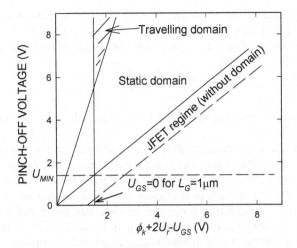

Fig. 5.1 Diagram of transistor stability. U_{MIN}, minimum pinch-off voltage for domain formation

$$v(E) = \frac{\mu E + v_S \left(\frac{E}{E_0}\right)^4}{1 + \left(\frac{E}{E_0}\right)^4},$$

(5.3)

where E_0, v_S, and μ are some typical value, the saturation velocity, and the mobility, respectively.

The model includes Poisson and discontinuity equations:

$$\nabla^2 \varphi = -\frac{q}{\varepsilon}(N_D - n),$$

$$\frac{dn}{dt} = \nabla(n\vec{v} + D\vec{\nabla}n),$$

(5.4)

where φ is the potential, n is the electron density, and D is the diffusion coefficient.

At the drain and source ohmic contacts, $n = N_D$ is given. At lateral sides, the normal component of the electric field and diffusion flow is equal to zero. At Schottky barrier, the electrical potential is $\varphi_G = \varphi_K + U_{GS}$ and electron density $n = N_D \exp\left(-\frac{q\varphi_G}{kT}\right)$ [125, 128].

In the simplest case, the breakdown can be detected by the ionization integral $\int g dx$ calculated along various strength lines of longitudinal field. The coefficient of impact ionization $g(E)$ can be approximated by exponential dependence [129]:

$$g = A \exp\left[-\left(B/E\right)^2\right],$$

(5.5)

where A and B are some constants. The breakdown voltage U_{BR} is obtained from the equation $\int g dx \approx 1$.

Due to nonlocal dependence of the electron drift velocity on the electric field, in general, the drift-diffusion model is not an appropriate description for solution of the breakdown problems in MESFETs and MODFETs with submicrometer dimensions. The typical dimension of electron velocity relaxation is ~ 0.6 μm [130] and thus similar to the structure dimension. In particular, the model does not allow calculation of the negative differential conductivity in MESFETs. However, the model still can be rather useful to obtain some qualitative estimation using the advantage in simplicity.

Using DDM, an analytical expression for the gate breakdown U_{GSBR} is obtained in [129]:

$$U_{GSBR} = \frac{4.4 \cdot 10^{13}\, \text{V} \cdot \text{cm}^{-2}}{\int_0^a N_D(y)\,dy},$$

(5.6)

where N_D is the donor concentration. The structure (Fig. 5.2) corresponds to the planar gate at the epilayer of thickness a. The depletion region boundary at b is assumed vertical.

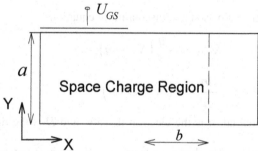

Fig. 5.2 Transistor structure for gate breakdown voltage U_{GSBR} calculation

A ratio similar to (5.6) has been obtained empirically in [131] with the constant 5.3×10^{13} V/cm. From (5.6), U_{GSBR} is a function of the total charge in space charge region and, therefore, does not depend on the doping profile in channel. Results obtained in [129] hold for gate recess configuration too.

Gate recess effect

The gate recess structures provide a more uniform electric field distribution in comparison with the planar structures [132].

Only when the spacing between the gate and the recess drain edge is more than 0.7 μm U_{DGBR} [132] the breakdown voltage corresponds to planar structure.

Surface depletion layer effect

Surface states and passivation play an important role in the gate-drain breakdown [132–134]. The states increase the recombination rate and this results in depletion layer formation. Negative surface space charge usually creates an embedded potential in the range of 0.6–0.8 V. Possible surface states are aperiodic order of atoms, Ga oxidation, defects of As, and broken bonds.

Due to captured electrons, the surface depletion layer decreases the surface field and thus increases the breakdown voltage and small channel current. The surface states provide impact on the U_{DGBR} scattering. The surface states are the reason why passivation impacts on the MESFET breakdown characteristics due to change in surface potential. Similarly, some difference in the maximum power capability can be explained by time delay effects related to recharging of the surface states [133].

In numerical simulation, a physical presence of the depletion region can be completed by adding a thin space charge region at the surface with potential $V_S \approx$ 0.7 V [134]. The negative space charge at the surface provides the field reduction in the space charge region of Schottky barrier and thus compensates the positive space charge of ionized donors. This results in increase of the field near the drain (Fig. 5.3). With V_S increase, the peak of electric field moves from the gate to the drain region. The maximum value E_{MAX} decreases (Fig. 5.3b) and U_{DGBR} increases significantly, for example from 18 up to 31 V in [134]. In the gate recess structure, additional maximum of the electric field near the recess edge appears (Fig. 5.3c). Moreover, an additional increase of U_{DGBR} in the recess structure is observed in comparison with the planar structure.

The calculated U_{DGBR} at various V_S values is in agreement with the experimental values at $V_S \cong 0.65$ V.

Quasi-hydrodynamic (QHD) model

Taking into account nonlocal dependence of the drift velocity (or velocity overshoot) is critical for MESFET and MODFET models. To address this issue, various hydrodynamic models are used [135–138] for correct prediction. A detailed description of one of the QHDs is presented in Section 5.2.4. The model has been widely used for consistent numerical simulation in the original studies of isothermal current instability and current filamentation.

From numerical analysis of the field distribution in the prebreakdown regime, four types of static domain states can be determined for an arbitrary gate bias and MESFET structure design (Fig. 5.4a). The domain structure is determined by average

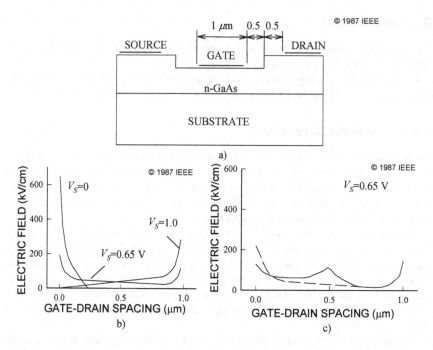

Fig. 5.3 Calculation of the depletion layer effect under gate-drain breakdown: (a) simplified cross section for the planar MESFET structure; (b) electric field distribution in the structure at various levels of the surface potential V_S; (c) the distribution in gate recess structure [134]

electron energy at the drain gate edge and the channel current flows through the narrowest region in the channel. In this problem, the ratio of the energy value ε_n to the intraband energy interval ΔE is important, rather than the absolute ε_n value.

Using current discontinuity equation, the physical interpretation of the domain types can be given as follows [136]. At the gate edge, the field increase and the drift velocity is close to its maximum value v_{MAX}. In this location, the channel thickness has a minimum thickness y_{MIN}. The linear current density j is proportional to $N_D y_{MIN} v_{MAX}$. At the same time, in the domain $j \sim n v_S (a - a_0)$, where a is the epilayer thickness, a_0 is the surface depletion layer thickness, and $v_S = v (E \gg E_t)$.

By comparison of $v_{MAX} y_{MIN}$ and $v_S (a - a_0)$, either a depletion or an accumulation process for electrons can be determined in the high-electric-field region ΔW. If $W > \Delta W$, then the carrier temperature of electrons in the gate region is increased enough to transit in high valley. The first domain type is the gate domain with $n < N_D$. If $n \approx N_D$, then the second type of domain with trapezoidal shape is formed (Fig. 5.4a, curve 2). If an adequate high energy is gained by electrons near the drain electrode, then triangular domains of the third or fourth types are formed in the structure: $W < \Delta W$, $n >_D N$ (Fig. 5.4a).

The drain static domain cannot be reproduced with DDM [125, 128]. Under experimental conditions, a visible light emission of the breakdown [139] provides visualization of field domain of a sharp form.

A physically correct model can clarify that the electrons are leaving the gate region with relatively low energy and that energy adequate for transit in high valley is obtained only in the drain region.

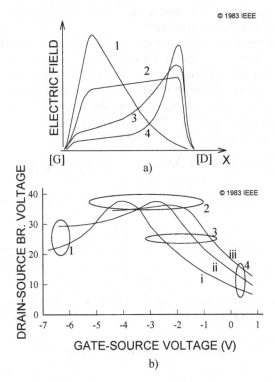

Fig. 5.4 Types of the static domains (a) and calculated dependence $U_{DS\,BR}(U_{GS})$ (b) for (i) $N_D = 2 \times 10^{17} cm^{-3}$, (ii) $N_D = 1.2 \times 10^{17} cm^{-3}$, (iii) $N_D = 8 \times 10^{16} cm^{-3}$; the regions of existence of the corresponding domain types "1," "2," "3," and "4" are marked by the ellipses [136]

The breakdown voltage value U_{DSBR} depends on the domain type. The domain structure and localization depend on operation regime and MESFET structure. The maximum value U_{DSBR} is obtained for the second domain type and does not depend on the active layer doping level (Fig. 5.4b). The gate bias increase results in U_{DSBR} decrease and first type domain formation. With the gate bias decrease (on-state), the second domain type is transformed to a sharp domain of either third or fourth type followed by U_{DSBR} decrease. Increase of the active layer doping level results in U_{DSBR} decrease at high current (Fig. 5.4b) due to increase of the domain of accumulation layer with electrons and resultant increase of maximum electric field.

On the basis of numerical simulation [136], the dependence of U_{DSBR} on the spacing between electrodes, gate recess geometry, active layer thickness, and doping level has been studied. The breakdown voltage does not depend on the gate-drain spacing for the first, third, or fourth type domains. For the second type domain, U_{DSBR} is proportional to the drain-gate spacing. Unfortunately, the second type domain is observed in the n-GaAs films with rather low level of the donor doping and therefore cannot be used to obtain high I_{DSS} and output power in practical cases.

With the gate length increase, the electron heating is more efficient. This results in transformation of the fourth or third domain types into the second type and is accompanied by corresponding breakdown voltage increase. Unfortunately, this effect results in reduction of the cutoff frequency.

The breakdown voltage in the on-state can be enhanced by introduction of gate recess structure. The recess provides a transformation of the sharp fourth type domain into the third or second types. Increase of U_{DSBR} depends on the angle of recess edge (Fig. 5.5a). Maximum value of U_{DSBR} is obtained at $\beta = 15°$ and does not depend on the active layer thickness (from the experiment $\beta = 16°$ [140]). For the vertical edge recess (Fig. 5.5b), the optimal ratio a_r/a is between 1.76 and 2.25 for a wide range of N_D and active layer thickness.

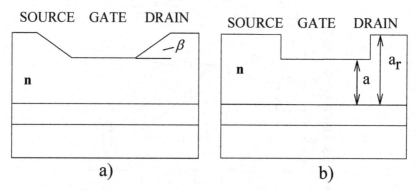

a)   b)

Fig. 5.5 Gate recess MESFET structure

Substrate deep levels effect

A number of peculiarities for MESFET operation are the result of deep levels in semi-insulating buffer layer [141–145]. Deep levels are widely presented both in pure and compensated GaAs. For example, level EL2 can be presented in pure GaAs, and Cr can be used for compensation of shallow donors. As a deep acceptor, Cr creates level with an activation energy of ~ 0.76 eV.

Epitaxial growth and anneal steps result in Cr diffusion from substrate into buffer region. Cr traps are able to capture free electrons and create the space charge region along the active layer and buffer layer boundary. This negative space charge acts as an intrinsic gate thus providing the backgating effect [142, 143]. Presence of the additional space charge region in the buffer and active film interface may even result in avalanche multiplication [142].

A more exact interpretation is suggested on the basis of the impact ionization of the traps [144]. A typical *I–V* characteristic with "kink effect" is shown in Fig. 5.6. The current increase is observed due to decrease of the buffer space charge region via release of the negative space charge of the traps.

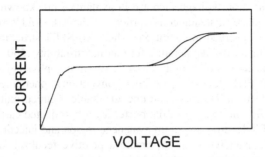

Fig. 5.6 *I–V* characteristic at breakdown with deep levels

5.2 Drain-Source Avalanche-Injection Instability and Filamentation

Irreversible breakdown of GaAs MESFETs is usually accompanied by an S-shape feature in the output *I–V* characteristic. The formed channels and local melting in the MESFET structure are evidence of filamentation processes before irreversible failure [139, 146, 147].

In applications, MESFETs can experience short-time electrical overloads. Even if a protection circuit against electrical overload is implemented in the system, for example amplifier, its reaction time might be insufficient (~1–10 ns) [148] to protect the device. This duration is adequate for catastrophic failure. Therefore, in [148–153] the breakdown characteristic under microwave signal has been studied.

In most of the papers, the external prebreakdown peculiarities are studied. However, the instability mechanism at high microwave frequency is not of a thermal nature since the critical power does not depend on the substrate temperature up to 400°C [149], pulse duration, and frequency [148].

External signature of failure in a large-signal operation mode is damage of the structure between the gate and source [148, 151], a formation [153] channel of local melting near the drain [149].

A detailed physical understanding of the electrical burnout in MESFETs was obtained from the reversible pulsed measurements of S-shape output *I–V* characteristics, electroluminescence [154, 155] followed by numerical simulation of the instability [156, 157] and filamentation [158, 159].

5.2.1 Experimental Observation of Current Instability and Filamentation in GaAs MESFETs

The drain breakdown, as a major limitation of safe operating area parameter drain voltage, output power, and reliability of GaAs MESFETs, has been studied in [160, 161]. Typically, the drain voltage increase in avalanche breakdown mode finally results in an immediate, instantaneous burnout of the MESFET's structure [146, 162]. This effect is observed in conditions where MESFET structure temperature is much less than the critical temperature for thermal instability (500–550°C) [149]. Thus, an electrical mechanism of burnout is the major physical limitation for typical GaAs MESFET devices. The most conventional phenomenological explanation of the electrical burnout is an uncontrollable drain current increase as a result of conductivity modulation of the buffer layer by injected carriers. The avalanche injection of holes from the drain avalanche region and injection of electrons from the source contact region [146] create a positive feedback due to mutual electron and hole space charge neutralization.

Essentially, this avalanche injection process is similar to the avalanche-injection effect in silicon MOSFETs. However, it has specific features related to the particular process and device architecture. Nevertheless, a physical nature of the effect can be explained by a parasitic n-i-n bipolar device with an open base indeed. The collector and emitter of this device are formed by corresponding drain and source contact regions. The floating base region of the device presents itself the buffer layer as a compensated semiconductor. Due to the low carrier lifetime in GaAs (~1 ns), the diffusion length for electrons is less than the length of this parasitic device base. Thus, a very low current gain condition is provided.

This phenomenological description is rather helpful for a quick start understanding of the major regularities in these devices and the similarities with already discussed phenomena in silicon devices. A detailed description of the processes is presented below.

Experimental setup for current instability study in MESFETs

The principle of operation of the experimental setup for investigation of the isothermal instability in GaAs devices is very similar to the commercial transmission line pulse (TLP) testers available from several vendors, for example Barth Electronics or Thermo Fisher Inc. At the same time, the equipment used in [155] provided several essential features that are not as yet implemented in the commercial testers. First, the pulse generation has been built using avalanche bipolar transistors, rather than mercury relay. This allows obtaining the pulses with rather high duty factor that was required to achieve a respectively high efficiency of the light emission. A long-term functional test between the pulses was disengaged. The pulse waveform had a lower rise time and pulse duration ~ 1 ns and 20 ns. The setup provided an analog output directly from the stroboscopic oscilloscope to the plotter to avoid the long-term procedure of uploading the digital data from the oscilloscope. These features allow adjustment of the voltage amplitude in real time toward reversible capturing of the beginning of the instability before the destruction of the sample.

The electrical instability and reversible observation of negative differential conductivity (NDC) region and S-shape drain current–voltage (I_D–V_{DS}) characteristics of MESFETs were first reported in [154]. The effects were observed at the 20-ns pulse duration of drain voltage at different constant gate biases. Similar effect was later obtained in HEMTs [155].

The measurement setup (Fig. 5.7) is based on the measurement method presented in [154, 155, 163, 164]: the pulse generator on avalanche bipolar transistors "G" provides rectangular pulses with up to 200-V maximum amplitude through the 50-Ω coaxial delay line Z. At point "0," a small part of the main signal is branched into the coaxial line Z_I through 1/200 divider R_I. The main portion of the pulse propagates through a coaxial 200-ns-delay line Z_T ($Z_T = Z_I$), drain limiting resistor R_T and is reflected from the drain of the measured device under test (DUT) structure.

After reverse propagation through line Z_T, a part of the reflected pulse was split through divider R_I at point "0" again. At point "I," this split signal part is summed up with the pulse previously reflected from the short circuit end of line Z_I.

In study [154, 155], the drain limiting resistor R_T was another specific feature that is usually not used with standard TLP measurements for ESD development. The resistor was required to limit the current through the device after snapback and has been later used for reversible measurement of the pulsed safe operating area. To avoid an inconvenient recalculation of the DUT voltage, the voltage is measured directly through the divider (R_U) and the coaxial line (Z_U) = (Z_T) = (Z_I).

The resulting signals from points "I" and "U" are connected to the separate channels of stroboscopic oscilloscope "SO." On the oscilloscope, the measurement moment is chosen at the end of the pulse. The signals of the oscilloscope beam deviation for the current and voltage pulses are proportional to the current and voltage amplitude of the DUT. These signals are connected directly to the plotter "P1." In [154], due to rather high duty factor, which was chosen in the range 100–1000, the TLP I_D–V_{DS} characteristics were taken in real time by plotting the I–V dependence on the generator "G" voltage control.

The light emission intensity distribution during the breakdown can be measured using a simple scanner setup (Fig. 5.7). In [154, 155], the DUT structure was attached to the platform of an optical microscope "M," which is connected with a shift sensor "S." In the image plane, a narrow gap "g" was located in order to cut a part of the corresponding image area ~1 μm of active region surface. The sensitive photomultiplier tube "PM" with S-20 photocathode type is located behind the gap. During movement of the sample, the signals from the sensor "S" and photomultiplier "PM" are sent to the plotter "P2." For a given point of a DC I_D–V_{DS} characteristic with high limiting resistor or a high-duty-factor pulsed I_D–V_{DS} characteristic, the light emission distribution is measured.

Typical pulsed I_D–V_{DS} characteristics of the GaAs MESFET (Fig. 5.8) structures for various constant gate biases are presented in Fig. 5.9. After saturation region, the following breakdown peculiarities were observed: the region of avalanche

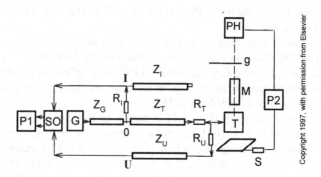

Fig. 5.7 The experimental setup for measurements of the drain pulsed I_D–V_{DS} characteristics and the light emission distribution from a planar FET surface [155]

breakdown, the NDC switching region, and the high-conductivity region with low holding voltage.

A reversible switching into a high-conductivity state is observed (Fig. 5.9) according to the load characteristic $R = (Z_T + R_T)$ with the switching time less than 1 ns. After the snapback, a typical avalanche-injection scenario of both the conductivity modulation and the filament formation has been confirmed.

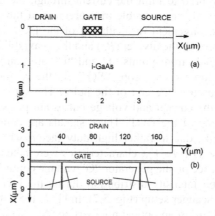

Fig. 5.8 Simplified cross section (a) and topology (b) of the MESFETs

The major components of avalanche-injection mechanism can be demonstrated from a comparison of drain and gate current characteristics. Under on-state conditions, the additional drain current increase before breakdown is mainly provided by the drain-source component. In the off-state, the drain current increase is provided by the gate current component. However, an additional slight drain-source current increase is observed immediately before switching. In the NDC region, the drain current is provided by the drain-source component at all gate bias values [154] thus pointing out the avalanche injection nature of the current instability.

Spectral and spatial light emission distributions provided further insight into the nature of current at breakdown. Similarly to the silicon devices, the broadband light emission spectrum pointed toward the recombination of carriers with energy higher than the band gap value. This emission is produced by avalanche carriers. The band gap light emission on the other hand indicated the recombination of injected carriers at relatively low electric field. Similarly, the localization of the carriers with the band gap radiation provided direct evidence of the injection process from the source junction, while a short-wavelength light emission from the avalanche region provided evidence of intensive impact ionization.

Indeed, before snapback the light emission is distributed uniformly along the drain or gate contacts (Fig. 5.10a,b, curves α). At the same time, the spectrum of electroluminescence remains unchanged and a typical uniform distribution for avalanche breakdown is observed along the drain contact prior to snapback.

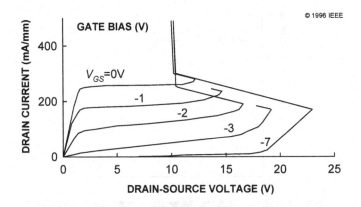

© 1996 IEEE

Fig. 5.9 Pulsed drain-source I_D–V_{DS} characteristics of the GaAs MESFET [156]

The current filament has been formed in the structure after switching. After the instability, the change in distribution of the light emission intensity has been observed both along the drain (Fig. 5.10a, curve β) and in the drain-source direction. Because of direct band gap recombination and high band gap value of GaAs, the light emission is typically easy to observe in an optical microscope. Due to high intensity and visual spectrum range, the current filament was observed in the optical microscope as a bright light strip of a few micrometers in the drain-gate spacing (Fig. 5.10).

A strong injection from the source region has been detected within the filament region. The light emission distribution originates from an intensive infrared radiation in the drain-source spacing (Fig. 5.10b, curve β). This radiation changes the light emission spectrum, in which the intensity of 0.88-μm wavelength increases. Thus, the band gap electroluminescence from the source region has changed the whole spectrum of electroluminescence after snapback.

With drain current increase in the filament state, the spectrum remains unchanged. In the structure with rather lengthy source and drain N^+-regions, the current filament dimension is increased up to the whole structure width ~50 μm.

Combined pulsed I_D–V_{DS} characteristics of infrared electroluminescence from the source region provided evidence that NDC region formation is caused by the "parasitic" avalanche-injection conductivity modulation of the drain-source spacing [155]. Thus, similarly to the silicon devices, NDC formation is the result of a positive feedback between the avalanche breakdown region near the drain in i-GaAs layer and the injection level from the source contact. This becomes possible due to electric field redistribution in i-GaAs buffer region as a result of mutual space neutralization of nonequilibrium injected carriers of opposite charge.

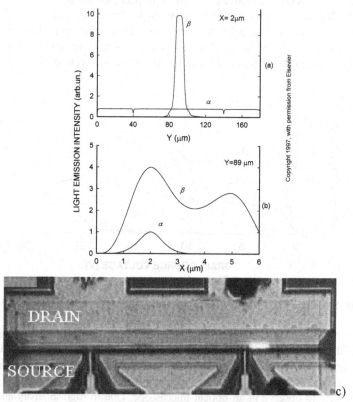

Fig. 5.10 Light emission intensity distribution of the active surface region of GaAs MESFET structure along the drain (a) and in the drain-source direction (b) [155]. The curves correspond to the states in the I_D–V_{DS} characteristics (Fig. 5.9). (c) Microphotograph of avalanche-injection filament under pulsed conditions with high duty factor

In some special cases, the switching can be reversible even under DC conditions. For example, in the experimental studies [154, 155] the switching was reversible at a high limiting resistor value $R_T \sim 1000\ \Omega$. Decrease of the drain load resistance or increase of the duty factor or a pulse duration results in MODFET local burnout. As a result of the burnout, the active region of the structure locally melts at the location of current filament.

The large overheating of the MESFET structure in the filament region is the final cause of local burnout under operation at high drain voltage or pulsed overload. The short time of filament formation and high current density inside the filament explain the instantaneous feature of the burnout process. Observed current instability is not dependent on gate avalanche breakdown directly. In fact, the switching voltage corresponds to the drain avalanche breakdown and increases with gate bias from 0 V to the pinch-off condition despite a gate current increase.

Further details of the isothermal current instability and filamentation will be demonstrated by the numerical simulation in the following sections of this chapter where the numerical solutions for gate and drain conductivity modulation and current filamentation effects will be presented.

5.2.2 Specifics of Electrical Burnout in Large-Signal Operation

The validity of the avalanche-injection mechanism has been confirmed for large-signal operation too. The avalanche-injection mechanism of the local electrical burnout of the MESFETs in a large-signal operation has been demonstrated by correlation of the reversible measurements for critical drain voltage in pulsed regime versus the test up to failure results in the RF regime.

Test up to failure method

Several lots of the experimental commercial 1-W X-band MESFETs have been studied including the devices with structural defects. The device architecture includes the aluminum gate and Au-Ge drain and source metallization system, 1000 μm total gate width, 1 μm gate length, and 3 μm drain-source spacing. The purpose of the experiment was to establish correlation between the reversibly measured snapback voltage in the 20 ns pulsed regime and the burnout voltage in the RF regime.

Experimental samples with various critical voltage values have been selected according to the TLP results of 20 ns pulsed regime from a few hundred MESFETs. From the experimental data, the switching voltage dependence $V_{SW}(I_D)$ was determined reversibly. The switching voltage V_{SW} at fixed test current level $I_{TEST} = 125$ mA was studied. This figure of merit was further used to select the final group of experimental samples with diverse critical voltage values for final RF test up to failure experiments.

In addition to the snapback characteristics, the current filament localization in each MESFET was recorded too from the results of light emission observation using optical microscope.

The tests at microwave operation were performed in the amplifier-like circuit with an input and output matching network. Both pulsed drain and pulsed input microwave power regimes were realized to avoid lattice temperature effect and thermal current instability conditions in the SOA regimes. Such a regime could be for example an equivalent of the short-term pulsed electrical overload.

In the RF test up to failure, the operating drain voltage was modulated with pulse duration below a typical thermal heating time with ~ 2 μs pulse duration at

10% duty factor. During the tests, the 5-GHz 2.2-μs input power pulses were synchronized with operating drain voltage pulse. The V_{DS} pulse amplitude was increased in 1-V steps up to MESFET failure at the given current level I_{TEST}. The maximum output power was adjusted by the matching network at every step during the time of ~1 min. Thus, the end result of each test was both the value of burnout voltage V_{BURN} and the observed location of the burnout region.

For the standard uniform devices without significant structural defects, the typical $V_{SW}(I_D)$ dependence has a typical V-shape (Fig. 5.11a). It corresponds to an increase of the maximum drain voltage with the gate bias increase from 0 V to pinch-off voltage and additional increase of the drain current in the pinch-off condition due to gate breakdown component. The dispersion of the switching voltage V_{SW} at test current I_{TEST} is from 10 V to 32 V.

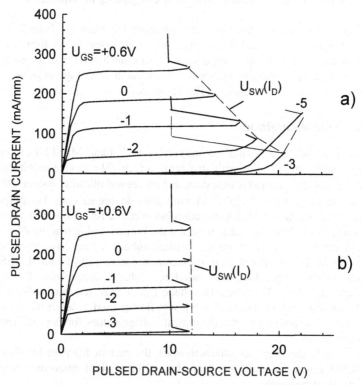

Fig. 5.11 Typical I_D–U_{DS} characteristics of 1-W MESFETs (AP602) at 20-ns pulse duration of the drain voltage (a) and MESFETs with local uncontrollable channel region (b)

The diversity in the critical voltage has been provided by several different factors. Among the major factors are lot-to-lot and wafer location variation. However, the major factor for a significant difference in SOA and the critical voltage was the presence of local defects.

In particular, in the case of MESFET structure with an intrinsic structural defect of local gate metallization break, the pulsed SOA has a vertical shape (Fig. 5.11b). Due to the presence of a small region with the characteristics uncontrolled by the gate depletion region, the snapback voltage of the whole device remains unchanged and independent of the gate bias. Such view of the dependence is the result of a local defect channel region which is not controlled by the gate bias (Fig. 5.11b). There are several reasons for the defect region formation, such as the gap in the gate metallization due to over etching or photolithography defects, the gate recess etching, or the deviation in local doping profiles. Usually the latter provides an intermediate dependence between the cases presented in Fig. 5.11. After the switching, the filament is usually formed in the region of structural defect.

Similarly to the previously described single finger devices, in these array power devices the visible electroluminescence along all 14 drain edges (Fig. 5.12b) has

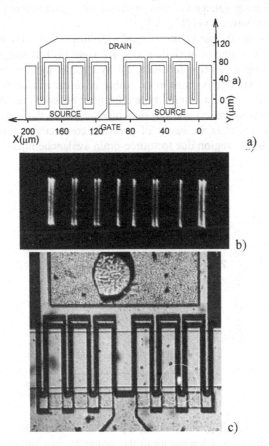

Fig. 5.12 Topology of the 1-W X-band MESFET (a) and photographs of MESFET's active region at 20-ns pulse duration of the drain voltage and gate bias $U_{GS} = -2$ V for the drain avalanche breakdown state (b) and the filament state (c)

been observed before the switching in the drain avalanche breakdown state. After the switching, the uniform distribution was substituted by a local region of bright light emission (Fig. 5.12c) from the filament.

No degradation of the electrical characteristics has been observed until final test step. After burnout, the gate-drain and gate-source spacing are short circuited. A melted metallization and semiconductor region of ~10 μm of and semiconductor has been observed in the failed devices. A coincidence in most cases of the localization of the melted region with the previously observed filament location has been reported. The median burnout voltage was found coincident with the median switching voltage (Fig. 5.13).

Since no change in the switching voltage was observed at the pulse duration from 20 ns to 2 μs (Fig. 5.13), the avalanche-injection current instability limits the drain voltage up to DC operation. The maximum value of the drain voltage is practically equal to the switching voltage at 20-ns pulse duration. The burnout voltage is independent of output power level (Fig. 5.13).

Practically the results can be understood as follows. In 5-GHz-frequency large-signal operation, the time of the half-period RF signal is less than the typical time required to develop a positive feedback for avalanche-injection instability. Only when the median voltage is close to the value of the snapback voltage in pulsed regime do electrical current instability and burnout become possible. Thus, in the pulsed large-signal operation, the physical mechanism of the burnout is the same as in pulsed regime. It is the result of parasitic conductivity modulation of the semi-insulating buffer region due to source-drain avalanche injection and filamentation [126, 129].

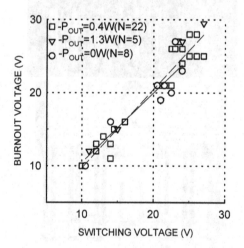

Fig. 5.13 Burnout drain voltage at various input power levels versus the measured switching drain voltage and the plotted dependence of the median burnout voltage on the switching voltage at linear function approximation for 0 (solid line) and 1-W output power (dashed line)

5.2.3 Current Instability and Filamentation in MODFET Structures and Short-Channel MESFETs

Monolithic millimeter-wave integrated circuits based on pseudomorphic GaInAs channel MODFETs (or MODFETs) took over leading positions for power applications [165]. Similarly to power GaAs MESFETs, the breakdown of a MODFET is observed under conditions of high-power and high-drain-voltage operation, however. The avalanche breakdown of the MODFET is studied experimentally and on the basis of analytic or numerical modeling [166–172] for various types and structures.

Practically, the instantaneous burnout and catastrophic failure mechanism of the MODFET under DC or large-signal operation is similar to the effects in GaAs MESFETs. It presents itself this mechanism involves the drain-source avalanche-injection conductivity modulation of the semi-insulating buffer regions followed by the current filamentation and local burnout [168]. The view of destruction after burnout is rather complex, but has a typical nonuniform local nature. At increased drain voltages, the power capability of state-of-the-art power MODFETs is limited by the gate avalanche breakdown [171]; therefore, the burnout is not a major limiting phenomenon in application.

As in the most commercial GaAs MOSFETs, in MODFETs the critical temperature for thermal instability is not achieved in standard operation regimes. Therefore, the dominant burnout mechanism is practically always of electrical nature.

The parasitic conductivity modulation of the undoped layers in pseudomorphic GaInAs/GaAlAs MODFET, GaAs/GaAlAs MODFET has been demonstrated by similar measurements of pulsed I_D–V_{DS} characteristics and light emission distributions [155]. MODFETs and MESFETs of similar topology have been compared in [155]. All structures have the same topology and recess structure: 180-mm gate width, 0.3-μm gate length, 3-μm drain-source spacing. The cross section and topology of experimental pseudomorphic GaInAs/GaAlAs MODFETs and GaAs/GaAlAs MODFETs are shown in Fig. 5.14a,b [156].

Typical pulsed I_D–V_{DS} characteristics of the GaInAs/GaAs MODFET structure and GaAs/GaAlAs MODFET are similar to the MESFET structure (Fig. 5.15 a,c,d). Both the avalanche breakdown and high-conductivity regions are observed. However, the NDC switching region in the MODFETs is at a gate bias corresponding to $I_D < 0.5 \times I_{DSS}$. From a similar comparison of the drain (Fig. 5.15a) and gate (Fig. 5.15b) dependence, the relationships between the gate and the source current component are similar to those of the GaAs MESFET discussed in the previous section.

This conclusion is also supported by the similar spatial and spectral distribution of the light emission at avalanche breakdown (Fig. 5.16, curve α) and the conductivity modulation mode (Fig. 5.16, curve β).

After the switching (at $I_D < 0.5 \times I_{DSS}$), in the optical microscope the filament is observed as a bright light strip of a few micrometers in the drain-gate spacing. Within the filament, the light emission distribution originates from an intensive infrared radiation from the gate-source spacing. This radiation changes the light emission spectrum, in which the intensity of 0.88-μm wavelength increases (Fig. 5.16).

With current increase at the filament the spectrum does not change and the filament increases up to the total gate width. A decrease of the drain load resistance, an increase of the duty factor or pulse duration results in MODFET local burnout. As a result of the burnout, the active region of the structure locally melts at places of current filament formation.

With decreasing gate bias from the pinch-off voltage to 0 V, the NDC region decreases, degenerates, and a transition into a high-conductivity region is observed at a positive differential conductivity (Fig. 5.15a) without filamentation. However, the light emission distribution in the drain-source direction and the light emission spectra are similar to the distribution in the filament state (Fig. 5.16, curve β).

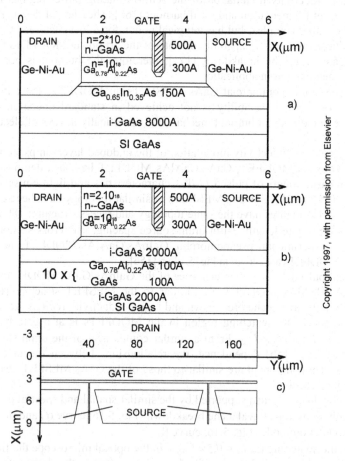

Fig. 5.14 Simplified cross sections (a, b) and topology (c) for the pseudomorphic GaInAs/GaAlAs MODFET and GaAs/GaAlAs MODFET, respectively [155]

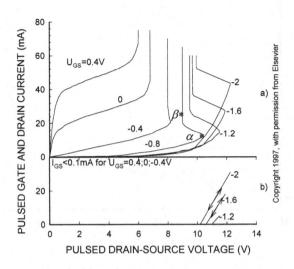

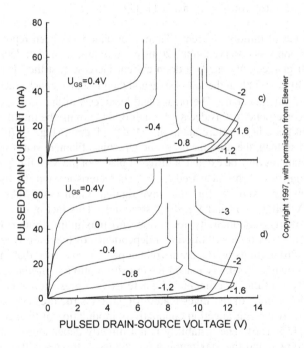

Fig. 5.15 The drain pulsed I_D–V_{DS} characteristics of the pseudomorphic GaInAs/GaAlAs MODFET (a) and GaAs/GaAlAs MODFET (c) and GaAs MESFET (d) of the same topology. (b) The corresponding dependence of gate current during the measurements [155]

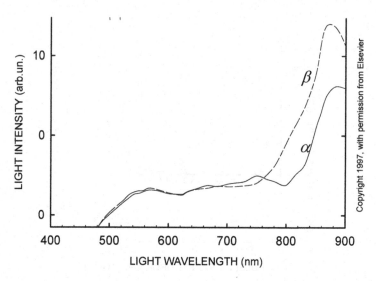

Fig. 5.16 Light emission spectrum of a GaInAs/GaAlAs MODFET. The states "α" before switching and "β" after switching are marked in Fig. 5.15a [155]

A new effect of multiple current filament formation has been reported in [156]. The phenomenon was observed in the 20-ns pulse regime, at a high duty factor ~ 5% and pinch-off gate conditions. At a drain current increase first the filament expands up to 20- to 30 μm width in the initial state and then instantaneously decreases to two new 10- to 15-μm filaments. The breaks and hysteresis in the I_D–V_{DS} characteristic (Fig. 5.17a) correspond to this decay. At high drain current, the multifilament states have a spatial periodic structure along the MODFET drain (Fig. 5.17b). Depending on the electrical load, the duty factor, or the gate bias, filament states with the same number of filaments may be excited at different places of the MODFET structure.

The peculiarities of the measured I_D–V_{DS} characteristics and the light emission characteristics are similar for the investigated GaInAs/GaAlAs MODFET, GaAs/GaAlAs MODFET, and MESFET structures. This fact points to a common mechanism of high-conductivity formation and maximum drain voltage limitation. The light emission distribution and dependence of the drain (Fig. 5.15a) and gate (Fig. 5.15b) currents on the drain voltage provide evidence that the drain avalanche breakdown is a primary reason for transition into a high-conductivity state. Detailed analysis of the experimental characteristics is presented in [155].

Thus, NDC regions of the investigated MODFET structures are caused by the same parasitic avalanche-injection conductivity modulation of the semi-insulating GaAs layer [155, 156]. The instability is not directly dependent on the gate avalanche breakdown, and the maximum drain voltage increases with gate bias from 0 V to the pinch-off condition despite the gate current increase. In the on-state, the high-conductivity region is achieved at the drain avalanche breakdown values at positive differential conductivity due to the same parasitic modulation.

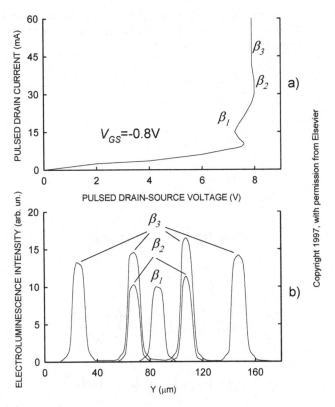

Fig. 5.17 Formation and evolution of multiple filaments under a 20-ns pulse regime with 1% duty factor. Experimental I_D–V_{DS} characteristic (a) and light emission intensity distribution along the drain (b) of the GaInAs/GaAlAs MODFET structure [155]

5.2.4 Numerical Simulation of the Avalanche-Injection Instability in MESFET Structures

The NDC, avalanche-injection conductivity modulation has been demonstrated in detail for GaAs MESFETs using 2D numerical simulation [157].

A proper quasi-hydrodynamic (QHD) model is required to take into account all necessary effects. A version of QHD model was chosen for simulation in [157]. This model is reproduced below.

The model included the balance equations for carrier densities (5.7) and (5.8), the mean carrier momentum in the simplified form (5.9) and (5.10), and the mean carrier energy in the form (5.12) and (5.14). Finally, the equation for total current is used in the form (5.17). A nonflux boundary condition has been used at the lateral sides. Dirichlet boundary condition has been assumed at the contacts for the electrostatic potential, carrier, and the mean energy density.

For carrier transport, specification equations of a QHD model were used [173, 174]. In addition, the electron and hole gas thermal conductivity has been taken into account in the form proposed in [175]:

$$\frac{\partial n}{\partial t} - \frac{1}{q} \operatorname{div} \bar{j}_n = G - R, \tag{5.7}$$

$$\frac{\partial p}{\partial t} - \frac{1}{q} \operatorname{div} \bar{j}_p = G - R, \tag{5.8}$$

$$\bar{j}_n = q n \mu_n (\varepsilon_n) \bar{E} + q D_n \operatorname{grad}(n), \tag{5.9}$$

$$\bar{j}_p = q p \mu_p (\varepsilon_p) \bar{E} - q D_p \operatorname{grad}(p), \tag{5.10}$$

$$\bar{j} = \bar{j}_n + \bar{j}_p, \tag{5.11}$$

$$\frac{\partial (n \cdot \varepsilon_n)}{\partial t} + \operatorname{div} \bar{j}_{\varepsilon n} = \bar{j}_n \cdot \bar{E} - \frac{n(\varepsilon_n - \varepsilon_0)}{\tau_{\varepsilon n}(\varepsilon_n)}, \tag{5.12}$$

$$\bar{j}_{\square n} = -\frac{1}{q} \gamma_n (\varepsilon_n) \cdot \varepsilon_n \cdot \bar{j}_n - \kappa_n (\varepsilon_n) \cdot \operatorname{grad}(\varepsilon_n), \tag{5.13}$$

$$\frac{\partial (p \cdot \varepsilon_p)}{\partial t} + \operatorname{div} \bar{j}_p = \bar{j}_p \cdot \bar{E} - \frac{p(\varepsilon_p - \varepsilon_0)}{\tau_{\varepsilon p}(\varepsilon_p)}, \tag{5.14}$$

$$\bar{j}_{\varepsilon p} = -\frac{1}{q} \gamma_p (\varepsilon_p) \cdot \varepsilon_p \cdot \bar{j}_p - \kappa_p (\varepsilon_p) \cdot \operatorname{grad}(\varepsilon_p), \tag{5.15}$$

$$\operatorname{div} \bar{J} = 0, \tag{5.16}$$

$$\bar{J} = \frac{\partial (\varepsilon_0 \varepsilon_d \bar{E})}{\partial t} + \bar{j}, \tag{5.17}$$

where $\bar{j}_n, \bar{j}_p, \bar{J}, \bar{j}_{\varepsilon n}, \bar{j}_{\varepsilon p}$ are the current density of electrons and holes, the total current density, and the energy flow density of electrons and holes; n, p, ε_n, ε_p are the electron and hole concentration and density of its mean energy;

$\mu_n, \mu_p, D_n, D_p, \gamma_n, \gamma_p, \kappa_n, \kappa_p, \tau_{\varepsilon n}, \tau_{\varepsilon p}$ are the mobility, the diffusion co-efficient, the differential thermal EMF, the thermal conductivity, and the electron and hole energy relaxation time; $\overline{E}$ is the electric field, t is the time; q is the elementary electron charge, ε_0 is the universal dielectric constant and ε_d is the dielectric constant of GaAs. The generation rate G of electron–hole pairs is expressed by the impact ionization coefficient and current density:

$$G = \alpha_n |\overline{j}_n|/q + \alpha_p |\overline{j}_p|/q.$$

The ionization coefficients are given by [176]

$$\alpha_n = A_n \exp\left[-(B_n/E)^2\right],$$
$$\alpha_p = A_p \exp\left[-(B_p/E)^2\right],$$

where

$$A_n = 1.83 \times 10^5 \text{ cm}^{-1}, A_p = 2.215 \times 10^5 \text{ cm}^{-1}$$
$$B_n = 5.79 \times 10^5 \text{ V/cm}, B_p = 6.57 \times 10^5 \text{ V/cm}.$$

For the recombination rate the Shockley–Read–Hall model is used [177]:

$$R = \frac{np - n_i^2}{\tau_n(p + n_i) + \tau_p(n + n_i)},$$

where n_i is the intrinsic carrier concentration and $\tau_n = 2 \times 10^{-10}$ s, $\tau_p = 6 \times 10^{-10}$ s. The dependence of the coefficients $\mu_n(\varepsilon_n), D_n(\varepsilon_n), \tau_{\varepsilon n}(\varepsilon_n)$ is defined using method [178]. This method assumes that dependence of kinetic coefficients for a uniform sample is the same as dependence for a nonuniform sample. The dependence of the mean energy density of electrons on the constant electric field (E_{SS}) in a uniform sample is defined according to a Monte Carlo program [179]. Using the dependence of E_{SS} on ε_n, the equation for a 1D, uniformly doped sample in a constant electric field is transformed to the equation

$$qE_{SS}(\varepsilon_n)v_n[E_{SS}(\varepsilon_n)] = \frac{\varepsilon_n - \varepsilon_{n0}}{\tau_{\varepsilon n}(\varepsilon_n)}. \tag{5.18}$$

The dependence $\tau_{\varepsilon n}(\varepsilon_n)$ is obtained by using the dependence of v_n on E_{SS}. The following expression is used [180]:

$$v_n(E_{SS}) = \frac{\tilde{\mu}_{0n} E_{SS} + v_m \left(\dfrac{E_{SS}}{E_{0n}}\right)^4}{1 + \left(\dfrac{E_{SS}}{E_{0n}}\right)^4} \, ,$$

where

$$v_m = 10^7 \, \text{cm/s}; \; E_{0n} = 4 \, \text{kV/cm}; \; \tilde{\mu}_{0n} = \frac{2\mu_{0n}}{b_0};$$

$$b_0 = 1 + \sqrt{\frac{N_d + N_a}{1 \times 10^{17} \, \text{cm}^{-3}}}; \; \mu_{0n} = 4000 \, \frac{\text{cm}^2}{\text{V} \cdot \text{s}}$$

where N_d, N_a are the donor and acceptor concentration. In the above, for the dependence $D_n(\varepsilon_n)$ the Einstein relation is used.

The hole energy relaxation time $\tau_{\varepsilon p}$ is assumed constant and equal to 1 ps. For a uniform sample in a constant electric field, $E_{SS}(\varepsilon_p)$ and $v_p(\varepsilon_p)$ may be found according to the equation

$$qE_{SS}(\varepsilon_p) v_p \left[E_{SS}(\varepsilon_p) \right] = \frac{\varepsilon_p - \varepsilon_{p0}}{\tau_{\varepsilon p}} \, . \tag{5.19}$$

The dependence $v_p(E_{SS})$ is defined by

$$v_p(E_{SS}) = \frac{\tilde{\mu}_{0p} \cdot E_{SS}}{1 + \left|\dfrac{E_{SS}}{E_{0p}}\right|} \, ,$$

where

$$\tilde{\mu}_{0p} = \frac{2\mu_{0p}}{b_0}; \; \mu_{0p} = 500 \, \frac{\text{cm}^2}{\text{V} \cdot \text{s}}; \; E_{0p} = 16 \, \frac{\text{kV}}{\text{cm}}.$$

Differential thermal EMFs γ_n, γ_p are supposed to be constant: $\gamma_n = \gamma_p = \dfrac{5}{3}$ [181]. The thermal conductivity coefficient is defined by [179]

$$\kappa_z = \frac{2k_z \varepsilon}{3q} \mu_z \left(\frac{5}{2} + m_z \right), z = n, p,$$

where k is the Boltzmann constant and $m_n = m_p = -0.5$ [182].

For numerical solution of the equation system (5.7)–(5.17), an absolutely stable, completely conservative finite-difference scheme has been used [183]. The scheme provided an accurate solution on the rough spatial and temporal grids.

To simulate MESFET's characteristics in the region with NDC, the equation system (5.7)–(5.17) was solved with "mixed" boundary conditions: fixed source and gate electrical potentials and fixed total drain current. To demonstrate the major simulation results, the breakdown characteristics of a simplified planar structure (Fig. 5.18) were calculated toward understanding of the NDC formation mechanism. The following structure parameters were introduced: drain-gate spacing 1 μm, gate-source spacing 0.5 μm, gate length 0.5 μm, length of the source and drain n^+-contact regions 0.3 μm, thickness of the active layer 0.18 μm, donor concentration in the active and contact layers 10^{17} and 10^{18}cm^{-3}, i-layer thickness 1 μm.

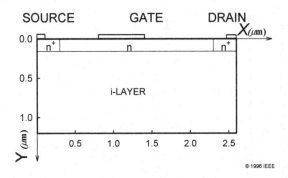

Fig. 5.18 Simplified GaAs MESFET structure for numerical analysis [156]

The calculated quasi-static drain I–V characteristics of the device at different gate biases are presented in Fig. 5.19a. The NDC regions are observed at the gate bias increase from 0 V up to the pinch-off voltage while the maximum drain voltage of NDC formation increases. Near the pinch-off conditions, a region of additional drain current corresponds to a region of the gate current I_G increase up to the maximum drain voltage value (Fig. 5.19). In this condition, the additional drain current is provided by the gate current component (Fig. 5.19b).

The module of electric field and carrier density depth profiles for two states of the MESFET are presented in Fig. 5.20 for comparison. The first state corresponds to point A on the drain current saturation region and the second to point B on the region with NDC of the drain I–V characteristic at the gate bias $U_G = -2$ V (Fig. 5.19a). At state A, two regions of the high electric field are formed in the MESFET: one near the drain n^+-edge and another near the gate edge (Fig. 5.19a). In the semi-insulating (SI) layer, the electric field is increased from the source to the drain due to the space charge limited current (Fig. 5.20b). The avalanche generation takes place in the high-electric-field region near the drain. In state A, holes appear in the SI layer (Fig. 5.20c). The hole density increased in the source n^+-contact direction and the holes are accumulated near the source (Fig. 5.20c).

In the high-conductivity state (B) (Fig. 5.19a), the electric field and carrier density depth profiles in the SI buffer layer are changed significantly (Fig. 5.20d–f). The injection level of the electrons and holes in the SI layer in comparison with state A is greatly increased. In state B, the electrons, injected from the source n^+-contact, and the holes created the quasi-neutral electron–hole plasma in the SI buffer near the source n^+-contact. In state B, the drain voltage is lower than in state A, but the electric field near the drain n^+-contact is slightly higher (Fig. 5.20d). In the SI layer near the source n^+-contact, the lengthy quasi-neutral region with carrier density up to $n = p = 10^{17} \text{cm}^{-3}$ is formed. In these conditions, practically equal density of the holes and electrons is observed in the source half of the structure (Fig. 5.20e,f).

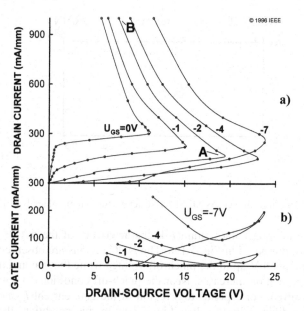

Fig. 5.19 (a) Calculated drain I_D–V_{DS} characteristics of the GaAs MESFET and (b) corresponding gate current-drain voltage dependencies [156]

The calculated I_D–V_{DS} characteristics of the GaAs MESFET model are mainly in agreement with the experimental data. The major difference in the NDC region of the calculated I_D–V_{DS} characteristics and the vertical NDC region of the experimental I_D–V_{DS} characteristics has been explained by the redistribution of the drain current along drain edge in the wide 3D device as a result of current filament formation.

With the simulation results, a perfect visualization of the avalanche-injection current instability and the positive feedback is provided directly. The high-electric-field region near the drain n$^+$-contact in the SI layer is formed due to the space charge limited current (Fig. 5.20a). A portion of holes that were generated near the drain is accumulated in the SI buffer near the source n$^+$-contact (Fig. 5.20c). The hole positive space charge lowered the barrier for electrons on the active film–SI junction and the additional electrons are injected in the SI layer from the source (Fig. 5.20b). The injected electrons create the additional negative space charge in the high-electric-field region near the drain (Fig. 5.20d). This results in the avalanche generation and hole injection in the SI layer increase and electron injection increase from the source n$^+$-contact. This positive feedback results in formation of the lengthy quasi-neutral area with a high density of electron–hole plasma (Fig. 5.20e,f) and electric field reconstruction in the SI layer (Fig. 5.20d). The electric field is being increased in the region of avalanche multiplication near the drain and decreased in another part of the SI.

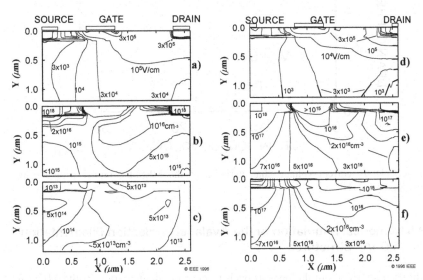

Fig. 5.20 Electric field (a, d), electron (b, e) and hole (c, f) density depth profiles of the MESFET in the avalanche breakdown state "A" (I_D = 150 mA/mm, V_{DS} = 18 V) and in the conductivity modulation state "B" (I_D = 1 A/mm, V_{DS} = 8 V) (Fig. 5.19a) at the gate bias V_{GS} = −2 V [156]

Particularly for the simulation results, it is directly evident that the avalanche-injection conductivity modulation rather than the avalanche drain-source breakdown itself limits the maximum drain voltage in this device type. Similarly to silicon devices. the electric field reconstruction in the SI layer provides NDC formation at relatively low multiplication factor M.

At high drain voltages, avalanche generation in the gate edge takes place and gate current increases (Fig. 5.19b). In the pinch-off condition at sufficiently high drain voltage, the reverse gate current significantly exceeds the saturation drain current (Fig. 5.19b). However, the holes from the gate avalanche region are not injected into the SI layer and thus do not impact the NDC formation.

Using numerical simulation, the principal role of the SI layer on the NDC formation of MESFET has been demonstrated using a model structure with no SI layer. In this case, no NDC is observed on the I_D–V_{DS} characteristics (Fig. 5.21) at any gate biases up to a high drain current level above the saturation drain current. Moreover, the maximum voltage of such a device is limited by the gate avalanche breakdown.

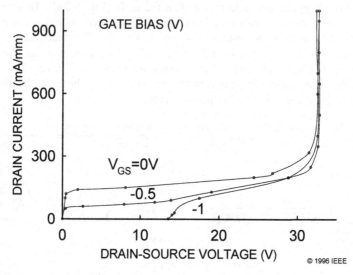

Fig. 5.21 Drain current–voltage characteristics of the MESFET model without semi-insulating layer [156]

5.2.5 Numerical Simulation of the Avalanche-Injection Filamentation in GaAs Structures

A direct numerical simulation approach to current filamentation in the MESFET structure is not possible for the 2D case. It requires the third dimension be taken into account to accommodate the gate electrode and the device features. However, 2D simulation of a simple n-i-n diode structure is rather informative. It provides important understanding of the major regularities related to the current filamentation effect in GaAs MESFETs. In this section the original simulation approach and results are presented.

According to experimental and simulation results in Sections 5.2.1–5.2.4, the uncontrollable drain current increase is observed due to the conductivity modulation of the buffer region by injected holes from the avalanche region near the n$^+$-drain and electrons from the n$^+$-source contact. However, the major peculiarities of operation in negative conductivity modulation mode are similar for different gate biase. This conclusion provides certain justification for reduction of the current filamentation problem dimension down to the current filamentation problem in the corresponding distributed parasitic n^+-i-n^+ structure.

Moreover, in the first approximation the current filamentation in the buffer region and the channel current impact in the on-state condition can be treated independently. At this point, it is assumed that the channel drain current component mainly influences the beginning of the conductivity modulation, rather than the states with high current density due to in-depth injection current flow.

The first 2D simulation demonstrates that the breakdown of GaAs n^+-i-n^+ structures (Fig. 5.22a) with the parameters typical for the contact and buffer layer of power MESFETs is actually accompanied by NDC filamentation.

For comparison, the calculated I–V characteristics of the 1D n^+-i-n^+ structures are presented in Fig. 5.22b. Due to low diffusion length in comparison with the i-region length L_I, the I–V characteristics are essentially similar to space charge limited (SCL) diodes discussed in the previous sections, namely, a linear increase of current from the beginning; small abrupt current increase due to avalanche multiplication; and finally NDC region formation. Higher maximum voltages, lower current of NDC formation, and higher holding voltages at the NDC states correspond to the larger values of L_I [157].

Redistribution of electric field, electron density, and hole density in the n^+-i-n^+ structure (L_I = 1.6 μm) is observed typical for conductivity modulation process in SCL diode. In Fig. 5.23, the distributions are presented for three states: the avalanche breakdown state at positive differential conductance (α), the small-current NDC state (β), and the high-conductivity NDC state (γ). Similarly to Si n-p-n structures, when drain (anode) current increases, electric field near the drain n^+-i junction increases, but that inside the i-region decreases (Fig. 5.23a) due to formation of quasi-neutral region with $n \approx p$ near the source n^+-i junction (Fig. 5.23b,c). The expansion of the quasi-neutral region is observed with the current density increase.

Unlike the case of n^+-i-n^+ structure, the NDC region formation for p^+-i-p^+ structure is not seen due to the reversed ratio of the electron and hole mobilities [157].

After this initial 1D simulation part, the numerical solution for 2D simulation problem is presented below. First, in the 2D case the shape of the NDC region of the wide n^+-i-n^+ structure (W = 50 μm, L_I = 1.6 μm) is already rather different than in the 1D case (Fig. 5.22b). The NDC region provides much lower holding voltage in comparison with the 1D device.

This change in the holding voltage is directly connected to a transient process that results in a principal nonuniform distribution of current along the drain. Calculated time–voltage dependence for fixed drain current is presented in Fig. 5.24a. At first, a quasi-stable state with uniform current distribution ("uniform state") along the drain contact is formed after a fast initial transient process. In this state, the voltage value, electric field, and carrier distributions coincided with those of the 1D structure (L_I = 1.6 μm) under the same current density (Fig. 5.23b).

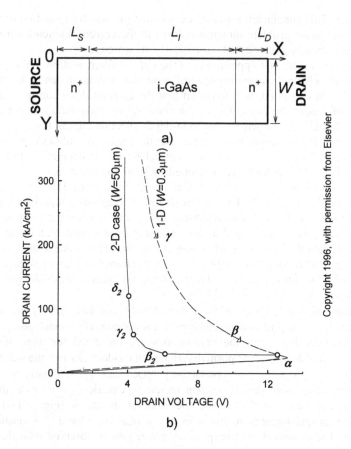

Fig. 5.22 Calculated GaAs n⁺-i-n⁺ structure (a) and comparison of *I–V* characteristics (b) for its 1D version with $W = 0.3$-μm drain width and truly distributed 2D n⁺-i-n⁺ structure with $W = 50$ μm at the i-region length $L_I = 1.6$ μm [158]

However, this uniform state remains stable only for a fraction of a nanosecond. After this, a new transient process results in formation of a filament state during a time duration of 300 ps.

From simple analysis of the 2D carrier density and electric field distributions, it can be concluded that inside the filament (Fig. 5.24b–d) the distributions are similar to those in the NDC state of the 1D model, while outside the filament the distribution corresponds to the low-current prebreakdown state.

Maximum carrier density in the filament region exceeds 10^{17}cm⁻³, but it is limited by the saturation in the n⁺-regions. Due to this limitation, the filament is expanded with drain current increase. Value of maximum voltage, NDC formation of 2D GaAs n⁺-i-n⁺ structures corresponds to the experimental *I–V* characteristics of GaAs MESFETs [155] after subtracting the channel component of the drain current, gate current and considering an additional hole injection from channel to the drain avalanche region. At the same time, filament dimension, electron–hole plasma distribution in the 2D GaAs n^+-i-n^+ structure correspond to the experimental spectral and intensity distributions of electroluminescence in GaAs MESFET [155].

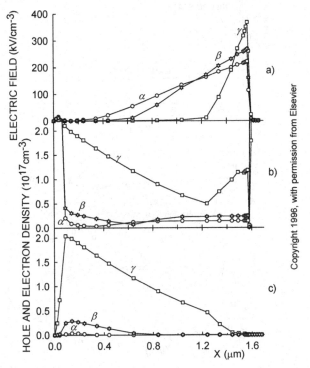

Fig. 5.23 Distribution of electric field (a), electron density (b), and hole density (c) in the GaAs n^+-i-n^+ structure with $L = 1.6 \, \mu m_I$, $W = 0.3 \, \mu m$ for three states: the avalanche breakdown state (α), NDC state (β), and high-conductivity state (γ). The states correspond to points α, β, and γ in Fig. 5.22b [158]

Simulation of the spatial instability using the model without any specific source of fluctuations is possible due to the presence of small numerical fluctuations in the finite-difference scheme. The transient numerical simulation of the current filaments is completed at the finite accuracy of the simulation with given numerical error. Thus, this numerical noise acts like a fluctuation in real device and allows the phase transition of the structure on different solutions in bifurcation points.

Similar results have been obtained for the 2D equivalent structures in the case of InAlAs/InGaAs and InAlAs/InP MODFETs. In spite of the absence of i-GaAs buffer in the InAlAs/InGaAs MODFETs on InP substrate, the light emission due to electron–hole recombination in the gate-source region was observed too [186]. Taking these experimental data into account, the assumption of a similar avalanche-injection conductivity modulation mechanism has been made and further confirmed by the numerical simulation.

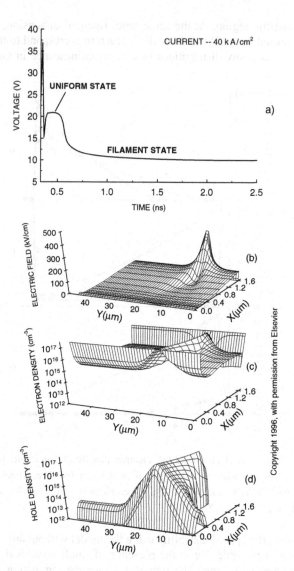

Fig. 5.24 Time dependence of the voltage after a current step to the structure with $W = 50$ μm (a); distribution of electric field (b), electron (c), and hole (d) densities for state β2 (Fig. 5.22b) [158]

The breakdown of $Ga_{0.47}In_{0.53}As$ n^+-i-n^+ and $GaAs/Ga_{0.47}In_{0.53}As$ heterostructure has been studied using a version of QHD model similar to [157]. To enable the simulation of the heterostructure, the transport equations of the model [157] required the following modification using quasi-electric vectors for electrons and holes [187]:

$$\vec{j}_n = -qn\mu_n(\varepsilon_n)\vec{E}_n - \frac{2}{3}\tilde{D}_n \, \mathrm{grad}(n\varepsilon_n),$$

$$\vec{j}_p = qn\mu_p(\varepsilon_p)\vec{E}_p - \frac{2}{3}\tilde{D}_p \, \mathrm{grad}(p\varepsilon_p),$$

$$\frac{\partial(n\varepsilon_n)}{\partial t} + \mathrm{div}\,\vec{j}_{\varepsilon n} = -\left(j_n\vec{E}_n\right) - \frac{n(\varepsilon_{n-} - \varepsilon_0)}{\tau_{\varepsilon n}},$$

$$\frac{\partial(p\varepsilon_p)}{\partial t} + \mathrm{div}\,\vec{j}_{\varepsilon p} = \left(j_p\vec{E}_p\right) - \frac{n(\varepsilon_{p-} - \varepsilon_0)}{\tau_{\varepsilon p}},$$

where $\vec{E}_n = \vec{E} + \nabla\chi, \vec{E}_p = \vec{E} + \nabla(\chi + E_G)$,

The breakdown of the GaAs, $Ga_{0.7}Al_{0.3}As$, $In_{0.52}Al_{0.47}As$, and InP n^+-i-n^+ structures has been studied in [188].

5.2.6 The Role of Contact n^+-Regions. Multiple Current Filament Formation

From the above sections, both the conductivity modulation and the final filament state in the MESFETs and MODFETs are determined by the parameters of a "built-in" n^+-i-n^+ structure formed by the source and drain n^+- and buffer i-layer regions.

From application case studies and ESD devices, it is well known that the n^+-contact region provides not only a better breakdown voltage, but also important current limitation and current ballasting features. Due to a better uniformity of breakdown current along the drain edge, and enhanced stability of MESFETs to overloads, n^+-regions are usually used in MESFET and MODFET devices.

At the same time, a distributed contact region considerably influences the filamentation in semiconductor structures. In particular, instead of a single filament in a simple structure having the S-shape current–voltage characteristic, in structures with a distributed contact region, static, pulsed, traveling, and periodic filaments are observed [189–194]. These phenomena are based on a separation between spatial (or temporal) parameters of current instability in the active region and damping in the contact region of structure. The multiple filaments were observed in MOD-FETs and MESFETs at 20-ns pulsed regime [158] (Section 5.2.3).

In addition to an interesting nonlinear physical effect, the understanding of filamentation in the GaAs structures is helpful for optimization, CAD, and reliability of MODFETs and MESFETs as well as other especially ESD silicon devices. Usually, multifilament states in semiconductor structures are analyzed and calcu-

lated on the basis of 1D reaction-diffusion equation for phenomenological simplified instability mechanisms [194]. Therefore, verification of this result by a QHD
model is of particular interest.

This section is devoted to the 2D simulation of the spatial instability in GaAs
structures with distributed contact regions and the investigation of multiple filament properties [159]. Two-dimensional QHD model and calculation procedure is
similar to [155] (Section 5.2.4).

Isothermal I–V characteristics and spatial distribution were calculated at a
given current value up to the time when a stationary spatial distribution of carriers
was formed. Any fluctuation source term is not included in the equations [155].
Spatial instability of uniform states is caused by numerical fluctuation in the finite-
difference scheme. The basic GaAs n^+-i-n^+ structure (Fig. 5.25) has anode and
cathode contact regions $L_C = 0.02$ μm and $L_A = 11$ μm, i-region length $L_I = 1$ μm,
and 10^{18} cm^{-3} donor concentration in the anode and cathode contact regions. The
acceptor (donor) concentration in the p- (n-) regions of the n^+-p-n^+ (n^+-n-n^+) structure is 10^{16} cm^{-3}. The calculation was made for a wide n^+-i-n^+ structure with $W =$
58 μm and a narrow structure with $W = 1.2$ μm contact width (as a 1D model).

Before breakdown, the I–V characteristics of the 1D n^+-i -n^+, n^+-p-n^+, and n^+-n-
n^+ structures (Fig. 5.26a) have a linear region SCL current (ohmic and saturation
for n^+-n-n^+). After the avalanche breakdown, the NDR region and high-
conductivity region with positive differential resistance are observed. The distribution of the electric field, and the electron and hole density at breakdown is the
same as the 1D n^+-i-n^+ structure. With positive differential conductivity at high
conductivity, the voltage drop across the anode n^+-region provides the positive
differential slope of the I–V characteristic. Finally, the maximum current density
is limited by saturation in the anode n^+-region.

Avalanche breakdown in the anode n^+-p junction of the n^+-p-n^+ structure is observed at a small current (Fig. 5.26a). In the n^+-n-n^+ structure, avalanche breakdown in the anode static domain is observed after saturation in the n-region (Fig.
5.26a). However, in these structures the current instability due to p- (or n-) region
conductivity modulation, NDR region, and current saturation in the anode n^+-
region are similar to the n^+-i-n^+ structure (Fig. 5.26a).

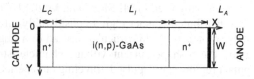

Fig. 5.25 Distributed 2D GaAs n-i-n structure for current filamentation study

Calculated I–V characteristic of the wide n^+-i-n^+ structure ($W = 58$ μm) is
shown in Fig. 5.26b. The shape of the NDR region of the I–V characteristic of this
structure differs from that for the narrow n^+-i-n^+ structure ($W = 1.2$ μm) (Fig.
5.26a). The NDR region is represented by different branches. Each branch
corresponds to a state with a fixed number of filaments.

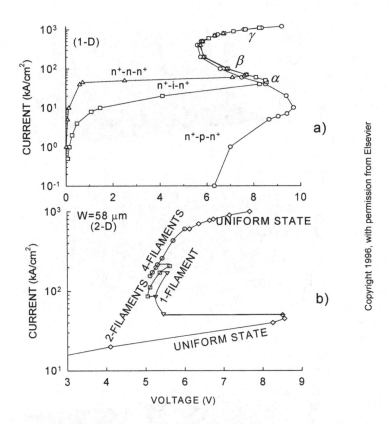

Fig. 5.26 Calculated I–V characteristics for the 1D ($W = 1.2$ μm) n^+-i-n^+, n^+-p-n^+, and n^+-n-n^+ structures (a) and the 2D ($W = 58$ μm) n^+-i-n^+ structure (b) [159]

For the current step over the critical current, a stable filament state is formed during the time interval of 0.5–5 ns depending on the simulation parameters. Two-dimensional distribution of the field, electron, and hole density is shown in Fig. 5.27 for the state with four filaments (three filaments and two half-filaments).

Inside the filaments, the field and carrier density distributions along current lines does not correspond exactly to the distributions in the high-conductivity state of the narrow (1D) structure calculated at the same voltage. Due to a current spreading in the n^+-region, the filament amplitude exceeds the maximum level in the 1D structure.

After current increase, the filaments are broadened to some critical width. At some critical current density, a new state with the largest number of filaments is formed (Fig. 5.28). The increased number of filaments is relaized with current increase is due to splitting of filaments in an initial state (Fig. 5.29). The decreased number of filaments with current decrease is due to suppression of some filaments in an initial state and their restructuring into a new state. The new state attains an exact spatial periodicity at large numbers of filaments.

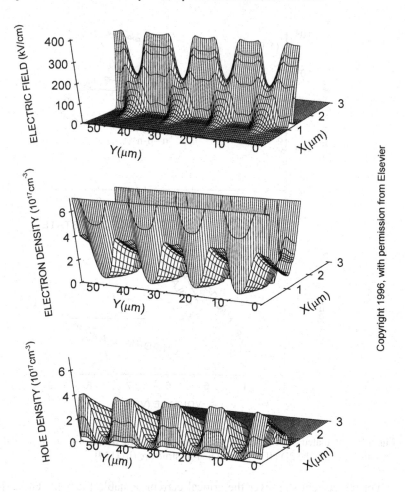

Fig. 5.27 The modules of field (a), electron (b), and hole (c) density distribution in 2D GaAs n⁺-i-n⁺ structure at current density $I = 450$ kA/cm² [159]

Decrease of the donor concentration in the n⁺-region results in decrease of the filament amplitude. The reduction of the length L_A (up to 0.2–0.3 µm) leads to an increase of the period (distance between filaments) and broadening of filaments up to a single filament. Subsequent decrease of L_A results in a transition to a narrow high-density filament. In this case, the amplitude of the filament is limited by optical and Auger recombination.

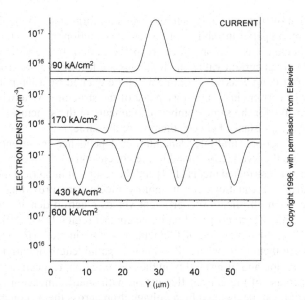

Fig. 5.28 Multiple filament states. Final stationary distribution of electron density in 2D structure (in the $X = 0.56$ μm plane) at different current values [159]

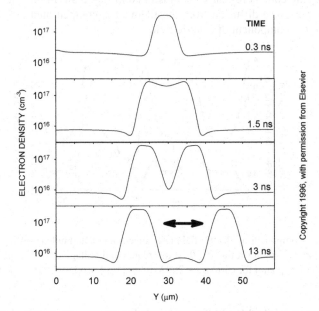

Fig. 5.29 Kinetics of the solitary filament splitting. Distribution of electron density versus time (in the $X = 0.56$ μm plane) after an abrupt step current increase from 90 up to 170 kA/cm^2 [159]

Formation of filaments in structures from a uniform state (along the contacts) is explained by a spatial instability of current distribution along contacts due to increase of fluctuations with an allocated spatial period. In the various calculations at the same given current value, the filaments are localized at an arbitrary place of the structure (along Y axis). A filament can be excited by a temporal local inhomogeneity of carriers in an arbitrary place of the structure. Meanwhile, at a given current density (for the same number of filaments in the structure), the amplitude of filaments and distance between them were the same.

The spatial periodic multiple filamentation, dependence on current and structural parameters qualitatively correspond to the conclusion of the theory of self-organization in active distributed media [194] and the results of numerical modeling of 1D reaction-diffusion equations for instability mechanism providing an increasing dependence of multiplication coefficient versus electron density at a high characteristic width of current spreading in the contact region [193, 194].

From [194], the dimension of filaments is approximately $(Ww)^{-1/2}$ and the distance between them is W, where W, w are the spatial length of current spreading in the contact region and of instability in the active region, respectively. In a uniform state, an increase of the current fluctuation with small dimension ($w \ll W$) in the active i-region can be damped by a voltage drop across the n$^+$-contact region only on the spatial width W. Therefore, this fluctuation is increased. The effect of voltage across the contact region V_A is illustrated in Fig. 5.30. In the edges of the filaments, a transverse diffusion from the filament is compensated by a drift current due to the E_Y-component of electric field.

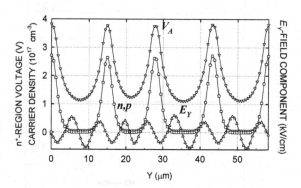

Fig. 5.30 E_Y-component of electric field (E_Y), electron (n), and hole (p) density distribution in the avalanche region (in the $X = 0.56$ µm plane) and voltage drop (V_A) on the n$^+$-region (in the $X = 1$ µm plane) for current $I = 200$ kA/cm^2 [159]. Copyright 1996, with permission from Elsevier

In epitaxial GaAs MESFETs, a typical donor concentration in the n$^+$-region is $1-2 \times 10^{18}$ cm^{-3}, and length and thickness of the n$^+$-region are 0.1–0.5 µm and 0.05–0.2 µm. To study the filamentation in MESFETs, the dependence of the filament amplitude on L_A in the n$^+$-i-n$^+$ structure is shown in Fig. 5.31. This dependence reveals that the limitation of filament amplitude is effective at $L_A > 0.3$ µm.

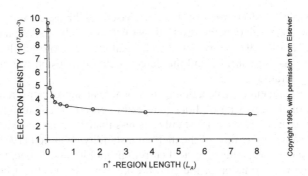

Fig. 5.31 Dependence of the filament amplitude (maximum electron density at $X = 0.56$ μm) on the length L_A for a single filament state ($I = 140$ kA/cm^2) [159]

Summarizing the above, one can conclude that the lengthy contact n$^+$-regions in GaAs n$^+$-i-n$^+$ (n$^+$-p-n$^+$, n$^+$-n-n$^+$) structures have two important effects: (1) a spontaneous formation of multiple current filament states and (2) a limitation of current density in the filaments on the current saturation level in the n$^+$-regions. Apparently, limitation of filament amplitude by the anode n$^+$-region is more effective than limitation by the cathode region due to local effect in the case of the drain (anode) breakdown. This simple conclusion is widely used in silicon grounded gate snapback NMOS devices for ESD applications, where a silicide blocked n$^+$-region ~ 2–6 μm long is added at the drain side.

In the structure without n$^+$-region, current instability results in a solitary large-amplitude current filament due to the same spatial dimension of the inhibitor and activator parameters. In this case, increase of the filament amplitude is practically unlimited with real device parameters before burnout due to high injection from the metal contact. A different scenario for current filament evolution is realized in the device with lengthy drain n$^+$-region. Depending on the distributed n$^+$-region parameter, the filament amplitude is limited. Multiple filament formation provides a uniform distribution of the dissipated power across the structure. With current increase, the distance between filaments is decreased down to a minimum value and following uniform high-conductivity state. Only when the current density becomes over saturation level in the n$^+$-region or under conditions where the in-depth current transport results in current flow redistribution closer to contact metallization is the secondary filament formed and result in device burnout in the pulse regime.

5.3 Double Avalanche-Injection Instability and Filamentation at Schottky Gate Breakdown in MESFETs

In the case of significant gate overload, an instantaneous local gate burnout is usually observed. One of the typical failure modes is gate burnout of small-signal MESFETs under conditions of overload by input microwave power [195]. The burnout takes place in a wide range of pulse duration from 1 ns up to DC regime [195]. Several thermal models were developed for prediction of the irreversible breakdown [196, 197]. Most of them are based on the assumption that the structure melting and failure is the result of thermal overheating inside a certain "defective" region [196, 197]. However, the physical mechanism of the local heat source formation was not clarified and the defect region dimension was used in models as a fitting parameter.

However, a natural explanation of the local heat source appearance is the electric or thermal current instabilities that lead to current filamentation. The characteristic time of the electrical instability is less than 1 ns, i.e., considerably smaller than the characteristic time of the thermal instability evolution [197]. Therefore, at short pulse overstress, the electrical current instability and filamentation cause the "defect" region formation at the gate breakdown. The electrical burnout mechanism is clarified on the basis of the numerical simulation [198–200].

In this section, the solution to the problem of MESFET gate burnout is presented by means of pulsed measurements and 2D simulation in order to demonstrate that the current instability and filamentation occur in MESFET structures and thus may result in burnout at some critical level of the gate avalanche breakdown current.

Pulse gate-drain I–V characteristics of the commercial GaAs MESFETs were measured at 10-ns pulse duration and 0.0001% duty factor using the technique of Section 5.2.1. The MESFETs have the following parameters: 3-μm drain-source spacing, 0.7-μm recess length, 0.5-μm gate length, 280-μm total gate width, 1-μm buffer layer, 0.15-μm active layer of 1.5×10^{17} cm^{-3} donor concentration, 0.2-μm contact layer of 1.5×10^{18} cm^{-3} donor concentration, Ti-Pt-Au gate and Au-Ge-Au drain and source metallization. The pulsed gate-drain characteristic is plotted in Fig. 5.33 as the solid line.

At some critical gate-drain voltage, the avalanche breakdown localized in the gate-drain spacing near the gate metallization edge. Under high duty factor, the evolution of avalanche breakdown may be observed by optical microscopy as a uniform light strip near the gate edge.

At some critical gate avalanche current, the reversible switching into a new state is observed (Fig. 5.33). However, current filament after the switching in this regime is rather difficult to detect. The light emission intensity in the new state after the switching is usually too small for the optical light intensity distribution measurements. A greater overload by increase in current, duty factor, or pulse duration results in irreversible failure and short-circuiting of the gate-drain spacing.

Typically, visible destruction in the gate-drain region of the burned out MESFETs is hard to find and special deprocessing and FA tools are required to visualize the damage area. The other way to indicate the short-circuit region localization in the failed MESFET is to apply a positive bias in the gate-drain circuit. This measure leads to gradual melting of the gate metallization around a region thus exposing the previously damaged region of ~1 μm width.

Thus, a typical cause of the short-circuiting is formation of conducting channel at a certain depth under the semiconductor surface. According to the experimental I–V characteristic (Fig. 5.33), current density inside this region is larger than 10^7 A/cm^2. The dissipated power in the channel is sufficient to cause the meltdown within a very short time period. Apparently, the short-circuit channel could be formed by the direct transport of AuGe-GaAs eutectic or metal-GaAs mutual diffusion [198] between the drain and gate electrodes. The pulsed drain-source I–V characteristics and burnout peculiarities were similar to the drain-gate characteristics.

Based on understanding presented in the previous sections, one can assume that for the case of gate-drain breakdown an absence of the current limitation by an n^+-region at the gate side must lead to unlimited current density in the filament region and therefore cause the appearance of rather narrow filaments with much greater amplitude in comparison with the case of avalanche injection in the n^+-i-n^+ structure and drain-source avalanche injection in MESFETs.

The numerical simulation in support of the above conclusion and the experimental data have been obtained in [183] using the same QHD model (Section 5.2.4) and finite-difference scheme. The characteristics of a simple planar structure were simulated with the following structure parameters: drain-gate and gate-source spacing $L_{DG} = L_{GS} = 1$ μm, gate length $L_G = 0.5$ μm, lengths of the source and drain n^+-contact region $L_S = L_D = 0.3$ μm, thickness of the active layer $W_{AL} = 0.17$ μm, donor concentrations in the active and contact layers 1×10^{17} and $1 \times 10^{18} cm^{-3}$, respectively, and thickness of the semi-insulating layer $W_{SI} = 1$ μm.

For the current filament solution, a model for M-i-n^+ structure with Schottky contact has been used to obtain the solution for quasi-1D case with width $\Omega = 0.2$ μm and the case of filamentation with width $\Omega = 10$ μm and $\Omega = 2.55$ μm (Fig. 5.32b). The i-region length of the M-i-n^+ structure is $L_I = 1$ μm, the n^+-region length is $L_n = 0.3$ μm, and the donor concentration in the n^+-region is $10^{18} cm^{-3}$.

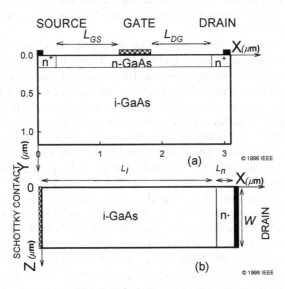

Fig. 5.32 Simulated cross section of GaAs MESFET (a) and GaAs Schottky M-i-n^+ (b) structures [199]

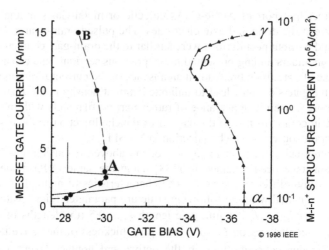

Fig. 5.33 Experimental pulsed gate-drain I–V characteristics of the GaAs MESFET for 10-ns pulse duration and 0.0001% duty factor (solid line), the calculated gate-source I_G–V_{GS} characteristics under $V_{DS} = 4$ V drain bias (dashed line), and the I–V characteristic of quasi-1D M-i-n$^+$ structure with contact width $\Omega = 0.2$ μm (dashed-dotted line) [199]

The calculated MESFET gate-source I_G–V_{GS} characteristic for the drain-source bias $V_{DS} = 4$ V is plotted in Fig. 5.33 as the dashed line. In comparison with the case of drain-source conductivity modulation, the minimum holding voltage at NDC is rather high.

Similarly to the simulation methodology used in the previous sections, the conductivity modulation mechanism can be derived from the analysis of the electric field, electron, and hole density depth profiles (Fig. 5.34) for the states before and after NDC region "A" and "B." In state "A," the impact ionization results in flow of nonequilibrium carriers through the semi-insulating layer. In this state, the picture of avalanche breakdown corresponds to the data presented in [135]. In state "B," the density of the nonequilibrium carriers in the semi-insulating layer is larger and exceeds 10^{17} cm^{-3}. The electrons and holes form a quasi-neutral region in the gate-drain spacing (Fig. 5.34e,f) and an additional region of avalanche multiplication is formed near the drain n$^+$-i junction (Fig. 5.34d). Further increase of the carrier injection from the avalanche regions leads to the expansion of the quasi-neutral region and decreasing of the drain-gate voltage, i.e., to NDC.

Thus, the conductivity modulation mechanism differs from the drain-source avalanche injection discussed in the previous sections. Essentially, it is similar to the double avalanche injection under breakdown of the reverse-biased p-i-n diodes. This physical mechanism is discussed in the following two sections.

The conductivity modulation of the drain-gate region is observed in the MESFET with $L_{GS} \approx L_{DG}$. In the MESFET structure with $L_{GS} = 0.5$ μm and $L_{DG} = 1$ μm, the modulation in the gate-source region is dominant up to $V_{DS} \approx 10$ V (Fig. 5.35).

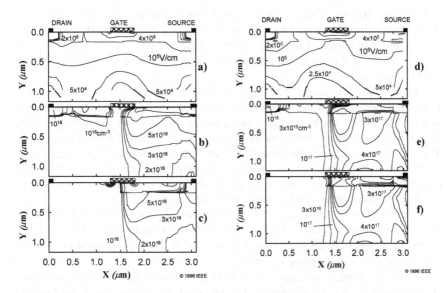

Fig. 5.34 Electric field (a, d), electron (b, e), and hole (c, f) density depth profiles of GaAs MESFET for the avalanche breakdown "A" and NDC states "B" [199]

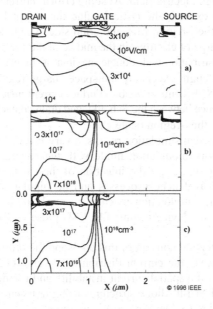

Fig. 5.35 Electric field (a), electron (b), and hole (c) density depth profiles in NDC state of the MESFET with L_{GS} = 0.5 μm and L_{DG} = 1 μm [199]

From the analysis of the cutlines of the density depth profiles, the similarity of the drain-gate breakdown to processes in the Schottky contact M-i-n$^+$ structure has been established. Indeed, at the drain-gate breakdown the gate contact, pinched active layer, i-region, and drain n$^+$-i junction form a parasitic "built-in" Schottky M-i-n$^+$ diode structure.

The 1D M-i-n$^+$ structure with negligibly small $\Omega = 0.2$ μm was simulated in order to demonstrate the scenario of the conductivity modulation (Fig. 5.32b). For consistency in this diode structure, the positive biased n$^+$-anode contact will be further referred to as a "drain" and the negative biased Schottky contact as a "gate." The I–V characteristic of the Schottky M-i-n$^+$ structure has an NDC region (Fig. 5.33). The NDC is the result of the conductivity modulation of i-GaAs layer due to partial spatial neutralization of the avalanche injected electrons and holes from the drain and the gate multiplication regions, respectively.

The distributions of electric field and carrier density are presented in Fig. 5.36. The understanding of the conductivity modulation is easy to derive from the comparative analysis of three states: the beginning of avalanche breakdown (α), NDC regime (β), and high-conductivity regime (γ).

As the drain bias increases, the avalanche multiplication at the gate space charge region begins. In this state (α), the generated electrons are drifting toward the n$^+$-i junction Accumulation of the electrons at the drain junction region creates a second region of high electric field. At some critical current density, the electric field at this junction exceeds the critical value for avalanche multiplication and generates holes into the i-region from the drain side. The hole space charge compensates the space charge of electrons in the middle of the i-region that finally results in the electric field distribution with double maxima of the electric field (Fig. 5.36a, state β). At i-region length exceeding the space charge regions near the n$^+$-i and M-i junctions, the NDC becomes possible. Further current increase leads to decrease of the potential difference between the Schottky and drain contacts due to the decrease of the electric field in the i-region (state γ).

According to numerical simulation results, a very fast transient process accompanies this conductivity modulation mode with the characteristic below 10 ps. This fact is not surprising since the injection of the carriers into the i-region is provided by the fast avalanche processes and no bipolar injection is involved. Thus, this conductivity modulation mechanism might be an interesting solution for the protection of the device against fast ESD pulse realized in charge device model (CDM).

According to the general concept of this book and the problem of the drain-gate burnout discussed above, the current distribution in an NDC state is unstable in general. As a result of the spatial current instability and filamentation, it is easy to obtain the distributed M-i-n$^+$ diode structures using transient simulation approach similar to the simulation problem in n$^+$-i-n$^+$ structures.

For current-controlled NDC, spatial instability resulting in current filamentation was first presented in [199] for the case of spatial instability in M-i-n$^+$ structure with width $\Omega = 10$ μm. After a step current increase, only ~ 300 ps was required

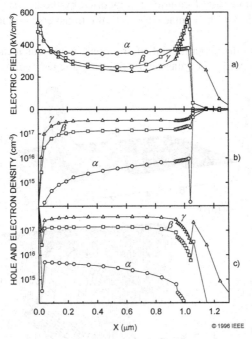

Fig. 5.36 Field (a), electron (b), and hole (c) density distributions in the narrow M-i-n$^+$ structure ($\Omega = 0.2$ μm), for the states: avalanche breakdown (α), NDC (β), and high conductivity (γ) (the states are so denoted in Fig. 5.33) [199]

to provide the transition into a new spatially nonuniform state with very narrow current filament.

Two-dimensional distribution of field and carrier density for the filament state is presented in Fig. 5.37. Typically for the current filament solutions, the electric field and carrier density distribution inside the filament region is close to that for the narrow (1D) structure in state "γ." At the same time, farther away from the filament the distributions correspond to the state of 1D structure before avalanche multiplication.

An important fact revealed by the simulation results is that the maximum current density inside the filament is not limited below the current saturation in the n$^+$-region (Fig. 5.38). This effect is the result of current spreading the n$^+$-region that essentially supplies more total current into the filament region than the corresponding saturation current density in the n$^+$-region. This provides an explanation for the very narrow and experimentally hard-to-observe filament realized in the case of drain-gate breakdown.

The effect of distributed drain n$^+$-region has already been discussed. It results in possibility of formation of solitary or multiple filaments [194, 201, 202]. Since for typical MESFET layers the typical size of the double avalanche-injection filament is much smaller than the avalanche-injection filament, the effect of spontaneous multiple state formation is supported by the larger ratio of the length of current spreading in the contact region to the transverse dimension of filament [195].

Respectively, in the experimental environment, numerous local melting regions are observed after MESFET catastrophic failure in the DC regime.

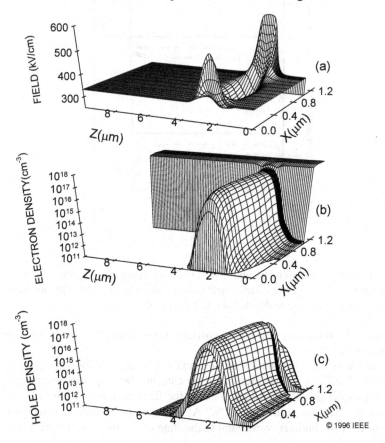

Fig. 5.37 Current filament in the wide M-i-n$^+$ structure ($\Omega = 10$ μm gate width), $L_n = 0.3$ μm. Field (a), electron (b), and hole (c) density distribution at 1×10^5 A/cm^2 current density [199]

Numerical solution for multiple filaments has been obtained for the case of wide M-i-n$^+$ structure with length of contact n$^+$-region $L_n = 9$ μm ($\Omega = 2.55$ μm). I–V characteristics of the structure have a small positive slope after the NDC region. The positive differential conductivity is the result of the additional voltage drop across the extended n$^+$-region. At some current level, the multiple filaments are formed spontaneously during ~2 ns and have the spatial-periodic structure (Fig. 5.39). The spatial period of filaments (distance between filaments) is decreased with the current increase populating more filaments in the device.

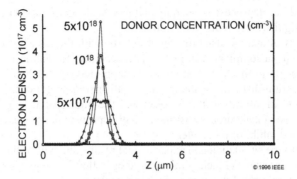

Fig. 5.38 Distribution of electron density in the M-i-n$^+$ structure ($\Omega = 10$ μm, in the $X = 0.54$ μm plane) for varying n$^+$-region doping level at 1×10^5 A/cm^2 current density [199]

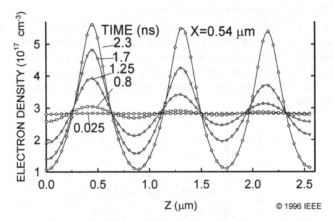

Fig. 5.39 Multiple filaments in the M-i-n$^+$ structure with long n$^+$-region ($L_n = 9$ μm, $\Omega = 2.55$ μm). Field (a), electron (b), and hole (c) density distribution at 1×10^6 A/cm^2 current density. (d) Kinetics of multiple filamentation from the uniform state [199]

The results of Schottky diode M-i-n$^+$ structure simulation agree with earlier studies [203–205]. In general, there are two "built-in" structures in the case of the GaAs MESFET structures connected in parallel. The first M(n)-i-n$^+$ structure is formed by the gate, (active layer $W = 0.17$ μm), i-buffer layer, and drain n$^+$-contact. The second M-n-n$^+$ structure is formed by the gate, active n-layer, and drain n$^+$-contact. The critical current density of NDC formation for M-i-n$^+$ is much lower than for the M-n-n$^+$ structure due to additional current required to compensate the space charge of the ionized donors in the n-layer. Moreover, the potential gap for holes in the n-i interface provides an escape of holes into the buffer layer from the n-channel penetrated. Thus, current instability and filamentation in the case of drain-gate burnout in the MESFETs are observed in the buffer layer.

This conclusion in particular explains why the gate-drain burnout usually takes place first in the MESFET structure's depth and therefore it is hard to observe both the filaments in pulse regime and the failure consequence using just a visual inspection without deprocessing etching steps. Visual light emission absorption in the n-region significantly reduces the intensity.

The analysis of spatial instability in structures with the distributed contact resistive layer was accomplished in [194]. Theoretically, formation of the periodic multiple filaments in the M-i-n$^+$ structure is supported by the theory of autosolitons in the reaction-diffusion systems. The small spatial length of the double avalanche-injection instability in the i-region and the high spatial length of current spreading and the voltage drop across the n$^+$-region provide the proper ratio between the activator and inhibitor parameters.

According to the simulation and experimental results, the typical time of filamentation is ~ 1 ns. At high-frequency microwave operation in the gigahertz range, the duration of the gate breakdown state is less than 10 ps. This time is less than the typical time of spatial instability evolution in the structure. Thus, the critical level of avalanche current is provided by some integral value of avalanche current at some critical level of pulsed microwave power.

For off-state breakdown of conventional power MESFETs, the maximum drain-source voltage is limited by the drain-source current instability of an avalanche-injection nature due to modulation of the buffer layer (Sections 5.2.1–5.2.4). The critical drain and consequent gate current for drain-source current instability in the real off-state operation are considerably lower than the critical current for the gate-drain instability.

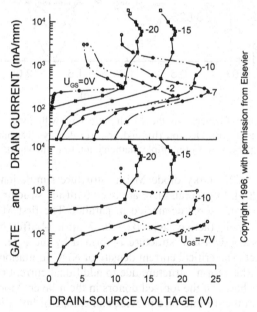

Fig. 5.40 Total calculated drain-source $I–V$ characteristics and gate current of the MESFET with $L_{DG} = 2 \times L_{GS}$ [198]

For off-state breakdown of conventional power MESFETs, the maximum drain-source voltage is limited by the drain-source avalanche-injection instability in the buffer layer. The critical drain and consequent gate current for the drain-source current instability in the pinch-off condition are considerably lower than the critical current for the gate-drain instability (Fig. 5.40).

For given simulation parameters at V_{GS} = 0, −2, −7, and −10 V the maximum drain voltage is still limited by the NDC due to drain-source avalanche injection (Fig. 5.40, dashed lines). Only at V_{GS} = −15 and −20 V is the maximum drain voltage limited by the NDC due to double avalanche injection in the gate-drain spacing (Fig. 5.40, solid lines).

Summarizing the above, one can conclude that electrical current instability and filamentation are always realized prior to the local thermal melting under MESFET burnout at some critical gate avalanche current. Namely, the instability leads into NDC in the drain-gate circuit and switching of the MESFET structure into the state with high-density filament located in the buffer layer. Only after these events the filaments act like a heat source for subsequent structure local melting and formation of Au-Ge-GaAs eutectic leakage channel that short circuit the gate with the drain in the buffer layer inside the filament.

Both the drain-source and the drain-gate current instability mechanisms are realized due to the "parasitic" modulation of the buffer layer conductivity. However, proper use of the n^+-regions considerably influences the filament. Maximum current density in the filament is limited by the saturation in the n^+-region, while the lengthy n^+-region provides the multiple filamentation across the structure width.

5.4 Microplasma Effect at the MESFET Breakdown

Another phenomenon that has been widely observed under certain conditions in original MESFET devices is formation of multiple highly localized regions of the avalanche breakdown. This phenomenon in MESFETs is similar to microplasma formation in reverse-biased p-n junctions [208–211]. Unlike previously discussed drain-gate current instability and filamentation, the microplasma effect accompanied the avalanche breakdown. Thus, essentially the state "α" in Fig. 5.33 might not be ideally spatially uniform along the contacts but often has a nonuniform distribution with local spots.

As a rule, the term "microplasma" is used for a narrow avalanche current region with the transversal dimension significantly less than the total structure dimension.

At the same time, microplasmas of ~ 1 μm dimension are experimentally observed at the ohmic contact (drain) breakdown [208, 209] in the region of drain static field domain. This kind of microplasma is referred to as "lighting points." In uniform samples, the periodic pattern of the microplasma formation is observed. However formation of microplasma near the drain contact is observed only in MESFETs without the deep gate recess or n^+-region. In serial MESFETs with the drain n^+-region, the distribution of the avalanche current is uniform up to the avalanche injection instability.

There has never been agreement between different experts in the field as to the cause of the microplasma effect. Two major polar opinions are based on the local structural defect with high electric field and the self-organization phenomena at the avalanche breakdown. This section is devoted to clarification of the general physical mechanism of the microplasma breakdown in semiconductor structures.

5.4.1 Microplasma at the Schottky Gate Breakdown in MESFETs

Microplasma breakdown was observed in MESFETs in [208, 210] at Schottky gate avalanche breakdown. However, in commercial MESFETs this phenomenon is not so regular.

The microplasmas at the gate breakdown are of dimension less than 1 μm and correspond to classical microplasmas in reverse-biased p-n junctions [212]. The microplasma breakdown is often accompanied by small current jumps, flickering noise, and hysteresis of the $I–V$ characteristic of 10- to 150-μA amplitude [209, 210].

The estimated current density inside microplasma from its visual dimension and $I–V$ characteristics exceeds $\sim 10^5$ A/cm^2. However, the excitation of the microplasma is observed at a very small total current to structure width ratio ~ 1 A/cm^2.

Up to now, the microplasma formation mechanism has not been well clarified. There are two basic concepts in this regard. From the classical point of view [212], microplasma formation is directly related to local inhomogeneities of the doping profile or contacts resulting in a local increase of the field and corresponding multiplication coefficient.

At the same time, microplasma near the gate preserves its dimension, localization, and does not reach a confluence up to the burnout [213]. This identical dimension of microplasmas, its development of spatial periodic order, and hysteresis in I–V characteristic led to an alternative hypothesis that the microplasma exhibits a self-maintained high-amplitude solitary filament state is formed in slightly nonequilibrium distributed active system [213].

The simplest example of such active system is the semiconductor structure compounded from two regions. One of them has a negative differential resistance (NDR). This NDR is compensated by a positive differential resistance (PDR) of another region [214]. Experimental observation of such phenomena is reported in [215–217] and validation by numerical simulation of multiple filaments in such structures is completed in [218–223].

In addition to a fundamental interest to understanding of microplasma formation at Schottky gate breakdown, this phenomenon is important for long-term reliability of power MESFETs. The maximum output power P_{OUT} of MESFETs and MODFETs is limited by the drain-source voltage U_{DSBR} in the pinch-off conditions at gate biases above $U_{GS} = U_{PIN}$. In optimized MESFET structures, U_{DSBR} is limited already by avalanche breakdown in the gate-drain spacing. The breakdown provides current in the gate circuit. Large-signal operation with the gate breakdown current in nonlinear regime at maximum power or load mismatch may result in decrease of the middle time before failure (MTBF). The decrease is accelerated from some critical gate current amplitude. Therefore, commercial power MESFETs usually have maximum rating for the reverse gate current in large-signal operation ~ 0.1 mA/mm.

At microplasma gate breakdown, the peak current through the solitary microplasma ~ 10–100 μA cannot be detected in the RF regime. At the same time, total breakdown current limitation cannot guarantee absence of high current flow through a low-voltage solitary microplasma, especially in power MESFETs with a high total gate width ~ 5–20 mm.

There is no doubt that the microplasma as a local "defect" region of high avalanche multiplication with the current density $\sim 10^5$ A/cm^2 results in decrease of the device MTBF. This may be one of the reasons for accelerated failures with time. Therefore, understanding of the microplasma and uniform breakdown mechanisms is practically important too.

In this section, the microplasma and uniform breakdown mechanisms of the MESFET gate breakdown are discussed using both experimental and numerical simulation results.

As will be shown below, microplasma type of breakdown can be achieved in the simplified Schottky M-n-n$^+$ structures. The solution for the microplasma states in 2D M-n-n$^+$ structures provides an explanation for a universal microplasma mechanism in reverse-biased Schottky junctions.

Commercial MESFETs with 100 mW of output power at 18 GHz have been used for the experimental data collection. The devices have double recess structure (Fig. 5.41), initial drain current $I_{DSS} = 110$ mA, total gate width 280 μm, gate length $L_G = 0.5$ μm, channel length $L_{DS} = 3$ μm, 0.2-μm contact n$^+$-layer of 3×10^{-18}cm^{-3} donor concentration, and 0.18 μm active n-layer of 3×10^{17}cm^{-3} donor concentration.

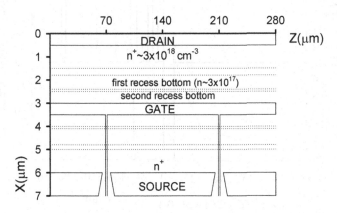

Fig. 5.41 Simplified topology of the experimental MESFETs

Maximum drain-source voltage at the pinch-off breakdown condition with good accuracy is equal to the drain-gate breakdown voltage at the floating source after adding the gate bias value $U_{GDBR} = U_{GS} + |U_{DSBR}|$. Therefore, for simplification, avalanche breakdown is studied in the floating source configuration. The DC drain-gate $I–V$ characteristic and corresponding light emission intensity have been measured using the scanner setup with photomultiplier.

The gate-drain breakdown DC $I–V$ characteristic and the electroluminescence distribution along the gate were measured in 400 randomly selected MESFETs from various production lots over a period of 1 year of production time. According to measurement results, all MESFETs were subdivided into three groups: the "uniform," the microplasma," and the "defect" groups. The subdivision was made based on the following criteria.

The "uniform" MESFETs ($\sim 60\%$ of the total amount) reveal a smooth $I–V$ dependence (Fig. 5.42a). The light emission distribution is uniform within the accuracy of measurements and solitary microplasma is not exited (Fig. 5.42b).

The devices with breaks and hysteresis in $I–V$ characteristics in the vicinity of the same breakdown voltage level as the uniform devices have been defined as

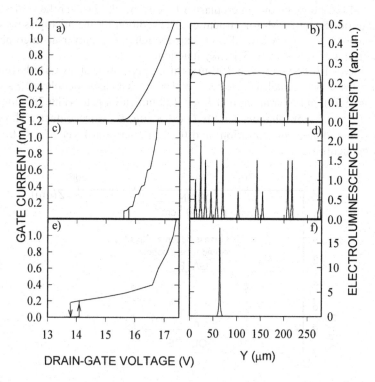

Fig. 5.42 Typical DC drain-gate $I–V$ characteristic with corresponding electroluminescence intensity distribution at the breakdown current of 0.4 mA for the "uniform" (a, b), "microplasma" (c, d), and "defect" (e, f) MESFETs

"microplasma" MESFETs (~ 30% of the total samples) (Fig. 5.42c). In these devices, each peculiarity in the *I–V* characteristics corresponds to formation of microplasma. The light intensity distribution is nonuniform and corresponds to the multiple high-amplitude microplasma states (Fig. 5.42d).

Finally, the MESFETs that showed a rather clear hysteresis region of the first microplasma switching with a break or hysteresis (Fig. 5.42e) at much lower breakdown voltage level have been defined as "defect" devices (~ 10%) assuming the presence of strong local inhomogeneity in the device structure. In this case, the light emission intensity distribution corresponds to a solitary microplasma state (Fig. 5.42f).

For a final set of 170 total "microplasma" and "defect" MESFETs, the distribution on the avalanche breakdown value (at $I = 3$ mA/mm) and the difference between the breakdown voltage and the first microplasma switching voltage are shown in Fig. 5.43.

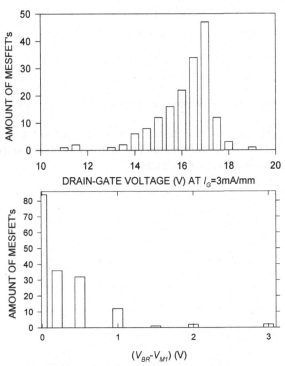

Fig. 5.43 Distribution of 170 "microplasma" MESFETs on the value of developed breakdown voltage and the first microplasma voltage difference

The numerical analysis employed a two-dimensional QHD model, boundary conditions, and calculation procedure similar to those for the model discussed in Section 5.2.4.

The recess MESFET structure with the surface depletion layer is calculated (Fig. 5.44a) for the following structure parameters: drain-gate and gate-source spacing $L_{DG} = L_{GS} = 0.5$ μm, gate length $L_G = 0.5$ μm, donor concentration in the active and contact layers 3×10^{17} and 3×10^{18} cm^{-3} respectively, and thickness of the semi-insulating layer $W_{SI} = 1$ μm.

The data points for I–V characteristics and spatial distribution were calculated from transient solutions for given current value up to steady-state condition. Similarly to the avalanche-injection current filament simulation problems, the numerical noise provided the source of fluctuations for physical solution of the nonlinear numerical problem.

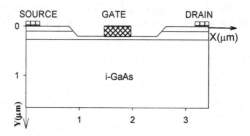

Fig. 5.44 Cross section for simulated MESFET structure

Calculated drain-gate I_G–V_{DG} characteristic of the MESFET structure is presented in Fig. 5.45. Positive differential conductivity is observed over the whole range up to the strong injection conditions.

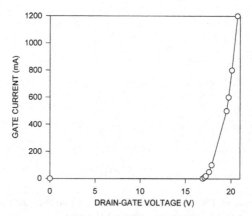

Fig. 5.45 Calculated drain-gate I–V characteristic of the MESFET structure

The electric field distribution in the drain-gate spacing at breakdown signify-cantly depends on the thin film effect and the presence of depletion layer and recess [224–227]. These factors result in smoothing of the electric field and formation before the avalanche breakdown of the state with two field domains near the gate edge and near the drain. The breakdown evolution is significantly different from the breakdown of uniformly doped Schottky diode presented below. In the 1D diode, the breakdown voltage is approximately twofold lower.

However, the positive feedback mechanism can be detected from the analysis of the distribution of electric field in MESFETs at different breakdown current levels (Fig. 5.46). At some critical current density, the breakdown in the gate and in the drain domain provide mutual neutralization of carrier space charge in the gate-drain spacing. This results in electric field decrease in the middle of the structure and total negative differential conductivity of the structure. These distributions are presented at a depth of 0.05 μm from the gate boundary in Fig. 5.46.

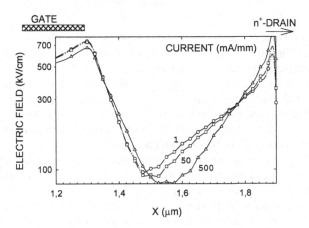

Fig. 5.46 Distribution of the electric field in the active layer in the MESFET channel ($Y = 0.2$ μm)

For some structure parameters, the compensation of the electric field in the central part can exceed the electric field increase in the gate and drain field domains. This results in the total NDR of the MESFET from the high current levels $\sim 10^4$ A/cm^2.

However, this high current NDR cannot explain the microplasma formation directly. The microplasma mechanism has been clarified using 2D numerical simulation of the breakdown of M-n-n$^+$ structures.

Calculations of the current instability and microplasma are made for the GaAs M-n-n$^+$ structure (Fig. 5.47). The structure parameters were similar to the MESFET parameters: length and donor concentration of the n$^+$- and n-regions are $L_{n+} = 1$ μm, $L_n = 0.5$ μm, $N_{n+} = 3 \times 10^{18}$ cm^{-3}, and $N_n = 3 \times 10^{17}$ cm^{-3}, respectively. The comparative calculation was made for a wide 2D structure with $W = 10$ μm and a narrow 1D structure with $W = 0.2$ μm contact width.

Apparently, MESFET breakdown for the 1D M-n-n$^+$ structure cannot fully reflect the real processes at gate breakdown due to significant influence of the surface on the electric field redistribution. In the gate-drain spacing, the positive space charge of ionized donors is compensated by the normal component of surface electric field.

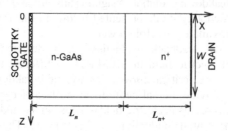

Fig. 5.47 Schottky M-n-n$^+$ structure model

From a general point of view, a more exact approximation of the breakdown in the gate-drain spacing in the Schottky M-i-n$^+$ structure can be provided taking into account scattering in the neutral donor atoms.

According to the calculated I–V characteristic of the 1D M-n-n$^+$ structure (Fig. 5.48a) and field distribution (Fig. 5.48b), the breakdown begins from the avalanche ionization in the gate domain at $V_{BR1} \sim 10$ V voltage value. Only at $V_{BR2} \sim 20$ V, when the space charge region attains the drain junction, does the double avalanche-injection mode of breakdown take place.

Since at high avalanche injection level the surface effect can be neglected, the dependence of the structure NDR on the n-region length can be obtained from the breakdown of the 1D M-n-n$^+$ structure (Fig. 5.48c).

Adding to the simulated 1D M-n-n$^+$ structure the n$^+$-region results in compensation of negative differential resistance of the M-n part of the structure by the positive differential ohmic and saturation resistance of the n$^+$-region at high current density. This 1D effect results in formation of the I–V dependence with total positive differential conductivity over the whole current range (Fig. 5.48a, dashed lines).

Unlike the case of drain-gate burnout discussed in the previous section, in this case the distributed contact region limits the current in the microplasma state locally in the n-region of the active film with the donor doping $\sim 10^{17}$ cm^3. Thus, the limitation of the current density is provided at a current level approximately one order of magnitude less than in the case of the drain-gate current filaments due to parasitic M-i-n$^+$ structure. Current saturation in the n-region makes the intrinsic NDC hidden and the whole structure has positive differential conductivity.

Major regularities for filamentation are similar to the problem of the drain-gate burnout with the exception of much more significant current limitation. In the calculated structures in the current range corresponding to NDR of the M-n part of the structure, the stationary solitary filaments are spontaneously formed. The simulated I–V characteristic is divided into different regions. Stationary distribution at the filament state of 2D M-n-n$^+$ structure with $W = 10$ μm is presented in Fig. 5.49.

The formation of filament is observed only if the parameters of the M-n-n$^+$ provide either total or intrinsic NDR. In the opposite case, the avalanche current distribution remains uniform and all numerical disturbances are dumped. In accordance with the simulation results (Fig. 5.48), the critical n-region length is ~ 0.4 μm at the donor concentration 3×10^{17} cm^{-3} and the current stratification is not observed.

In this heavily doped n-base region of M-n-n$^+$, the thin film effect is essential. In real MESFET 3D problem, the current density between filaments must be significantly lower according to the *I–V* characteristic (Fig. 5.45).

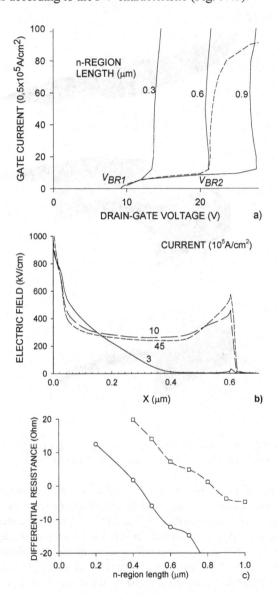

Fig. 5.48 Calculated *I–V* characteristic (a) and the electric field distribution (b) in the 1D M-n-n$^+$ structure with $L_n = 0.6$ μm and $L_{n+} = 0.1$ μm (a). The case $L_{n+} = 1$ μm is shown with dashed lines. (c) Dependence of the differential conductivity value on the n-region length (L_n)

For qualitative estimation of current stratification in the real device, a lightly doped base structure is calculated with the parameters: $L_{n+} = 5$ μm, $L_n = 1$ μm, $N_{n+} = 5 \times 10^{17}$cm^{-3}, $N_n = 5 \times 10^{15}$ cm^{-3}, $W = 26$ μm (Fig. 5.50).

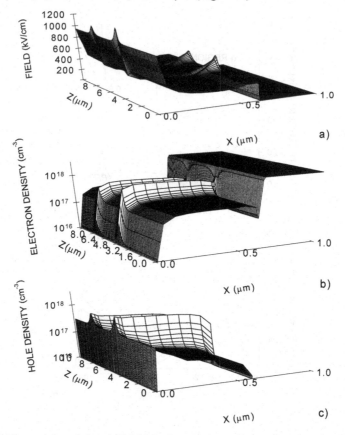

Fig. 5.49 The modules of electric field (a), electron (b), and hole (c) density distribution in 2D GaAs M-n-n$^+$ structure for four-filament state (at 100 kA/cm^2)

The scenario of filamentation is similar to the case of avalanche-injection filamentation in n-i-n structures. Each branch of I–V characteristics corresponds to some stable state with fixed number of filaments. After current increase, the filaments are broadened up to some critical width and become unstable at some critical current density. Then a new state with a large number of filaments is formed (Fig. 5.51). It is achieved by splitting of filaments in an initial state and repumping over into a new stable state. Respectively, the number of filaments decreases with current decrease due to suppression of some filaments in an initial state and their rearranging into a new state. The new state attains a spatial periodicity at small distance between filaments. For small number of filaments in the structure, the redistribution after transition to a new state did not result in periodic stationary states within the parameters of the simulation problem. This effect is most likely related to the limited accuracy due to finite grid dimension.

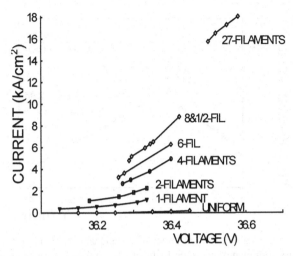

Fig. 5.50 Calculated I–V characteristics for 2D M-n-n$^+$ structure

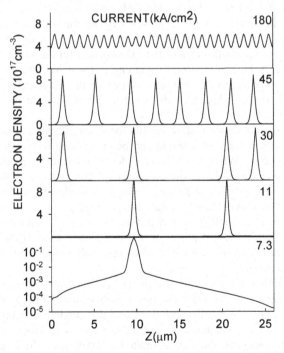

Fig. 5.51 Stationary distribution of electron density (in the $X = 0.54$ μm plane) for different current (I/W) values

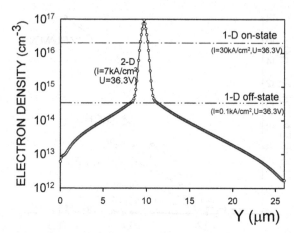

Fig. 5.52 Two-dimensional autosoliton effect. The stationary distribution of electron density (in the $X = 0.54$ μm plane) for single-filament state in the 2D structure (solid line) and for high- and low-conductivity states in the 1D structure at the same voltage (dashed lines)

In the structure with a small n^+-region, only the single filament is observed (L_{n^+} = 0.3 μm). In the structure with lengthy n^+-region, the field and carrier density distribution along current lines in the filament state did not coincide with the distributions in the on- and off-states of the 1D structure for the same corresponding voltage (Fig. 5.52). Due to a current spreading in the n^+-region, the filament amplitude significantly exceeded the possible level in the 1D structure on-state, but in the filament tails it is less than the level in the off-state at the same voltage level.

Thus, from the above simulation results the microplasma and uniform breakdown mechanisms in GaAs MESFETs can be understood within the same physical model as follows. When the MESFET design (structure, recess, and surface depletion layer) provides the positive differential conductivity of the part of the channel between the gate and drain recess edge, the avalanche current distribution is uniform and stable over the whole gate current range.

If the latent or total NDR is provided, then the microplasma can be exited and formed as a solitary current filament or self-maintained state. If the microplasma exited its minimum current turn-off level corresponds to the local current density 10^4 A/cm^2 at its dimension of ~ 0.5 μm.

The second important observation is the critical current of the microplasma excitation process and a role of local structure inhomogeneity. If the gate-drain breakdown voltage value is not fully optimized and the maximum drain voltage V_{BR2} (corresponding to the breakdown at the high current) is significantly higher than the breakdown voltage in the gate domain V_{BR1}, then the critical current of the MESFET can be obtained at a high total gate current.

On the contrary, at a sharp breakdown characteristic with $V_{BR1} \sim V_{BR2}$ the voltage overloads ~10 mV of small local inhomogeneity with the local high field result in microplasma excitation at a not high total gate value $\sim 10^{-6}$A. Moreover, this current does not depend on the total gate width, but with the gate width increase the probability of local defect region is increased.

According to the experimental results, the current density in microplasma region j_M can be estimated using its visual dimension ~ 0.1–0.5 μm and the current value after the switching ~100 μA: $j_M > 4 \times 10^4$ A/cm². The uniform current density before the switching j_U is less than 10 A/cm². High ratio between the maximum current density in the filament and its tales ($\sim 10^5$) provides a similarity between the filament and autosoliton states [194].

In the structure with latent NDR, a spontaneous formation of multiple filaments is not different from the presented structure [223]. Nonrestructuring of the multiple filaments in a spatial periodic pattern [207] at high distance between filaments is in correspondence with experimental data [202].

The calculated multiple filamentation in p^+-n-n^+ structures exhibits the same regularities. A number of physical mechanisms were suggested for avalanche current stratification structure, for example the mechanism of increase of short-wave fluctuation (of a multiplication region length) due to carrier space charge separation [228].

Employing an additional assumption, a mechanism both less sophisticated and more universal can be suggested for the case of microplasma breakdown in Schottky and in p-n junction diodes. This assumption concerns an additional submicrometer lightly doped or compensated region near the junction or metal–semiconductor interface. For example, if a lightly doped n^--region in p^+-n junctions is formed, then according to simulation results for microplasma formation the dimension of this layer can be ~ 0.5 μm or higher. In this case, even a tenfold lower doping level in this assumed region is sufficient to achieve microplasma solutions.

Existence of the NDR in the external $I–V$ characteristics is not necessary for microplasma filament formation. It is the only necessary that the parameters of the structure provide an internal NDR of the M-n (p^+-n) part of the structure. In this case, positive differential conductivity of the whole structure due to a voltage drop in the n^+-region can be observed.

The phenomenon of current filamentation in the structures with "latent" (internal) NDR offers a good explanation of the microplasma mechanism observed at PDR breakdown in the MESFET's gate or IMPATT diodes. Similarly to other current filamentation simulation problems, the instability in the structures with internal NDR is realized due an increase of the current fluctuation, i.e., due to numerical errors of finite-difference scheme solution. If a numerical error with a small dimension of $w \ll W$ in the active n-region is occur, then the current increase cannot be effectively suppressed by a voltage drop across the n^+-contact region due to a high spatial width $\sim W$ of current spreading [194].

According to the presented measurement, the control of microplasma in MESFET by measurement of the total gate current is not effective. The diagnostic of low-voltage microplasma is the most useful for the reliability problems.

5.4.2 Microplasma Formation in n⁺-i-n⁺ Structures

In the original studies, the formation of microplasma-like states was observed at the breakdown in the drain-source breakdown. The phenomenon has been observed as a formation of multiple periodic lighting dots at some critical drain-source breakdown voltage [208]. The necessary condition for microplasma formation was metal–semiconductor junction, while in the structure with the drain n⁺-region the microplasma states have never been confirmed. In commercial C- or X-band MESFETs with the typical n⁺-n recess structure, this phenomenon is observed usually only as a single lighting spot significantly localized on the structural defect. For millimeter-wave FETs which use alloy contacts, this phenomenon is of some interest.

The lighting points mechanism has been clarified by the investigation of the GaAs thin film test structures without gate contact with various drain-source spacings L_{DS} (from 1 to 25 μm) and the small width $W = 15$ μm with various thicknesses $d = 0.1$–0.6 μm (Fig. 5.53). In these structures, the distribution of the electroluminescence in the surface is investigated using the scanner setup with sensitive photomultiplier and spatial resolution ~ 1 μm.

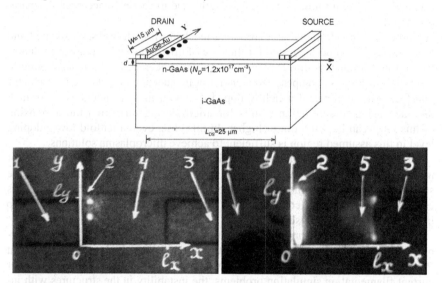

Fig. 5.53 Simplified drawing and microphotographs of GaAs planar test structure with lighting points (1, drain; 2, lighting points; 3, source; 4, thin film of n-GaAs; 5, electroluminescence from Gunn domain)

The breakdown in the structure is determined first by the GaAs film thickness. At the thick active layer films (0.4–0.6 μm, $I_{DSS} = 280$–400 mA/mm), the traveling Gunn domain is spontaneously formed. The Gunn oscillation is stable and its frequency is proportional to $(L_{DS})^{-1}$ (~ 4 GHz at $L_{DS} = 25$ μm). The N-shape I–V characteristic corresponds to the traveling domain formation (Fig. 5.54, curve 1).

The traveling domain amplitude is high. This resulted in uniform avalanche ionization in the film of the drain-source spacing [230]. The electroluminescence distribution along the drain contact remains uniform up to the burnout.

At the thickness value of 0.25–0.35 µm ($I_{DSS} \sim$ 200 mA/mm), Gunn oscillations became chaotic and "noise" type. Gunn oscillations are coherent at high avalanche level corresponds [127]. The N-shape region in the I–V characteristic is degenerated and part of the samples have the S-shape hysteresis at breakdown (Fig. 5.54, curve 2). A wide spectrum of oscillations is observed. The breakdown in the drain-source spacing is enhanced. In such structures, the drain static domain is formed. At some critical U_{DS} value, the multiple bright lighting points formation near the drain is observed in the optical microscope. The visual points dimension is ~1 µm.

At thickness value below 0.2 µm, the Gunn oscillations are suppressed. The electroluminescence far from the drain in the drain-source spacing is enhanced. At some critical U_{DS} value, the same multiple bright lighting points near the drain contact are formed.

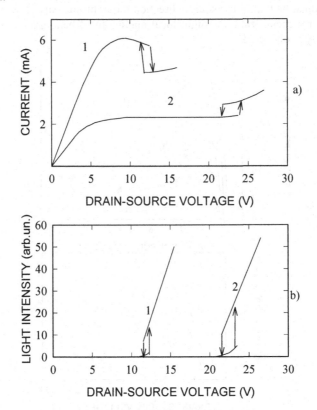

Fig. 5.54 I–V characteristic (a) and electroluminescence intensity (b) dependence on the n-GaAs film thickness $d \sim 0.6$ µm (1) and $d \sim 0.25$ µm (2) of the test structures

In the most uniform samples, the lighting points have a spatial periodic structure in the drain direction at the same dimension of each point [229].

The regularities of lighting points formation in the MESFETs without n^+-region and recess correspond to the phenomena in the test structures. The gate presence results in dumping of the Gunn domain at the highest drain saturation current in comparison with the test structures (Fig. 5.55a).

The second peculiarities of the lighting points formation are connected with the local inhomogeneity which plays an important role at a high total gate width.

The nonuniform breakdown is observed both at breakdown in the traveling Gunn domain and at breakdown in the static domain. This results in a complex view of the total light emission intensity dependence on U_{DS} ($I_\lambda(U_{DS})$) (Fig. 5.55b, $U_{GS} = 0$–1 V).

The values U_{SAT} and U_{BR} for nonuniform structures and the initial part of the $I_\lambda(U_{DS})$ dependence are lower than for the uniform ones. For the nonuniform structure, the excessive current before the burnout is very small (Fig. 5.56a, curve 2).

The mechanism of lighting points formation is a rather complex problem [209, 231]. However, additional measurements of electroluminescence intensity distribution demonstrate that in the source direction the light intensity is decreased, but near the source metallization a slight local maximum of band gap recombination light intensity is observed.

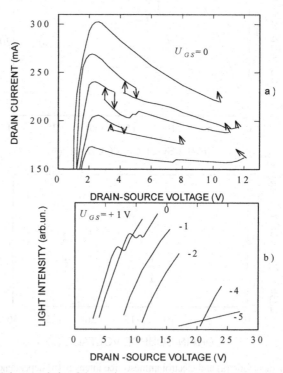

Fig. 5.55 (a) I–V characteristic of the various test structures and (b) electroluminescence intensity at various U_{GS} values for MESFET structure

Therefore, the injection from the source contact is important in this mechanism. This fact allows an assumption about similar basic avalanche-injection mechanism of breakdown and filamentation (Sections 5.2.1–5.2.4). The second basic fact is the presence of a sharp static domain due to the metal–semiconductor interface junction and corresponding local inhomogeneity due to a nonuniform edge of the drain metallization or local irregularities of the metal–semiconductor interface (Fig. 5.56). In particular, due to these factors no lighting points formation has been observed in the structure with n$^+$-region under the drain contact metallization [149].

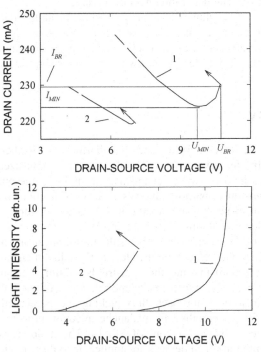

Fig. 5.56 Experimental I–V characteristic (a) and electroluminescence intensity (b) for uniform MESFET structure "1" and MESFET with local inhomogeneity "2"

In general, the measure of breakdown voltage increase by itself can provide higher robustness to overloads. However, the majority of typical technological means of U_{BR} increase result in trade-off for the major transistor capabilities: noise, frequency, amplification. The problem of improvement of breakdown characteristics requires a rather complex optimization approach. An optimal device-level solution is possible by variation of the doping level and thickness of the active layer, the buffer layer thickness, the gate-drain spacing, and the gate length [150]. The practical methodology of this optimization problem is calibrated physical process and device simulation.

Several technological solutions toward increased stability are listed below. The gate recess [129, 132, 139, 149, 231, 233] provides an essential advantage at operation of the transistor in large currents, as it prevents formation of the sharp electric field domain at the drain metallization. At small current levels and in the pinch-off condition, this measure provides no advantage in increase of the breakdown voltage. On the contrary, the breakdown voltage in planar structures is usually higher. However, at higher current level, increase of the breakdown voltage is realized.

Matching of n^+-region and recess was first suggested in [137, 149]. The n^+-region is formed as a part of n^+-epilayer in epitaxial MESFETs or can be obtained by ion implantation. The usual thickness of the layer is ~ 0.2 μm. With introduction of n^+-edge, the drain ohmic contact impact on the breakdown does not become a limiting factor. In this case, microplasmas, lighting points, and local heating effects are either locally ballasted or at least separated from the drain metallization edge. The long-term reliability of the devices with n^+-drain is significantly better [232].

5.5 Summary

The most important regularities and peculiarities of isothermal avalanche-injection instability for devices with Schottky gate are emphasized in this chapter. First, the devices involve two mechanisms: the drain-source and the gate-drain burnout. The drain-source burnout involves avalanche-injection conductivity modulation and current instability in parasitic n^+-i-n^+ structure formed by the contact regions and the semi-insulating buffer i-region.

The specifics of the carrier transport results in nonlinear and nonlocal dependence of the mobility and velocity on the electric field do not change the conductivity modulation mechanisms and the catastrophic failure scenario in principle, both for DC and large-signal microwave operation too, although they significantly complicate the analytical approach thus making the numerical simulation approach most useful. Essentially this avalanche injection process is similar to the avalanche-injection effect in silicon MOSFETs. The major regularities for isothermal SOA limitation for Si devices are applicable to these devices too.

In the case of drain-source burnout, the physical nature of the effect can be explained by phenomena in the parasitic n-i-n bipolar device with an open base with very low current gain in lengthy buffer layer with rather low carrier lifetime in GaAs ~ 1 ns. The case of gate-drain burnout involves the double avalanche-injection conductivity modulation mechanism. However, the numerical solution for the gate electrical instability and for current filament can be obtained using the same simulation methodology. Absence of the local level current limitation by a current saturation region at the gate side results in uncontrolled current density increase in the filament region providing for rather narrow filaments of much greater amplitude in comparison with the avalanche-injection filaments.

Finally, based on the presented results, the authors of this book believe that formation of microplasma is a reflection of the same double avalanche-injection physical phenomena, but realized on a different scale under conditions of local current limitation.

Chapter 6
Degradation Instabilities

Reliability of any semiconductor device is mainly determined by the physical conditions of its operation. If electrical, thermal, mechanical, or other loads are close to some limit or exceed it, then the probability of catastrophic failure is rather high. If transistor operation conditions guarantee a safe operation regime and the operating regime never exceeds SOA limits, then the failure probability is sufficiently low. However, even in this case it is never equal to zero at any real operation regime including storage. The main reason for this is continuous evolution of various degradation processes inside the transistor structure.

Degradation changes in the structure are not always available to direct observation or measurements, but a direct consequence is gradual change in transistor parameters and operation capability degradation.

However, some degradation relaxation processes have practically no influence on the datasheet transistor parameters until some threshold or critical value of the changes is reached. Then, attainment of the threshold results in drastic changes in transistor operation. Obviously, such processes usually do not provide the possibility of visual inspection unlike for example in the case of electromigration.

As a rule, the rate of degradation process evolution depends on operation conditions: device temperature, electric field level, and current density. Meanwhile, the time before failure is usually determined by some assumed ratings for deviation in the transistor parameters. Therefore, degradation failures are mostly conditional.

As mentioned in Chapter 1, in the case of new device types the portion of observed cumulative catastrophic failures usually exceeds degradation failures by a few times. While this is true for general statistics, at application some degradation failures result in major limitation of the time before failure, for example in the case of various autonomous systems with long-term operation. In any case, degradation failure is not a matter of secondary importance.

In spite of the high importance of degradation failures, this chapter is focused only on several limited aspects of these phenomena related to catastrophic failures. First, the methods of degradation failures are widely studied. For this reason, results obtained for the degradation mechanism have been the topic of numerous books. A variety of basic degradation processes were studied on the fundamental level [234–237]. A particular realization of these processes in transistor structures corresponds to common regularities and, therefore, does not require detailed presentation.

A limiting set of published studies are involved below with the sole purpose being to provide brief introductory material required for understanding of the basic degradation mechanisms. Then the rest of the chapter is focused on discussion of the catastrophic failures related to nonlinear effects, for example degradation instability phenomena.

6.1 Electromigration

Electromigration is the phenomenon related to transport of a material, usually metallization, under conditions of high current density [238, 239]. For observation of the phenomenon during real time, a critical current density level 10^5–10^6 A/cm^2 can usually be easily achieved in thin metal films without considerable Joule heating. Therefore, electromigration is not an exotic phenomenon in microelectronic applications.

Metal transport along the current lines is usually finalized by creation of a metallization break, or a short circuit event due to creation of a pile of metal accumulations. Electromigration failure becomes a dominant degradation mechanism at current densities $j > 10^5$ A/cm^2 and temperatures $T > 125°C$.

Metal atom flux is the result of electron "wind" effect. Therefore, the electromigration theory is based on quantum mechanics formalism [234]. For ion flux with the effective charge Z^*_{EFF}, the following simple formula holds [234]:

$$\vec{\varphi} = N\vec{v} = N\mu Z^*_{EFF}\rho\vec{j} , \qquad (6.1)$$

where N, v, j, ρ, μ are the concentration of metal atoms, the velocity of directed atom drift, the current density in metal, the specific resistance, and the ion mobility, respectively. The mobility can be expressed by the Einstein relation: $\mu = D/kT$, where D is the diffusion coefficient for metal ions.

In a simple case, the mass transport is described by the one-dimension continuity equation

$$\frac{dN}{dt} + \mathrm{div}\,\vec{v} = 0 . \qquad (6.2)$$

Since usually $\mathrm{div}\,\vec{v} = 0$, the equation can be presented in the form [240]

$$\frac{dN}{dt} = -N\vec{j}\,\mathrm{grad}\left[\left(D/kT\right)Z^*_{EFF}\rho\right] = 0 . \qquad (6.3)$$

From (6.3), it follows that if the parameters D, ρ, T are constant along the current direction, then $dN/dT = 0$, and the break of the structure is not realized even at variable cross section of the film.

If $\mathrm{grad}\left[\left(D/kT\right)Z^*_{EFF}\rho\right] > 0$, then in a corresponding region the matter accumulation takes place ($dN/dT = 0$); if $\mathrm{grad}\left[\left(D/kT\right)Z^*_{EFF}\rho\right] > 0$, then a void formation is observed under the assumption of positive current flow direction.

There are two types of gradients in metallization lines of real transistors: the microscopic due to grains of metallization film and the macroscopic due to various reasons connected with operation regime, design structure, and technological peculiarities of the device.

Electromigration in the thin films with microscopic gradients results in uniform structure damage only. As a result, the metal line is covered by cells of voids and pileup of metal material accumulation. These peculiarities grow with time in chaotic order [241, 242]. The failure occurs when the voids cross the whole line.

Macroscopic gradients result in dominant accumulation or, on the contrary, the voids grow on lengthy regions of metallization. The view of electromigration failures in real transistors (Fig. 6.1) demonstrates the presence of regions with high value of macroscopic gradients in transistor structures. Apparently, the latter determines the value of middle time before failure. The microscopic gradients are present as a statistical background and result in scattering in the time before failure [240, 241].

Influence of macroscopic gradients on the time before failure can be estimated using the following simple considerations [240]. The probability P of metallization line intersection by the region of voids during time period Δt is proportional to Δt and the speed of atom density decrease in given section dN/dt. It is inversely proportional to N and the line section S:

$$P \sim \frac{\Delta t \left(dN/dT \right)}{SN}.$$

The probability of the absence of line damage during Δt is $P_1 = 1 - P$, and during the time period $t = \Sigma \Delta t$: $P_{1t} = \prod (1 - P) \approx 1 - \Sigma P$.

The line is damaged if $P_{1t} = 0$. By replacing the sum with the integral value, the failure condition is given by: $1 - \dfrac{1}{C} \displaystyle\int_0^{t_{MTBF}} \dfrac{dN/dT}{SN} \, dT = 0$,

where t_{MTBF} is the middle operation time before failure; $1/C$ is the proportionality coefficient. Substituting dN/dt from (6.3) and assuming $\mathrm{div}\,\vec{v}$ to be constant over time, the expression for t_{MTBF} is given by [241]

$$t_{MTBF} = \frac{CS}{j} \left[\mathrm{grad} \left(D/_{kT} Z^*_{EFF} \rho \right) \right]^{-1}. \tag{6.4}$$

Obviously, for t_{MTBF} estimation. the maximum possible value of gradient is used.

The influence of microscopic gradients and film microstructure can be adequately reflected by presentation of atom flow in i-section $\varphi(x_i)$ in the form of casual value [241, 242]:

$$\varphi(x_i) = \sum_{r=1}^{n(x_i)} NZ^*_{EFF} \frac{D}{kT(x_i)} \cos \varphi_r \rho h \delta j, \tag{6.5}$$

where h is the film thickness, $n(x_i)$ is the number of grain boundaries in i-section, δ is the median width of the grain boundary, D_r is the diffusion coefficient along r-grain boundary, and ϕ_r is the angle between the transport direction and r- boundary. The diffusion coefficient D_r is given by

$$D_r = D_0 \sin\left(\frac{\theta_r}{2}\right)\exp[-E(\theta_r)/kT],\qquad(6.6)$$

where θ_r is the angle of disorientation between the adjacent grains and $E(\theta_r)$ is the activation energy of the electromigration processes.

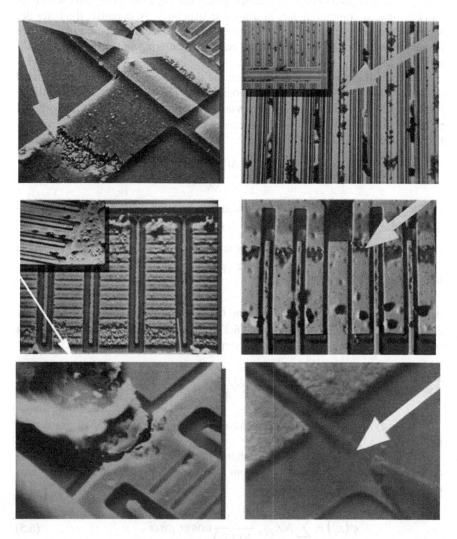

Fig. 6.1 Various types of electromigration failure of bipolar transistor

Distribution of the casual value $\varphi(x_i)$ is determined by casual values $n(x_i)$, θ_r, φ_r and other factors with scattering.

Application of (6.5) for $\varphi(x_i)$ in the continuity equations allows one to transit in t_{MTBF} scattering estimation. This was made in a number of studies using numerical simulation, for example [242].

Numerical simulation of electromigration process in transistor structure is rather effective for reliability estimation. The basis of such simulation consists in modeling of flow of aluminum atoms in some section x_i of contact metallization. As a failure criterion, the following condition is assumed: propagation of the damage region in some distance l_{cr} of a few grain diameters. The possible casual values are performed by the numerical generator of casual values according to a particular distribution law for the given parameter. For example, the grain dimension has a logarithmic-normal distribution law, while the angles θ and φ have a uniform distribution law.

Traditionally, it is assumed that the temperature dependence of major electromigration process parameters obeys Arrhenius's law. First, this is true for the diffusion coefficient (6.6). Besides, the amount of activated ions N is exponentially increased with temperature too [238, 243]: $N \sim \exp(-E_f/kT)$, where E_f is the energy of defect formation for the following elementary transport process.

Approximation of $t_{MTBF}(T)$ dependence subject to Arrhenius's law is a matter of serious study. It is not always the case that the line $\ln t_{MTBF}(T^{-1})$ drawn through three experimental points is a straight line. One of the possible reasons is change in the dominant activation factors with change in ambient temperature.

The fact that activation energy values E_a in polycrystalline films have rather wide scatter (0.3–1.2 eV [243]) is evidence of electromigration process sensitivity to various factors. Obviously, E_a depends on the grain structure, film growth technological process and its chemical composition as well as presence of various impurities. For example, use of Cu slows the electromigration process [244, 245].

Such strong E_a dependence on the film structure affects the t_{MTBF} value: a number of experimental and theoretical studies [246, 247] reported a nontrivial t_{MTBF} dependence on the mean grain dimension and its ratio to the film dimension. For example, the lowest values of t_{MTBF} are obtained in films with approximately two times ratio between the width and the average grain size [246]. Increase of the grain dimension dispersion results in t_{MTBF} reduction [241, 246, 247].

Bipolar transistors

Particular transistor design and its operation peculiarities have an effect on the mode of electromigration failure. A few typical failures of power microwave transistors due to electromigration [240] are discussed below. These failures provide a good illustration of electromigration specific to bipolar transistors.

The failure of n-p-n transistor due to void formation near the bonding thermocompression site in the emitter contact pad region is depicted in Fig. 6.1a. The current density in contact metallization achieved a value of $3–5 \times 10^5$ A/cm^2 and exceeded by more than one order of magnitude the current density in bonding wires. Due to Joule heating, the film temperature differs from the wire temperature at the contact site by 5–15°C, which provides a rather high temperature gradient.

According to (6.4), the dependence for middle time before failure is given by

$$t_{MTBF} = \frac{CS}{Z^*_{EFF}\,\rho j}\left(\frac{d\mu}{dT}\frac{dT}{dx}\right)^{-1},$$

(6.7)

where $\mu = D/kT$.

The expression for film temperature can be assumed in the form: $T(x) = T_0 + R_T(x)j^2\rho$, where $R_T(x)$ is the thermal resistance per area unit, and $j^2\rho$ is the Joule power. Then the expression for t_{MTBF} can be rewritten as

$$t_{MTBF} = \frac{CS}{Z^*_{EFF}\,\rho j}\left[\frac{d\mu}{dT}\left(j^2\rho\frac{dR_T}{dx}+2j\rho R_T\frac{dj}{dx}\right)\right]^{-1}.$$

(6.8)

An additional factor facilitating failure evolution is the difference in material structure of film and the bonding wires. The gradient of diffusion coefficient at the contact site due to low grain dimension in aluminum films is ten times higher than in the bonding wires

The consequence of aluminum electromigration directly along the emitter line is shown in Fig. 6.1b. The transistor structure has been tested in the regime with stationary hot spot formed due to emitter current filamentation. This resulted in void formation in the regions with $dT/dx > 0$, and aluminum accumulation in regions with $dT/dx < 0$ (positive direction x is assumed along the electron current lines). Importantly, dT/dx depends on the transistor operation regime: with U_C increase at $P_C = I_C U_{CB} = $ const, the filament is narrowing down and dT/dx is increased. Respectively, t_{MTBF} is decreased.

An example of void formation at the site of contact between the aluminum film and Mo (or NiCr) emitter balancing resistors is given in Fig. 6.1c. In this case, the major accelerating factor is the gradient of diffusion coefficient, since $D_{Al-Al} \gg D_{Mo-Mo}$ and t_{MTBF} is inversely proportional to the diffusion coefficient gradient:

$$t_{MTBF} = \frac{CS}{Z^*_{EFF}\,\rho j}\left[\frac{1}{kT}\left(\frac{dD}{dx}\right)\right]^{-1}.$$

(6.9)

In the above expression, the temperature T is assumed uniform along x. However, at high current densities the Joule heating of the emitter balancing resistors cannot be neglected. Therefore, t_{MTBF} dependence on j becomes stronger. Metallization break frequently takes place in the oxide step with highest current density (and heating) due to the lowest film thickness. Besides, the film structures on silicon substrate and on SiO_2 are different and have various D values.

MOSFETs

Test of MOSFETs has been done at elevated heat sink temperature (from 60 up to 125°C) in the active regime with maximum current density in metallization of $3-5 \times 10^5 A/cm^2$. The electromigration has an activation energy $E_a \sim 0.7$ eV and is observed both for the drain and source lines (Fig. 6.2) [248]. On the source finger, the voids are formed near the feeder only, but Al accumulation is observed along the whole length. On the contrary, on the drain finger the voids are formed along the whole length, but accumulation takes place in the common metallization region connected to all drain lines (Fig. 6.2). Accumulation of Al as pileup and crystals leads to multiple breaks in the dielectric passivation layer. In some cases, the pileup connects to other metallization lines, producing short circuit damage.

Since voids and Al accumulation regions exist in the lengthy regions of source and drain metallization, it can be assumed that along the source lines $\mathrm{div}\varphi < 0$, while along the drain lines $\mathrm{div}\varphi > 0$. The variation of flux along lines can be explained by a linear increase of current density through the transversal film (i.e., j_x current component parallel to the boundary Si-Al) from the end to the beginning of a line, since Al transport in structure depth is impossible (i.e., possible only along the axis X):

$$\mathrm{div}\,\varphi \sim \left(\frac{dj_x}{dx}\right)\left\{Z_{EFF}^* \rho N D_0 \exp\left(-\frac{E_a}{kT}\right)\right\}\left[\frac{1}{kT}\right]. \tag{6.10}$$

Fig. 6.2 Electromigration in MOSFETs

In the source lines, $\dfrac{dj_x}{dx} < 0$ and therefore aluminum is accumulated (x is the direction of electron flux), but on the drain, $\dfrac{dj_x}{dx} > 0$ and voids are formed.

Estimation of $\dfrac{dj_x}{dx}$ can be performed using the following assumptions. The total current in a single finger is $j_{max}dl$, where j_{max} is the maximum current density at the beginning of the finger, d is the thickness, and l is the line width. Assuming that the current density flow in the structure is uniformly distributed along the line of L length and $j_{max}dl/Ll = j_{max}d/L$:

$$\frac{dj_x}{dx} \approx \left(j_{max} - j_{max}\frac{d}{L} \right)\bigg/ L = \frac{j_{max}}{L}\left(1 - \frac{d}{L} \right). \tag{6.11}$$

It is easy to see that for a long finger with $d \ll L$, the result is obvious: $\dfrac{dj_x}{dx} = \dfrac{j_{max}}{L}$. Since $t_{MTBF} \sim (\mathrm{div}\varphi)^{-1}$, but $\dfrac{dj_x}{dx} \sim j_{max}$, then $t_{MTBF} \sim \left[\exp\left(E_a/kT\right)\right]\big/ j_{max}$, which corresponds to the experimental data.

GaAs MESFETs

Electromigration in GaAs MESFETs is reported in [249–251]. The most typical scenario is depletion of the metallization edge of drain ohmic contact and metallization material accumulation on the source contact [250]. This phenomenon is the result of divergence in flux of migrated atoms in the contact regions of semiconductor–metallization material.

Frequently, the voids are growing at the beginning of source lines (Fig. 1.1c), where metallization comes into the active n-layer step. At the end of these lines, gold is accumulated. This feature of damage justifies the assumption that the main cause is the temperature gradient in the source line edges. Indeed, in the case of MESFETs the hot spots are not formed, the center of the active region is usually heated uniformly, and temperature decrease is observed near the edges. The activation energy of Au electromigration along the source and drain ohmic contacts is 0.9–1.0 eV according to the estimation presented in [248, 250].

Electromigration can be a cause of aluminum gate metallization degradation too [249, 252]. Indeed, the gate bias in active operation regime provides an accelerating factor for the gate degradation. Voids are formed and accumulate first in those places of gate metallization where the current density has a maximum according to the particular device's architecture. At zero bias conditions, the voids in these places are not formed. Current value through reverse-biased Schottky gate at 250°C is one- to twofold higher than the corresponding current at room temperature. However, according to estimation [249], j_{max} is less than 10^4 A/cm^2. This value is not adequate for significant electromigration at 250°C over hundreds of hours. Apparently, an exact value of current density in real gate metallization is determined by the surface inhomogeneities especially at steps and mesa structure edges.

6.2 Mutual Diffusion of Materials

The most natural degradation process in transistor structure is a mutual diffusion of heterogeneous materials in the region of its contact.

The mutual diffusion takes place at any real temperature and is accelerated by Joule heating. It can be accelerated by electric field too in the electrical regime.

According to the thermodynamic law, the mutual diffusion in principle is an irreversible phenomenon. Therefore, it is a genuine degradation mechanism in comparison with the electromigration that can be reversed by changing of the current flow direction.

It is established that in transistor structures, the mutual diffusion is present practically everywhere with varied speed and intensity. However, this fact does not always impact the reliability parameters significantly. Active electrical regions are the most critical from this point of view: p-n junctions, induced channel region, ohmic contacts. Even negligible breaks of structure perfection in these regions can result in parametric failures of transistor.

Mutual diffusion on Al-Si boundary

This degradation mechanism is the most appreciable in power bipolar microwave transistors. In these devices, the mutual diffusion in the system Al-Si results in increase of the base and emitter metallization resistance and emitter junction degradation [246, 253].

Mutual diffusion consists in dissolution of Si in Al since the partial diffusion coefficient of silicon in aluminum D_{Si-Al} is higher than that of aluminum in silicon D_{Al-Si}. At elevated temperature, flows of aluminum and silicon atoms through the interface boundary are in opposite directions. This results in formation of both excessive nonequilibrium vacancies and corresponding vacancy gradient into the silicon boundary region. Vacancy diffusion or its occupation by aluminum atoms practically results in the flow through the boundary Si-Al [253]. Finally, the speed of mutual diffusion in the initial stage is determined by some effective diffusion coefficient $D \gg D_{Al-Si}$. However, over some period of time a saturated solid compound of silicon in aluminum is formed. This results in practical end of the mutual diffusion process.

With transistor operation in the active regime, the mutual diffusion process can be prolonged and accelerated by the current through the boundary. This happens if "electron wind" direction results in taking away of silicon atoms from the Si-Al into contact pads. That is why the accelerated degradation is experimentally observed in emitter lines of n-p-n transistors. The highest depth and linear dimensions of the etched holes are observed in the emitter metallization regions with maximum current density and temperature.

Discussed mechanism limits the middle time before failure in transistors with a small depth of emitter junction (~ 1 μm) and a very thin base. This mechanism with $E_a = 1$ eV becomes dominant at $T > 180°C$.

Peculiarity of the mutual diffusion mechanism in power microwave transistors is a trend to the final local thermal breakdown. With aluminum diffusion in depth of the emitter region in local places, the base thickness is reduced (Al overcompensates the base impurity), thus gain is increased and emitter current is redistributed. This results in local temperature increase and acceleration of the diffusion process. Such positive feedback provides a degradation process with instability features. The degradation instability is discussed in more detail in Section 6.5.

It is important to emphasize that hot spot presence in the structure accelerates evolution of the degradation process dramatically. Figure 6.3 illustrates test results for a few production lots of transistors at three U_{CB} values (I_C and T_0 were fixed). If $U_{CB} < U_{cri}$ (U_{cri} is the value of U_{CB} that corresponds to hot spot formation in i-transistor), then the dependence of accumulated failure percentage safe linearity (Fig. 6.3, curve 1). If for some transistors $U_{CB} > U_{cri}$, then a new region appears on the dependence. Such region corresponds to transistors with abnormally low t_{MTBF}.

In commercial devices, multilayer metallization systems are used on the basis of refractory materials: Mo-Al, NiCr-Al in order to suppress Si atom penetration into aluminum metal layer.

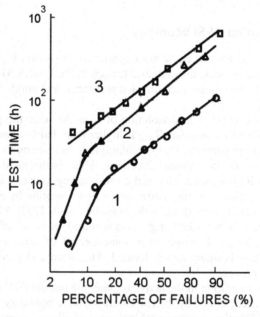

Fig. 6.3 Hot spot effect on the degradation failure as the accumulated percentage of failure after the long-term tests (1, $U_{CB} = 15$ V; 2, $U_{CB} = 9$ V; 3, $U_{CB} = 6$ V)

Formation of intermetal compound between gold and aluminum

Formation of intermetal compound between gold and aluminum (Au_5Al_2, Au_2Al_3, $AuAl_2$, and so on) was observed due to mutual diffusion of these materials in the region of bonding wire contact [254]. Formed intermetal clusters act like traps for atoms Al and Au. This results in formation of voids between the bond wires and contact pads up to complete loss in device control due to formed open circuit conditions.

Operation time before failure can be determined as the time of Au-Al mutual diffusion propagation through the contact layer thickness l_C, i.e., $t_{MTBF} \sim l_C/v_{dif}$, where v_{dif} is the speed of diffusion front propagation. Since $v_{dif} \sim D$ and $D \sim \exp(-E_a/kT)$, then $t_{MTBF} \sim \exp(E_a/kT)$ and, therefore, evolution of the degradation process obeys regularities according to the Arrhenius equation.

Activation energy of this degradation mechanism in GaAs transistors has high scatter: $E_a = 0.5–1.1$ eV [248, 254, 255]. Since a gradual change in any parameter is not observed, the failure can be considered as relaxation. A negligible increase in the noise coefficient is probable, but the major failure precursor is the presence of the traveling contacts with gate control lost. In this stage, the thermocompression contact reveals total degradation. This is confirmed by optical microscope observation (Fig. 6.4).

It is experimentally found that the speed of Au-Al contact degradation is determined by the contact temperature and that it is not related to the current flow or the effect of potential difference. Intermetal layer is formed as a rule at rather durable (within 100 hours) effect of high temperature. However, this process can be initiated in the production stage as a result of the deviation in the wire bonding thermocompression operation.

A complete removal of the degradation of intermetal Au-Al compound is, obviously, an exclusion of Au-Al contact itself in design and use of the uniform contact systems Al-Al or Au-Au. A practical way to avoid this problem is an additional surface layer on the contact pad, for example Au-Ti-Pt [255].

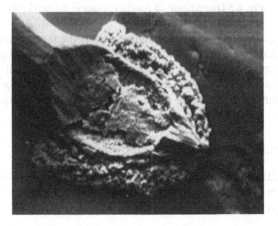

Fig. 6.4 Photograph of GaAs MESFETs after failure due to Au-Al intermetal compound

Degradation of Au-Au connections

The nature of this degradation process in MESFETs has not been completely clarified. The initial stage is characterized by formation of emptiness in the thermocompression point that leads to a sharp decrease of contact mechanical strength [256]. The most probable cause of damage is gold diffusion from the thermocompression point into disordered film of contact pads. Mechanical stress can support this process in Au film.

Diffusion of Au atoms in film metallization is accompanied by formation of Au pileup in its surface. In the initial stage, it circulates the contact as a halo and, then, fills all contact strips to some extent (Fig. 6.5). Simultaneously in thermocompression compound, the emptiness pinches the compound down to GaAs. Sometimes the contact break does not occur, the degradation process speed is decreased, and the distraction picture is fixed. However, not all Au-Au contacts in the transistor are damaged. This is evidence of the thermocompression effect on the degradation evolution.

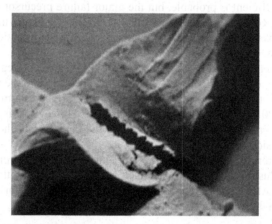

Fig. 6.5 Degradation of Au-Au contact in the bond pad

The major accelerated factor for given process is the temperature, since neither current nor voltage presence influences the degradation speed. According to high-temperature storage results (180, 200, and 220°C), the activation energy is $E_a =$ 1.2 eV. Usually, visually observable emptiness and "shells" formation are assumed as a failure criterion while no change in the electrical characteristics might be detected.

Mutual diffusion of GaAs and metallization material

According to [251], this mechanism is one of the most dominant degradation failure modes of GaAs MESFETs. It causes increase of contact resistance (I_{DSS} decrease), and degradation of frequency capabilities.

A conclusion of a number of studies [249, 255, 257] is that the increase of contact resistance is related to gallium diffusion in contact metallization at high temperatures. There are assumptions that vacancies accumulated in the substrate act as acceptors and compensate the donors in GaAs layer.

The process of Ga diffusion can be enhanced or suppressed depending on adsorption properties of metallization. In particular, gold layers are able to accumulate Ga in appreciable amount, but Ag and Ti-Pt layers present some barriers for Ga [249]. Characteristics of mutual diffusion processes of ohmic contact degradation apparently dependon the particular properties of the used metallization system. For example, in the case of system Au/Ni/Au-Ge, it is experimentally proven that ohmic contact degradation is the result of penetration of Ni atoms into GaAs material. For this degradation mechanism, a theoretical estimation of relative conductivity reduction at zero gate bias Δg is obtained in [257]:

$$\Delta g_D = \left\{ \left(1 + \frac{\Delta R_{DS}(T,t)}{R_{DS}(0)} \right)^{-1} - 1 \right\} 100\%,$$

where $R_{DS}(0)$ is the initial resistance between the drain and source electrodes; $\Delta R_{DS}(T,t)$ is the increase of R_{DS} due to annealing during the time t at the temperature T. If the thickness of high-resistance Ni diffusion layer is d and its area S, then

$$\Delta R_{DS}(T,t) = \rho \frac{d}{S} = \rho \sqrt{Dt} \Big/ S ,$$

where ρ is the specific resistance of diffusion layer and D is the diffusion coefficient; $D = D_0 \exp \left(-E_a \Big/ kT \right)$. Then, $\Delta g_D = \left\{ \left(1 + \sqrt{bt} \right)^{-1} \right\} 100\%$, where $b = [\rho / R_{DS}(0)S]^2 D$. Considering that $(bt) \ll 1$, finally $\Delta g_D = \sqrt{bt} \, 100\%$.

A good agreement was obtained between the experimentally measured and theoretically calculated Δg_D values [257]. The activation energy for the failure mechanism is 1 eV under the assumption of 10% reduction in g_D as the failure criterion.

6.3 Charge Instability

Typically under the charge instability in semiconductor devices, phenomena related to change in the charge state of the structure regions are assigned due to the generation and transport of various charged particles. Change in charge state is equivalent to change in the electrical state and thus frequently results in change of several parameters including a termination of the device operational capabilities.

The latter contains a practical sense of charge instability notion. One of the typical forms of the charge instability is hot carrier degradation.

In general, the charge instability, as a physical phenomenon, can be detected practically in every transistor structure under certain conditions. However, its consequences from the device reliability point of view can be rather different. Obviously, the appearance of failure due to charge instability depends on the amount of charged particles that are generated or traveled in the structure. Usually, it is characterized by the surface density N. The failure itself is determined by the change of the parameters. A third factor is the ratio of the typical time of charge instability evolution to the typical time of operation regime.

From the energy balance point of view, the instability processes are subdivided on activation and inactivation [258]. In the first case, particle participation in the process requires overcoming of some potential energy barrier E_a. Respectively, the density of activated particles, the relaxation time, and a number of other kinetic parameters are varied according to Boltzmann's law with factor $\exp(E_a/kT)$. For the inactivated processes, the temperature effect is negligible.

The activation process can be reversible and irreversible. For the reversible process, the typical time of direct process $\vec{\tau}$ (i.e., process in the presence of operation regime) is higher or comparable with the relaxation time to the initial structure state $\overleftarrow{\tau}$ after removing the DC regime or at influence in reverse direction. For the case of irreversible process, $\vec{\tau} \ll \overleftarrow{\tau}$ [258]. The reversible processes usually are the ion migration, the dipole polarization, and the charge carrier capture on traps and deep layers. Chemical and electrochemical interaction as a rule results in irreversible instability. The latter is true for diffusion of particles too.

Mobile ion migration

Processes of charge ion migration take place mainly in open surfaces of transistor structures, inside dielectric films, and at the interface boundary and surfaces between these films and semiconductor material. Positively charged ions, usually the ions of alkaline metals such as Na^+ and H^+ ions, play a most appreciable role. Apparently, participation of negatively charged ions in the migration is minimal [258, 259].

The "fast" and "slow" modes are distinguished in ion migration processes of Na^+ and H^+. The value $E_a = 0.7$–0.9 eV is typical for the fast ion migration, and $E_a = 1.0$–1.4 eV is typical for the slow mode. The activation energy of fast and slow drift of hydrogen ions is within the intervals 0.3–0.66 and 0.9–1.3 eV, respectively [258]. The variation in the migration speed is due to the difference between physical natures of the process that limits the migration. In one case, this can be emission of ions from traps, whereas in another case, the ion diffusion through the dielectric or ion migration in compound with other atoms. The fast processes are usually connected with migration of free ions along interstitial or on "channels": lengthy breaks of dielectric structure with higher ion mobility [258].

The highest positive charge density is achieved in the passivation SiO_2 films. In thermally grown oxide SiO_2, the surface density up to $N \approx 10^{12} cm^{-2}$ can be accumulated. This value is comparable with the subsurface layer inversion of p-type silicon with the acceptor concentration $N_D \approx 10^{16}\text{cm}^{-3}$. Formation of the depleted or inverted regions at the bipolar transistor surface is the major result of the charge instability. It results in significant increase of saturation current levels for emitter and collector p-n junctions.

Appearance of the subsurface depleted regions increases the total area of p-n junction and corresponding generation-recombination currents. Inverted layers of n-type can provide a leakage current path between the emitter and collector junctions that is a few orders of magnitude higher than the normal level. Obviously, silicon p-n-p transistors with high-resistance collector region ($N_A \approx 5 \times 10^{15}\text{cm}^{-3}$) are the most sensitive to the inversion channels. In order to reduce the charge instability, additional P-field implants or special guard rings of diffusion regions of p^+-type are used in the device design. The inversion in these regions is impossible due to high subsurface doping level.

In MOSFETs, the ion migration process is mainly concentrated in the subgate dielectric space inside the MOS structure. At positive gate bias, the ions drift to the semiconductor surface thus resulting in decrease of the threshold voltage U_T. In drain saturation current I_{DSS} conditions, the effect is opposite. At negative gate bias, the drift direction and the trend of U_T and I_{DSS} change are reversed.

In power MOSFET arrays with complex gate metallization, the mobile charge distribution in the oxide region can be nonuniform. Respectively, the regions of charge instability can be scattered in the structure. Nonuniform evolution of the charge instability might result in current redistribution and degradation of transistor transconductance. This process is discussed in Section 6.5 in greater detail. One of the most sensitive and informative characteristics from the charge instability point of view is the transconductance on gate bias at constant drain-source voltage $I_D(U_{GS})/U_{DS} = $ const. In the uniform drift scenario, shift in this characteristic is parallel to itself along the voltage axis. However, in the case of local drift the change of the slope dI_D/dU_{GS} is observed mainly in the range of small drain currents I_D (Fig. 6.6).

Ion drift process in the gate dielectric is frequently associated with other charge instability phenomena related to capture or emission of the electrons and holes from traps, deep layers, and generation of the surface states [260]. Sometimes this results in a complex U_T evolution and degradation. In the beginning, U_T can decrease due to positive charge drift to the Si surface, and then increase due to capture of electrons by traps and neutralization of the positive charge after current drift.

Deep layer recharging

The drift of parameters due to deep level recharging is mostly typical for field-effect transistors with an active region that interfaces various boundary surfaces (semiconductor–dielectric, different semiconductor and contact materials). An increased density of charge states occurs near such surfaces in the form of various electron and hole traps [237, 258]. For example, in silicon MOSFETs the traps are concentrated mainly at the silicon–dielectric interface (gate or thick field oxides, shallow trench isolation, passivation layers). In MESFETs, in addition to surface regions, the intensive electron and hole capturing processes are realized at the boundaries of active film–buffer, buffer–substrate, and in the substrate itself. Here, compensated Cr ions or other levels of GaAs, for example level EL2, are acting as capture centers [261].

In real transistors, the processes of deep level recharging are rather diverse. A few processes of different nature can be formed simultaneously. The practical difficulty relates to the fact that the processes, various in essence, might result in the same consequences. For example, a typical result of hot carrier degradation instability in low-noise MESFETs is the drift of the on-state drain resistance and a number of other microwave parameters.

According to the understanding [261, 262], the instability is the result of a slow change of the channel thickness at bias U_{DS} and U_{GS}. This effect can be produced by at least two basic mechanisms: the "surface" mechanism due to recharging and change in charge density on the active layer surface in space charge region and the "volume" mechanism due to deep level recharging in space charge region at the boundary of active layer–buffer (or –substrate) [263]. In the case of the surface mechanism, drift suppression can be provided by passivation [264]. In [264], the drift activation energy was scattered in the range from 0.8 to 0.16 eV thus providing evidence of a few involved mechanisms.

However, an exact identification of the particular degradation mechanism is a complex problem. There are several known methods that allow high probability clarification of what basic mechanism (surface or volume) is responsible for drift in I_D [262, 264]. One of them is the dependence $I_D(t)$ at negative bias of substrate application. Volume mechanism, as a rule, corresponds to slow-speed reduction of I_D due to traps recharging in the buffer and at the internal interface regions. For the surface mechanism, a small change in I_D is observed. Another identification of charge instability mechanism is possible with some probability using spectral and C–V measurements [262].

Hot carrier degradation

One of the most conventional charge instability effects in MOSFETs is the capture of hot electrons and holes on deep levels in oxide or Si-SiO$_2$ interface. This process has numerous peculiarities [265–268], but is usually referred to as hot carrier degradation. Capture of hot carriers is possible practically under any operation conditions of MOSFETs, but the effect is most visible in current saturation regime with maximum substrate current. In the prebreakdown regime, it can result in an appreciable degradation of the transconductance, threshold voltage, channel resistance, and other parameters. In published studies, a leading role is given to holes as a trap center in oxide. The holes are either injected from the semiconductor [265, 268] or generated in the oxide due to the avalanche ionization process [266].

Intensive hole injection in the oxide of n-channel transistor is observed at $U_{DS} > U_{GS}$ when the electric field near the drain provides positive direction for holes. Near the source, the field has an opposite direction and its value is rather low;

therefore, in this channel region there is no injection into oxide. Injected holes create in oxide the neutral trap levels of acceptor type [265].

Capture process for electrons and holes can take place simultaneously at creation in oxide by avalanche generation [266]. In this case, according to U_T measurements the process of hole capture is dominated in the first 50 s (U_T is decreased); then it is suppressed by the capture process for electrons. At $t > 350$ s, U_T exceeds the initial level and further increases [266].

A nonuniform hot carrier capture along the channel is mainly concentrated near the drain in the maximum electric field region. It results in reduction of the maximum transconductance and the maximum channel conductance. The capture process is only slightly activated by temperature since carrier injection in the oxide is provided mainly by tunneling. However, the relaxation process can be significantly accelerated by heating.

6.4 Degradation Phenomena in Dynamic Regimes

When degradation failure mechanism for the DC regime is well studied, then, as a rule, its evolution in a given dynamic regime is rather predictable. Practically for each mechanism there are a number of characteristic time parameters that determine the kinetics of the process. Ratio between the main time parameters of dynamic regime determines the possibility of degradation evolution.

For example, at electromigration failures a typical relaxation time of ion interaction process with electron wind is ~10 ps [234]. Therefore, in any real frequency range the electromigration evolution will be determined by instantaneous values for $j(t)$ from the continuity equation and the process kinetics is given by average value during the period τ [240].

For the temperature gradient case (6.6), the time before failure in dynamic regime is given by

$$\tilde{t}_{BF} = \frac{CS}{Z_{eff}^* \rho \bar{j}} \left[\frac{d\mu}{dT} \left(\overline{j^2} \rho \frac{dR_T}{dx} + 2\bar{j}\rho R_T \frac{\overline{dj}}{dx} \right) \right]^{-1}, \qquad (6.12)$$

where $\bar{j}, \overline{j^2}, \dfrac{\overline{dj}}{dx}$ are the median values across the period. Obviously, in the general case the value $\tilde{t}_{BF}$ is less than the corresponding value t_{BF} in the DC regime according to (6.8) at $j = \bar{j}$ and $P = \overline{P}$.

If electromigration failure is due to the diffusion coefficient gradient, then

$$\tilde{t}_{BF} = \frac{CS}{Z_{eff}^* \rho \bar{j}} \left[\frac{1}{kT} \left(\frac{dD}{dx} \right) \right]^{-1}. \qquad (6.13)$$

In given case, $t_{BF} \approx t_{BF}$. Thus, at transistor structure operation in dynamic regime the time before failure depends on current shape and frequency and can be significantly different from the time before failure in the DC regime with the same peak power level.

The visual consequences of electromigration failure are as a rule similar to the observed consequences in the DC regime [269]. However, damage localization can be different [270].

At analysis of the degradation conditions in dynamic regimes, the possibility of relaxation for a number of degradation phenomena must be considered. This is especially important for a pulse regime with low duty factor. For example, the processes of charge instability can totally relax between the pulses.

There are a number of specific degradation mechanisms, for dynamic regimes. These mechanisms are mainly connected with cycle effects (thermal, mechanical). For example, in pulse operation regimes the thermal cycling causes alternative mechanical strengths due to difference in the thermal characteristics of metallization films, semiconductor material, and isolated films. Relaxation of such strengths can be accompanied by breaks in the thin film conductors, local short circuiting, and metallization exfoliation [271]. At "slow" thermal cycling evolution, fatigue phenomena are possible.

Level of thermal stress at thermal cycling can be estimated by the simple formula [271]

$$\Delta\sigma = \left(\alpha_f - \alpha_s\right)\Delta TE, \qquad (6.14)$$

where $\Delta\sigma$ is the amplitude of stress, α_f, α_s are the mean coefficients of film thermal expansion and substrate, ΔT is the maximum difference between substrate temperature during the period, and E is the elasticity modules. If $\Delta\sigma$ achieves the fluidity limit of film, a plastic deformation and corresponding breaks are possible in the structure [270].

6.5 Degradation Instability

This final section essentially justifies why degradation phenomena have been included in this book.

It has been reported that a number of current redistribution phenomena are related to degradation processes. These phenomena are called degradation instabilities. In contrast to the above-discussed thermal and isothermal instabilities, the degradation instability can be developed under unchanged external conditions: current, voltage, ambient temperature. Process of current redistribution is initiated and supported by changes inside the structure due to a degradation process with variable speed during the instability evolution.

Two types of degradation instability scenario can be identified. The first is realized when the degradation process plays a triggering role for some thermal or isothermal instability. The second is realized when the degradation process is directly involved in current filamentation and its speed has some distribution in the structure too.

The first case can be easily illustrated by an example of bipolar transistor in the DC regime at $U_{CB} = U$, $I_C = I$, and the constant heat sink temperature $T_C = T_0$. It is assumed that the transistor is in a stable state, since the voltage $U < U_{CR}$, where U_{CR} is the critical voltage of emitter current thermal filamentation (at given $I_C = I$). In addition, it is assumed that due to some degradation process the bonding composite between semiconductor chip and the package is damaged with corresponding loss in thermal conductivity properties. Practically, this means that thermal resistance is increasing with time: $R_T = f(t)$. From (3.16), $U_{cr} \sim R_T^{-1}$, then after some delay time U_{cr} becomes equal to U and the hot spot is formed. If R_T degradation continues, then the filament is narrowing down thus creating conditions for the instantaneous thermal breakdown and catastrophic failure.

Similar phenomenon has been reported for GaAs transistors in the form of degradation of drain-source breakdown voltage due to chemical reaction at the chip surface [272]. Delay time for the first type of degradation instability is determined by the degradation process kinetics, its characteristics according to Arrhenius's law $t_D \sim \exp(E_a/kT)$.

The second type of degradation instability was observed in structures with charge instability [273, 274]. It can be illustrated in the example of power n-MOSFETs [274] with Mo-Al gate metallization. The drift and reduction of the threshold voltage were observed at test of MOSFET structures in thermo-field regime: source connected to the drain and voltage between the source and gate U_{GS} with polarity of induced channel formation.

Shift of transconductance characteristic of the structures $I_D(U_{GS})$ was observed to the left practically in parallel (Fig. 6.6, curves 1, 4). This provided evidence of charge instability in the gate dielectric of the MOSFETs. Most likely, the instability mechanism is related to migration of positively charged ions in electric field from the metal–SiO$_2$ interface to the Si–SiO$_2$ interface.

The test results for MOSFETs in active regime are different from those discussed above. At some power levels, a significant decrease of the slope of transconductance characteristic is observed together with U_T drift (Fig. 6.6, curves 2, 3). Thus, during the tests the threshold voltage U_T is decreased nonuniformly along with the induced channel width and channel current density redistribution.

This current density redistribution has been confirmed by direct measurements of the temperature distribution in the structure using IR microscopy. With decrease in the slope of the transconductance characteristic, formation of a region with local overheating is observed in the structure under corresponding increase in maximum temperature with time. At adequately high drain-source voltage U_{DS}, the process evolution is limited by thermal breakdown in the filament region when the peak temperature exceeds some critical level ~ 280–$320°C$. Depending on U_{DS} and I_D, the time of nonuniform state formation before irreversible breakdown is changed from a few minutes up to tens of hours. If the test terminated prior breakdown, then a fast restoration of initial U_T value and transconductance characteristic can be provided by thermal anneal.

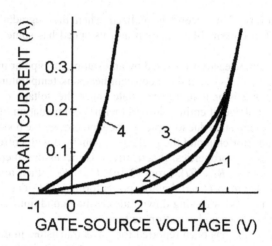

Fig. 6.6 Evolution of transconductance $I–V$ characteristic of MOSFET at charge instability evolution: initial characteristic "1," and the characteristics after 5 ("2"), 10 ("3"), and 24 hours ("4")

Physical mechanism and regularities of instability evolution

The above experimental results provide direct evidence of channel current filamentation in the MOSFETs in active operation regime. Mechanism of the phenomenon is connected with the process of U_T decrease due to charge instability of gate dielectric. Exponential U_T dependence on T greatly impacts the drift. Under these conditions, with given total current through the structure the nonuniform fluctuations $\delta j_{CH}(x)$, $\delta T(x)$, $\delta U_T(x)$ can grow.

Indeed, if a small fluctuation $\delta j_{CH}(x)$ appears, then the temperature in a local region with high j_{CH} is increased. This results in faster local decrease of U_T in comparison with other structure regions. This effect causes corresponding further j_{CH} and T increase. Since the total current is fixed, the process is accompanied by j_{CH} redistribution with corresponding cooling of other regions and decrease in speed of U_T drift. Thus, "stratification" of the drift speed U_T takes place. In the local region, U_T increase is accelerated while at the same time being decelerated in other regions. This positive feedback results in filamentation.

Apparently, this nonuniform state cannot be stable and it disappears after reaching the minimal U_T value in the filament region. Then U_T drift outside the filament region will lead to leveling of j_{CH} distribution on the channel width and dumping of the j_{CH} distribution nonuniformity. It is clear that the higher U_T drift is, the higher both the slope of the transconductance characteristic and the local temperature.

Analysis of the discussed instability mechanism appearance can be accomplished taking into account the fact that the distributions j_{CH} and T on area in the one-dimensional approach are given by the following equation system:

$$\begin{cases} \tau_T \dfrac{\partial T}{\partial t} = l_T^2 \dfrac{\partial^2 T}{\partial x^2} + j_{CH} U_{DS} R_T + (T_0 - T), \\[3mm] \tau_U(T)\dfrac{\partial U_{boun}}{\partial t} = l_U^2 \dfrac{\partial^2 U_{boun}}{\partial x^2} + (U_{boun}^0 - U_{boun}), \\[3mm] \displaystyle\int_0^{l_x} j_{CH}\, dx = I_{CH}. \end{cases} \qquad (6.15)$$

with boundary and initial conditions:

$$\left.\frac{\partial T}{\partial x}\right|_{x=0,l_x} = \left.\frac{\partial U_{boun}}{\partial x}\right|_{x=0,l_x} = 0,$$

$$T\big|_{t=0} = T_0; U_{boun}\big|_{t=0} = U_{boun}(0).$$

where τ_T, l_T are the typical time and dimension of temperature changing, R_T is the thermal impedance on the area unit, T_0 is the heatsink temperature, τ_U, l_U are the typical time and dimension of threshold voltage variation, U_T^0 is the stationary value of U_T, l_x is the width of induced channel region, and I_{CH} is the given total channel current. Temperature dependence of τ_U has a good approximation by exponential function: $\tau_U \sim \exp(E_a/kT)$ where E_a is the activation energy of drift process U_T.

By linearization of (6.15), relatively small disturbances of δT, δj_{CH}, $\delta U_T \sim \cos(\pi n x/l_x)\exp(\gamma t)$ ($n = 1,2,\dots$) and considering $\tau_T \ll \tau_U$, the relation for increase increment γ is given by

$$\gamma \tau_U = \frac{R_T U_{DS} \dfrac{\partial j_{CH}}{\partial U_{boun}} \dfrac{\partial \ln \tau_U}{\partial T}\left(U_{boun} - U_{boun}^0\right)}{1 + l_T^2 P_n^2 - R_T U_{DS}\,\partial j_{CH}\big/\partial T} - \left(1 + l_U^2 P_n^2\right), \qquad (6.16)$$

where $P_n = \pi n/l_x$. For simplicity, it is assumed that $j_{CH} = S(U_{GS} - U_T)$, where S is the transconductance. Taking into account that in the absence of thermal breakdown $R_T U_{DS}\dfrac{\partial j_{CH}}{\partial T} \ll 1$ from (6.16) the instability criterion ($\gamma > 0$) is given by

$$R_T U_{DS}\frac{E_a}{kT} S\left(U_{boun}(0) - U_{boun}^0\right) > \left(1 + l_T^2 P_n^2\right)\left(1 + l_U^2 P_n^2\right). \qquad (6.17)$$

Expression (6.17) is in good agreement with the experimental results. From (6.17) the instability boundary is determined by some dissipated power level. The higher S is, the higher U_T drifts in thermo-potential regime, and therefore the instability excitation is provided at a faster rate.

Unfortunately, this linear theory is not suitable for the analysis of nonuniform states and formation kinetics. For this purpose, numerical simulation is the most effective tool. The simulation code was developed and the distributions $j_{CH}(x,t)$,

$U_T(x,t)$, $T(x,t)$ were calculated at given regimes: I_{CH}, U_{DS}, T_0 using equation system (6.15) and the stationary heat equation, on the assumption that variation of the distribution $j_{CH}(x)$ instantaneously influences the temperature due to $\tau_T \ll \tau_U$. With the purpose of disturbance simulation in the initial moment, a small nonuniform deviation is applied to $U_T(x)$ distribution. For simplicity, the case of rectangular deviation at structure edge is used.

The simulation results (Fig. 6.7) validated the possibility of the discussed instability scenario. In addition, at unchanged I_{CH} and U_{DS} the nonuniform state exists only during a certain period of time (Fig. 6.7b). Observation of this process in practical cases is difficult since the temperature in the center of the filament usually rapidly becomes higher than the critical value for irreversible breakdown.

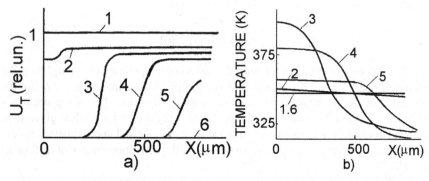

Fig. 6.7 Calculated boundary voltage (a) and temperature (b) distributions on the structure surface after 1 h ("1"), 4 h ("2"), 10 h ("3"), 15 h ("4"), 20 h ("5"), 22 h ("6")

6.6 Summary

In this chapter, a review of basic degradation mechanisms is provided to emphasize again the corresponding differences between previously described catastrophic phenomena followed by a discussion of the catastrophic failures related to nonlinear effects due to example degradation instability phenomena. The original concept of opposing the change in energy and matter is illustrated in greater detail.

In most cases, degradation changes in the structure are reflected by change in the transistor parameters. However, they are not always available for direct observation or measurements during operation. Similarly to catastrophic events, only the direct consequence of the gradual changes in transistor parameters can be observed.

The possibility of degradation instabilities significantly complicates failure analysis when degradation processes play a triggering role for some thermal or isothermal instability. This is the case when the degradation process is directly involved in current filamentation and its speed has some distribution in the structure too.

Chapter 7
Conductivity Modulation in ESD Devices

7.1 ESD Design

Various scenarios for current conduction under breakdown and instability conditions can be realized in the electrostatic discharge (ESD) pulse regime depending on the device design and biasing circuit parameters. Usually, the complexity and nonlinearity of the conductivity modulation processes in the case of ESD events limits the implementation of compact models for direct circuit simulation in ESD pulse conditions. Perhaps for this reason, numerous ESD devices and clamps have been proposed to emphasize a particular feature of the process, device architecture, or specification for the protected pins. Often, these solutions are based on a rather empirical and phenomenological design approach. It is not easy to navigate through numerous publications in the field where devices with the same operation principle have beedn given different names. For example, the same snapback BJT device from bipolar process can be presented in CMOS process under another name—field oxide (FOX) or thick field oxide (TFO) device. At the same time, even in the case of the snapback NPN BJT the device operation can be rather complex due to superposition of the lateral and vertical current conduction.

With this backdrop, the physical level of understanding of semiconductor structure operation at ESD pulse conditions is critical for successful design. Thus, in spite of the variety of ESD protection devices and clamps, the physical effects enabled during ESD events are similar to the phenomena discussed in the previous chapters. The purpose of this chapter is to demonstrate that high-current operation under ESD pulse not only obeys the same principles, but even is confined within a rather limited set of the same isothermal basic conductivity modulation mechanisms as those involved in physical limitation of electrical SOA in standard devices. This statement is true with the exception of the additional double injection conductivity modulation mechanism typical for p-n-p-n, thyristor, IGBT structures and also responsible for latch-up phenomenon in CMOS integrated circuit. Below, several introductory sections to the field are presented to aid understanding of the feature material.

7.1.1 The Field of ESD Protection

ESD protection devices have been often mentioned in the previous chapters. ESD protection became a very established field of knowledge due to integrated components development. Numerous studies have been done in the field. The EOS/ESD Association (www.esda.org) currently supports numerous activities including annual symposia. Practically every international conference on microelectronics includes some ESD-related papers. ESD standards have been created. The ESD community counts thousands of dedicated specialists in industry and higher education, and practically all steps of integrated product development involve ESD development and consideration from the process development up to the product release phases. ESD requirements can be treated simply as a conventional part of the standard integrated product and system-level specifications.

A strategic problem of ESD development can be formulated as implementation of the added capability to integrated products with regard to withstanding some standard ESD tests. From the circuit design point of view, this essentially means an added functionality of integrated product to function under some specific high current pulse conditions. The way to realize this is design and implementation of an additional peripheral ESD protection network connected in parallel to the original functional circuit blocks.

In a simplified manner, the ESD protection network is achieved through the connection of ESD devices and ESD clamps into a so-called ESD pad ring circuit. More advanced and sophisticated solutions may involve a shared functionality between ESD network and internal circuit, self-protection solution, distributed or multiple-stage ESD blocks and a variety of passive components from limiting poly- and current saturation resistors up to a sophisticated distributed RF network. At the top of the device-level capabilities, an important consideration at high current level is a certain requirement on the network resistance, current crowding effects, and dynamic coupling of the circuit components.

Thus, on the circuit design level, ESD solution could be considered an implementation of a pulsed power circuit on the top of the normal circuit. On the one hand, this secondary pulsed power circuit should provide pulsed power operation with some voltage waveform parameters in the ESD time domain. On the other hand, in the ideal case this circuit should not interfere with normal circuit operation. The major functionality of the ESD protection circuit is ability to turn-on into a high-conductivity state during ESD pulse; dissipate the energy of the ESD stress thus providing limitation of the pin voltage below the pulsed absolute maximum limits in pulsed regime of those devices that are directly or indirectly connected to the pin.

One of the most critical requirements of an ESD protection network with respect to competitive advantage of the product is that it occupy a space as small as possible on the chip. Therefore, the proper way to achieve such capability is implementation of small-footprint device-level solutions based on isothermal conductivity modulation mechanisms.

Thus, typical device-level ESD solutions are essentially pulsed power devices with some biasing components that are specifically designed to work in the breakdown and one of the isothermal conductivity modulation modes.

ESD development includes a rather complex set of cross-disciplinary tasks in the fields of circuit design, physics of semiconductor devices, physical ESD clamp design with nonlinear physics of operation, effects in interconnects, materials and topological array problems to balance the current density distribution.

However, the physical phenomena responsible for conductivity modulation are very similar to the diode, transistor semiconductor structures discussed in the previous chapter. The only addition to complete this statement is the double injection conductivity modulation mechanism mentioned above.

This chapter has no goal to provide a complete overview of the whole device-level ESD development area. This task is accomplished in numerous ESD books and review publications in the field; for example [275–285]. According to the purpose above, this chapter is focused on a very specific task, namely, to demonstrate the connection between previously discussed physical principles of SOA limitation in the standard devices and the physics of ESD device operation thus highlighting the universal nature of the physical principles and conclusion regarding mutual validity of the methodologies developed for both in areas of discrete components, integrated components, and ESD devices.

SOA and instability boundary in the case of ESD pulse regimes

On the device level, the problem of ESD design is somewhat opposite the traditional reliability problems of physical limitation of the device parameters and thermoelectrical SOA. In the case of standard devices, the SOA is first required to ensure sufficient area for reversible device operation regimes with confirmed long-term reliability parameters via the accelerated tests. If there is no major limitation of the long-term reliability parameters, then increase of the SOA is an admirable task. On the contrary, in the case of device-level ESD solutions, the goal is essentially subdivided into two major components: the ESD pulse operation of standard devices and the reversible operation of ESD devices in the conductivity modulation mode, while long-term reliability parameters are usually not an issue.

Thus, for ESD devices, reversible operation in the secondary breakdown mode presents the normal operation regime, while the physical limitation of the operation regime is related to pulsed thermal effects caused by high dissipated power in the rather small subsurface semiconductor region.

The problem of proper ESD device design is mainly confined within several tasks: provide proper energy balance inside the device, proper current balance on the clamp level, collect proper total current level, and finally deliver some triggering and holding voltage parameters for adequate voltage limitation of the protected pin.

The first mentioned component of the problem is to determine a specific characteristic of the standard devices in integrated process that are interfacing with the external pins under ESD pulse conditions. This pulsed ESD SOA is required to at least roughly identify the so-called ESD protection window. This window is usually defined as an absolute maximum voltage of the device in the ESD pulse domain. At the same time, in general the absolute maximum voltage is a function of the control electrode state, pulse width and rise time and temperature. In a simplified case, the minimum voltage range in the SOA can be assumed as the ESD protection window. Then this fact is confirmed by the product ESD testing. The finding of pulsed ESD SOA limits might be a very complex task that involves the specific output device topology such as in the case of power arrays.

The second component of the ESD task is related to proper ESD device design itself. This task can be treated as a design of a device that has at least some area of reversible operation in the current instability mode under conductivity modulation regime. The boundary of instability would normally correspond to pulsed SOA limitation for standard devices.

Thus, a primary goal is to deliver ESD device primitive as a component of the future ESD clamp. In a simplified version, the problem can be converged down to implementation of a scenario when operation of the internal circuit during ESD events is always realized inside the ESD protection window. In this case, the output device or clamp could be connected directly and independently to the pad and could be treated on macro level as a two-terminal "ESD diode."

Thus, on the device level this second component of the problem essentially consists in implementation of an array device with reversible high current operation in the conductivity modulation mode under proper characteristics of the turn-on voltage, the minimum holding voltage and high current operation (clamping) voltage, high level of pulsed current density, and several other secondary parameters related to normal operation regime. The last may include turn-off voltage, parasitic capacitance and leakage current, turn-on speed and of course the required space on chip, number of metal layers, and additional ESD implants required.

Thus, practically in a majority of cases the new semiconductor device should be designed with not just a reversible primary pulsed SOA, but with a particular pulsed instability boundary parameters. Moreover, it is desired that the device be designed as a small-footprint "free" device to the given process technology. This means that at the device level, the development goal must be achieved by applying topological solutions to the device architecture, rather than involving additional expensive mask and implant process steps.

In properly designed ESD devices, the real physical limitation is related to ESD pulse energy dissipation. In most cases, the irreversible SOA of the ESD device converges to a single point due to low control of the bulk lateral conductivity modulation in the high injection conditions by the surface or high resistive base electrodes. The physical limitation for irreversible SOA of ESD devices is a rather complex problem. However, the major regularities are the same as those highlighted in the Chapter 4 summary. Thus, in most cases this problem can be converged down to the conductivity modulation of the contact regions and achieving much higher injection from the metal electrodes, exceeding backend metallization limits, a localized heating and melting due to current filamentation.

ESD circuit design

Due to dynamic coupling in the circuit during ESD pulse test, the state of a particular output device strongly depends on the internal circuit block design. Since the absolute maximum voltage of the output device is usually a function of the control electrode potential, the failure point will depend on the dynamic coupling provided by the circuit.

In general, the internal circuit could be designed to withstand much higher voltage overstress level than the corresponding SOA limits for standard components. This could be achieved by providing smaller coupling effect to the control electrodes the devices interfacing with pins, by using stacked components, limiting resistors' so-called keep-off and ESD pulse detection circuits. The latter are implemented to ensure that the internal circuit blocks will operate at the highest possible triggering voltage when the control electrode remains at low potential or in order to drive the control electrode of the ESD device.

At the same time, one of the important considerations for ESD network design is added limitation of the absolute maximum voltage of the normal circuit operation by the triggering voltage of the ESD device. In some cases, this dilemma can be resolved using dV/dt triggering of the ESD devices. However, this approach is not suitable for high-speed and high-voltage power circuits, particularly for switching voltage regulators that operate in the same rise time domain as ESD.

ESD pulse specification

During the era of integrated product development, various specifications for ESD pulse were created and became industry or custom standards. A detailed description of the ESD pulse characteristics and the equivalent circuits can be found in the introductory chapters of numerous popular books in the field for example [275, 276], original publications [286], and standard documents for example [287–290]. The examples of standards for Human Body Model (HBM) pulse are: ESDA (ANSI) STM5.1-2001, JEDEC JESD22-A114-E, IEC 613240-3-1, AEC Q100-002 REV-D, and EIAJ ED-4701/304. Similar standards can be found for Machine Model (MM), Charge Device Model (CDM), and ESD system level [286]. A consolidated summary of the most commonly used ESD pulses is presented in Table 7.1. Typical non-system-level corporate requirements for HBM, MM, and CDM ESD passing level are 2 kV, 200 V, and 1 kV, respectively.

From the ESD device design challenges point of view, there are two major spec types: the nonsystem level and the system level. The non-system-level spec usually targets some minimal requirements to protect the integrated circuit component on the way to incorporating into the system, packaging, handling, electrical tests. For example, HBM ESD pulse spec for nonsystem level requires withstanding some pulsed current level ~ 1.33 A with a rise time ~ 2–10 ns (Table 7.1). In the test, the circuit is not powered.

In contrast to the nonsystem spec, the system-level spec targets protection of some circuit pins under normal operation conditions. In addition, usually system-level requirements target a much higher current level too. The complexity of the system-level protection problem concerns the possibility of transient latch-up that can occur if the ESD clamp provides a holding voltage lower than the power supply voltage.

Due to the fast rise time in most cases, the ESD pulse automatically provides the conditions for pure electrical turn-on. Electrical current instability is initiated in quasi-isothermal conditions and provides further triggering into high current conductivity modulation state. In most practical cases, the lattice temperature change can be neglected before the triggering due to uniform current distribution and short time before the triggering.

After the switching, heat dissipation becomes significant. However, the heat dissipation scenario is significantly different from DC operation. Due to the rather short pulse duration, the heat dissipation is realized in a rather small area of a few micrometers in the vicinity of the device active region. Thermal heat dissipation, as well as electromechanical stress and breakdown of dielectric, blocking junction robustness and backend limits provide physical limitations for ESD device operation.

Table 7.1 The most common examples of non-system-level ESD pulse parameters

ESD pulse	Peak current to pulse voltage ratio (A/kV)	Rise time/pulse width (ns)
Human Body Model	0.67	2–10/150
Machine Model	17.5	2–10/66–90
Charge Device Model	9	0.25/2
IEC 61000-4-2	3.75	0.2/50–150

7.1.2 ESD Devices

ESD design specification involves high power dissipation and high current density. In the majority of cases, small-footprint snapback ESD clamps are designed. The snapback devices involve specific design solutions to operate in conditions of the avalanche-injection and double injection conductivity modulation mechanisms under high power generation in the local regions.

The reversible operation in the conductivity modulation conditions is achieved both by a proper energy balance inside the device structure and by current ballasting in the conductivity modulation mode using the contact diffusion regions and backend metallization design. For contacts and metallization, usually ~40-fold higher current density limits in the case of ESD pulse are assumed in comparison with the electromigration ratings.

In this book, integrated ESD devices are classified according to the conductivity modulation mechanisms involved in their operation (Table 7.2). Since the maximum current density in the device usually does not exceed 1–10 mA/μm, the ESD clamp design usually involves implementation of arrays with a total contact width ~100–1000 μm. The specific advantages and disadvantages for each device are briefly discussed below.

Forward-biased and avalanche ESD diodes

Simplified cross section of the ESD diode is similar to that of typical integrated diodes (Fig. 7.1). Specific ESD diode application includes the ESD diode for forwardbias operation and application on the diode in avalanche breakdown conditions.

The major figures of merit for forward-bias diodesd are on-state resistance, leakage current at operation voltage, noise and small-signal parameters, and reverse voltage tolerance.

Table 7.2 Major conductivity modulation mechanisms realized in ESD devices for high-current density

Conductivity modulation mechanism	ESD devices	Typical lateral current density (mA/μm)
Avalanche breakdown	Avalanche diodes, blocking junctions, PMOS, PNP	0.01–0.1
Avalanche injection	Snapback NMOS, NPN, field oxide devices	0.1–1
Double avalanche injection	P-i-n, M-i-n diodes	0.1
Double injection	LVTSCR, SCR, bipolar SCR, LDMOS-SCR	10–100

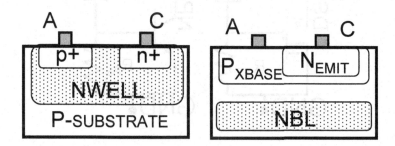

Fig. 7.1 Simplified cross sections for CMOS (a) and BJT (b) based ESD diodes

The avalanche diodes are designed for ESD pulse operation in the avalanche breakdown mode. For these devices, the on-state resistance and the critical current are the major figures of merit. In spite of relatively low critical avalanche current density, the avalanche diodes are very useful both as a reference voltage and as two-stage ESD protection components (Figs. 7.2, 7.3).

In BiCMOS and BCD processes, the avalanche devices can be optimized for the desired voltage range using combination of the BJT, CMOS, and DMOS device regions, different RESURF techniques, surface and bulk conductivity.

In spite of different physical mechanisms for ESD purpose in the first approximation Schottky diodes could be treated similarly to avalanche diodes.

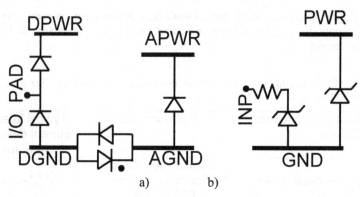

Fig. 7.2 Examples of major applications for forward conduction (a) and avalanche (b) diodes in ESD protection networks

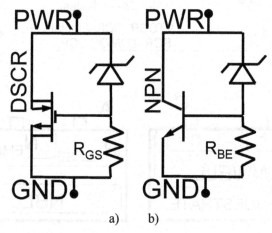

Fig. 7.3 Application of avalanche diode as reference voltage (a) and reference current (b) component in NLDMOS-SCR and NPN clamps

NPN BJT, snapback NMOS, FOX devices

Avalanche-injection conductivity modulation is implemented in snapback characteristics in NPN BJT, snapback NMOS, thick field oxide devices (TFO, FOX), and other lateral BJT-like devices. Simplified cross section for some of these devices is presented in Fig. 7.4. A common feature of these devices is the presence of a parasitic NPN structure that is able to provide short-term reversible operation under ESD pulse conditions.

The most important parameters of the NPN-based ESD devices are related to the waveforms of their ESD operation extracted from transmission line pulse (TLP) measurements and DC measurements: the triggering voltage and current, the minimum holding voltage, the on-state resistance, the pulsed maximum current and voltage, DC breakdown voltage, and leakage current. The parameters of introduced parasitic effect on the implemented circuit are critical too: noise, parasitic capacitance, and small-signal parameters in the case the device is used as a local clamp for input/output pins.

Due to bipolar current gain and space charge neutralization during conductivity modulation, these devices are superior to the avalanche diodes. They provide approximately tenfold higher current density. At the same time, the bipolar conductivity modulation has several side effects. First, it is certain difficult to

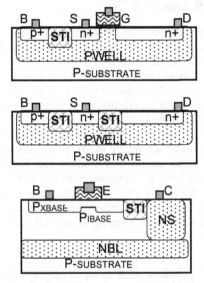

Fig. 7.4 Simplified cross sections of snapback NMOS, FOX, and snapback NPN BJT ESD devices

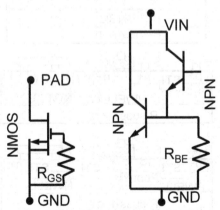

Fig. 7.5 Examples of typical implementation of the ESD clamps using avalanche-injection conductivity modulation in NPN devices: grounded gate snapback NMOS and BVCEO NPN clamp

realize all important parameters simultaneously within the same device. For example, the most conflicting parameters are low leakage and high breakdown voltage tolerance versus low dV/dt effect, proper turn-on characteristics under compatibility with given integrated process and device architecture. Therefore, these devices are usually used as a part of the ESD clamps that provide additional biasing or dynamic coupling circuit to adjust the triggering parameters (Figs. 7.3b and 7.5).

LVTSCR, FOX-SCR, bipolar SCR

From the power dissipation point of view, the most advanced ESD device can be achieved using double injection conductivity modulation. This is realized in different SCR structures. Device architecture involves implementation of parasitic NPN and PNP structures. In most cases for given NPN, NMOS, NLDMOS, the corresponding SCR structure can be implemented by addition of an isolated P^+-emitter region connected to the positive electrode. Simplified cross sections for some of these devices are presented in Fig. 7.6.

The most important parameters for SCR devices are similar to NPN-based ESD devices. However the major advantage in low power dissipation due to low holding voltage creates a major application problem too. Due to the holding voltage below the power supply level the application of the SCR's might create jeopardy of transient latch-up. Due to possible short term electrical overstress events during application or test conditions power supply latch up phenomenon is possible and should be thoroughly considered by obeying the absolute maximum voltage ratings in all pulsed regimes.

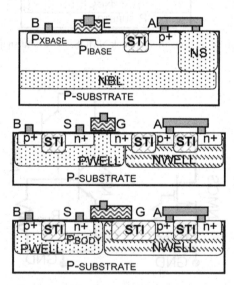

Fig. 7.6 Simplified cross sections for bipolar SCR, LVTSCR, and NLDMOS-SCR devices

This measure, however, is not possible in the case of system-level ESD tests and thus limiting SCR application. Several methods have been suggested to increase the holding voltage of the SCRs toward latch-up immunity in the case of power clamps [281, 291–294]. In spite of all the limitation SCR-type clamps, perhaps, the only small-footprint solution for high-voltage pins.

Similarly to bipolar ESD devices, a simultaneous design of the triggering, high current, and DC parameters presents an even bigger challenge in the case of SCRs. For similar reasons, the SCR-type devices are usually used as part of the ESD clamp circuit with some dynamically coupled or biasing components.

Other ESD devices

A large number of different names for ESD devices and clamp design appeared in ESD publications during the ESD era. We believe that was done for emphasis of the novelty of the particular free device or clamp biasing circuit architecture, specific features of the proposed solution or process technology.

However, excluding exotic nonsemiconductor devices (spark gaps, nano-electronic polymers, micromechanical devices), in the integrated semiconductor ESD devices one or more elementary devices discussed above can be determined, in spite of a possible complex interplay between positive and negative feedback in the device structure, alternative current path at different current levels and current balancing triggering techniques.

This emphasis here is not to trivialize the rather complex challenges of ESD development nor the significant progress that has been made by research groups in the ESD area. ESD development still brings many challenges and requires quite sophisticated solutions.

On the contrary, this is done to highlight that the same regularities are involved in the physics of ESD device operation in spite of this great variety of device- and clamp-level solutions for integrated circuits. One of the complex examples of such devices is dual-direction structures will be discussed in this chapter so as to demonstrate the general approach in particular.

Triggering techniques and clamp design

Despite our conventional circuit-level understanding of the ESD clamp building block as a diode, in most snapback devices the control electrode plays an important role. It is actively used to adjust the clamp characteristics inside the ESD protection window. This is achieved by control of the field electrode above the threshold voltage or by current injection through bipolar control electrode during ESD stress. High variety of the triggering techniques and sometimes sophisticated interplay between different current paths in the ESD devices can be subdivided on following major methodologies:

(i)   implementation of the reduced avalanche breakdown voltage of the internal blocking junctions;

(ii) the triggering initiated by the displacement current through the bipolar junctions;

(iii) dynamic coupling of the field or bipolar control electrode through the internal device parasitic capacitance;

(iv) active biasing of control electrode using additional driver circuit.

ESD device layout

As mentioned above, typical layout of the ESD cells involves a specific array type or multifinger design. The specific focus is both on simultaneous turn-on of the device into snapback mode and on balance of the high current density in particular avoiding device limitations due to the interconnects. The layout design for high-current conditions should take into account the current crowding, voltage drop on interconnects, and metal bus lines taking advantage in voltage drop on parasitic interconnect resistance for the internal circuit connection.

7.2 Spatial Thermal Runaway in ESD Devices

A complete scenario of conductivity modulation, current and spatial electrical and thermal instability for the case of ESD operation is revealed and demonstrated in [295] using 2D simulation TCAD tools [296]. It has been studied for the case of SiGe BJT devices in a 0.24-μm BiCMOS process.

The experimental TLP collector-emitter I–V characteristic for the device is presented in Fig. 7.7a. The final catastrophic failure mechanism was irreversible local burnout at some critical pulsed current (Fig. 7.7b). Numerical simulation analysis demonstrated that the major physical mechanism responsible for the processes in these devices is similar to the conductivity modulation in standard NPN bipolar devices discussed in Chapters 2–4.

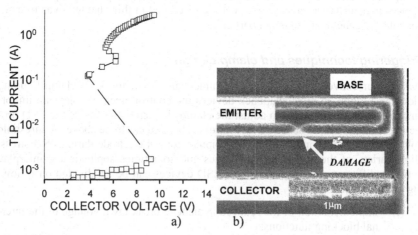

Fig. 7.7 TLP snapback curves for BJT pulse operation and SEM photographs of the damaged device

In [295], the analysis has been completed using thermal coupled mixed-mode simulation. The reference BJT structure was obtained from calibrated process simulation for a 0.24-μm BiCMOS process (Fig. 7.8a).

The quasi-3D simulation methodology for current filament solution is similar to that described in Chapter 4 and 5 in greater detail. The distributed multiple-cell structure has been created from segments of the original BJT cross section from the physical process simulation (Fig. 7.8b) thus providing a mathematical approximation for the general 3 D problem. The local lattice temperature has been taken into account for quasi-3D simulation filament for the first time.

For visualization of the current stratification effect in transient conditions, the maximum lattice temperature values have been extracted across the structure during transient ESD response. The dependencies of the values, T_1, ..., T_{10}, on time have been compared for the cut lines with Z_1, ..., $Z_{10} = 1$ μm, ...,10 μm, respectively. At the uniform temperature distribution, all of these dependencies were coincident. The appearance of the difference between them indicated formation of appearance of a spatially nonuniform solution.

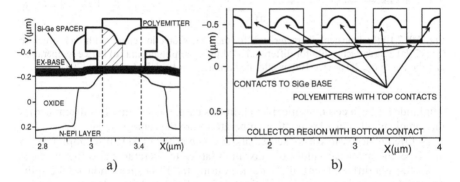

Fig. 7.8 Cross section of the original NPN BJT (a) and 9.6-μm distributed structure in order to obtain a 0.3 μm part (a) by multiple reflections. (b) The intermediate structure after three reflections [295]

A complete description of the processes inside during thermal coupled mixed-mode operation is achieved by superposition of both the 2D simulation for the device cross section (Fig. 7.9, 7.10) and the quasi-3D simulation for current filament solution (Fig. 7.11).

In the case of 2D simulation for BJT cross section, the device is triggered in the snapback regime after reaching a critical triggering voltage (Fig. 7.10). The maximum lattice temperature, extracted during the ESD stress, corresponds to the power generated during the stress.

The current filamentation process, taken into account, reveals that the real physical limitation is much lower. The quasi-3D BJT structure (Fig. 7.7b) in general provides a uniform response to ESD stress only in the range of HBM ESD pulse

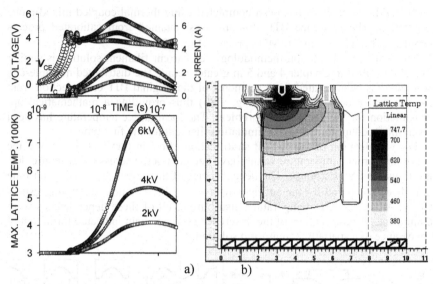

Fig. 7.9, 7.10 (a) Transient characteristics for the ESD protection clamp with BJT structure (Fig. 7.7a) at area factor 200 for different HBM pulse amplitude and (b) lattice temperature distribution in the 2D device 100 ns after beginning the 6-kV HBM pulse (b) [295]

amplitude below a certain critical level. In this case, the solution is uniform across the sample width and all extracted temperatures are the same $T_1 = T_2 = T_3 ... = T_{11}$.

A different scenario was observed above some critical ESD pulse amplitude. In this case, the extracted values of maximum lattice temperatures at different Z co-ordinates in a different part of the device during the ESD stress indicate the splitting of the uniform numerical solution. This splitting of the solution for lattice temperature exhibits current stratification and the formation of solitary hot spot.

The simulated temperature response to different levels of HBM stress is illustrated in Fig. 7.11. No spatial redistribution of the emitter current has been observed at any pulse amplitude below 2.7 kV and the maximum temperature in the device is below 700 K. Conversely, in the regime when the ESD stress exceeds this critical threshold, a catastrophic type of current stratification appears leading to hot spots with a maximum lattice temperature >1500 K (Fig. 7.11b).

The lattice temperature profiles demonstrated the casual process of the current filament and final hot spot formation in the ideal structure and the multiple filaments (Fig. 7.12). Similarly to the simulation problem [297, 298], in the numerical experiments the numerical fluctuations enabled transient solutions with stable spatially nonuniform current distribution realized in casual location across the structure width. The final spot dimension and other parameters were independent (Fig. 7.12a, b). The stable solutions for multiple hot spots have been obtained too in correlation with the experimental results for multiple local destructions (Fig. 7.12c).

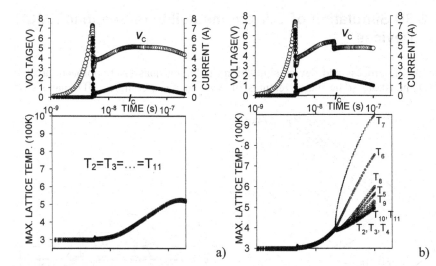

Fig. 7.11 Mixed-mode transient ESD characteristics for the protection clamp with the equivalent $W = 9.6$ µm distributed BJT structure at area factor 42 for 2 kV (a) and 2.8 kV HBM pulse (b)

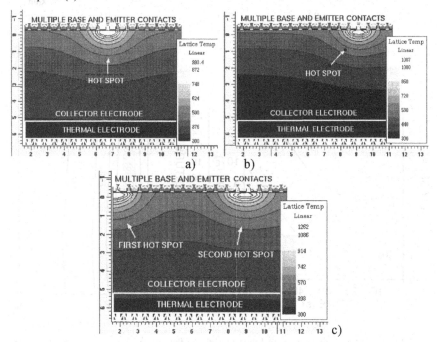

Fig. 7.12 The depth profiles for lattice temperature distribution at $Y = -0.15$ µm in the distributed BJT equivalent structure after 100 ns of 2.8 kV (a), 3 kV (b), and 4 kV (c) HBM ESD stress

7.3 3D Simulation of Current Instability in Snapbac NMOS Devices

While 2D process and device simulation, including quasi-3-D approach, areable to demonstrate some major features of phenomena, a complete 3D analysis of

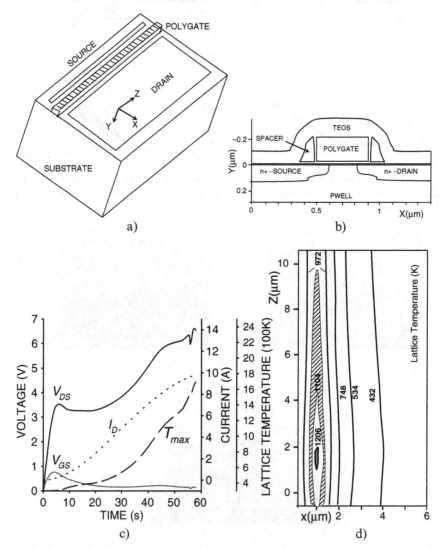

a)

b)

c)

d)

Fig. 7.13 3D NMOS structure (a) and the cut plane in the gate region vicinity (b). Transient ESD characteristics (the drain voltage, the drain current, and the maximum lattice temperature) of the 10 μm 3D NMOS structure with area factor 50 for 10 kV HBM (c) and the cut plane representing hot spot formation at Y = 2 μm (d)

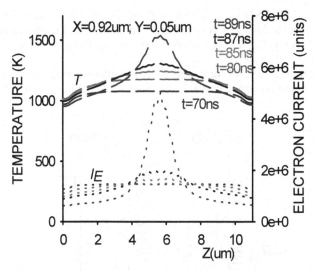

Fig. 7.14 Kinetics of current stratification obtained from 3D simulation solution

the problem is ongoing. Today, the solution for hot spot formation with a suitable level of accuracy and with a reasonable number of grid points can be obtained by available 3D tools [295].

The first 3D simulation results for this problem have been demonstrated in [295] for the case of a rather simplified 3D structure NMOS structure with a 10 μm contact width (Fig. 7.13a,b). The calculated ESD HBM transient characteristics of this simulation object are presented in Fig. 7.13c. Starting from HBM pulse amplitude ~ 6 kV, the maximum lattice temperature became nonuniform representing the current filament and corresponding hot spot formation phenomenon. A cut-line along the drain junction, under the gate demonstrates the spatial electrothermal current instability of the drain current evolving with time and the solitary hot spot in the structure (Fig. 7.13d). Kinetics of current stratification is presented in Fig.7.14.

Thus, in both BJT device and NMOS structure, the physical limitation of ESD device operation is related to rather fast thermoelectrical current filamentation. At the same time, before the critical level the avalanche-injection conductivity modulation is limited by either subcollector or drain ballasting region, thus limiting the current density below the critical level.

7.4 Conductivity Modulation in BJT and Bipolar SCRs

ESD operation of NPN BJT devices in the case of different triggering circuit is similar to the BJT operation in avalanche-injection conductivity modulation discussed in Chapters 3 and 4. An example of comparative analysis between transient triggered and voltage referenced ESD power clamps is presented in [300]. Different architectures of the ESD clamps with both the external and internal breakdown voltage reference techniques are compared. In particular, the grounded base or BVCER BJT clamp (Fig. 7.15a) has been compared to enhanced Zener clamp (Fig. 7.15b).

The turn-on voltage level for conductivity modulation mode of the first clamp in general is based on either internal collector-base blocking junction breakdown

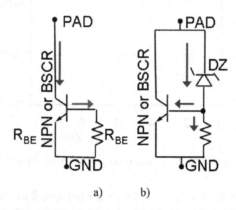

Fig. 7.15 Negative and positive base current direction in the case of the BVCER (a) and enhanced Zener (b) BJT ESD clamps

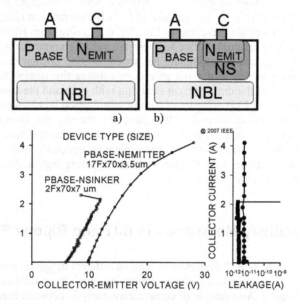

Fig. 7.16 Simplified cross sections and experimental TLP characteristics for the PBASE-NEMIT (a) and PBASE-NSINK (b) lateral avalanche diodes [300]

or dynamic effects. The latter can be called dV/dt effects and related displacement current in the collector-base junction.

The second clamp is specifically designed to be less sensitive to dV/dt effect due to the fast avalanche breakdown in the relatively small avalanche reference diode in the base-collector circuit. Thus, turn-on of the BJT into conductivity modulation mode occurs at negative and positive base current for the first and second cases, respectively.

The avalanche breakdown voltage reference can be achieved by using the avalanche diodes formed by the PBASE-NEMITTER and PBASE-NSINKER lateral junctions (Fig. 7.16).

In addition to the standard BJT clamps, Bipolar SCR [281] clamps have been studied in [300] (Fig. 7.17). The physical mechanism responsible for the snapback effect is discussed below.

For comparative analysis, the collector-emitter 100 ns TLP characteristics have been measured under the conditions of given initial base current. Even though a constant base bias was applied and only the initial base current value can be controlled at each pulsed measurement, this technique provided for a rather informative insight into the pulsed SOA of both devices. The ESD protection window limitations for standard BJT can be estimated from the instability boundary (Fig. 7.17b). The same characteristics can be used to implement corresponding voltage referenced clamps both the BJT-based (Fig. 7.17b) and Bipolar SCR based (Fig. 7.17c).

Another version of the ESD clamp device has been designed with the internal Zener diode inside. This internal junction has been formed by stretching the PBASE diffusion to create an overlap with the collector region, thus forming a surface junction between the base and the N-sinker regions (Fig. 7.18). Similarly, the internal Zener diode structure was implemented in the BSCR, where an additional floating N-EMIT region was added between the PBASE and the P-EMITTER.

Comparison of the TLP characteristics of different NPN BJT clamps is presented in Fig. 7.18. In the case of the external Zener clamp, the breakdown voltage of the device is significantly reduced, the critical current required for snapback is rather large, and the device operation is not optimal to provide the proper waveform for the desired voltage range (8–10 V for the particular application). In contrast, the device with the internal blocking junction provides excellent triggering characteristics that produce both the low triggering current and holding voltage. However, this device had a lower high current tolerance. The high current tolerance was improved by removing the N-sinker region (Fig. 7.19).

Similar snapback characteristics, but with better high current tolerance were observed in the BVCER and internal Zener BSCR clamps (Fig. 7.19); also, in the case of the base N-sinker overlap the corresponding blocking junction was formed with a lower breakdown voltage of ~ 7 V.

In spite of the similarities in the design and operation, the internal and external voltage reference clamps have quite different triggering mechanisms. These mechanisms are defined by the direction of the base current and the interplay between the avalanche and the injection effects. The ESD pulsed operation and the experimental results presented in this study can be interpreted within the same

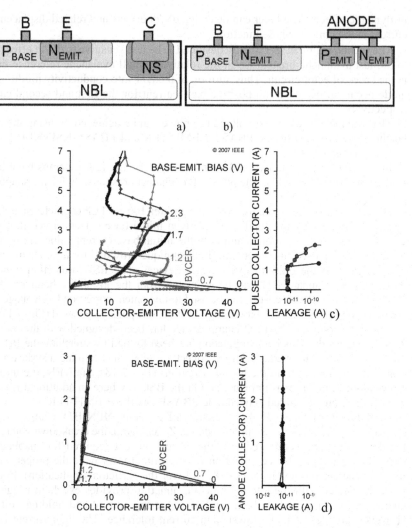

Fig. 7.17 Simplified cross sections for standard NPN BJT (a) and BSCR (b) devices and measured TLP collector-emitter characteristics at constant base emitter bias (c) and (d), respectively [300]

framework of the positive feedback induced by the impact ionization and the injection processes in the bipolar devices under normal operation conditions discussed in Chapter 4. Indeed, the universal criterion for avalanche-injection instability in the bipolar devices can be expressed as $\alpha M > 1$, where α and M are the current gain and the carrier multiplication factor, respectively.

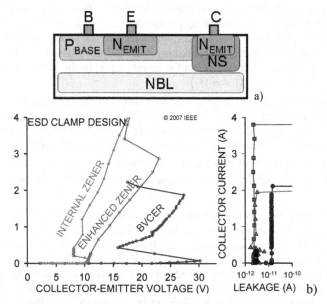

Fig. 7.18 Comparison of the measured TLP *I–V* characteristics for the BVCER, enhanced Zener clamp, and internal Zener clamp [300]

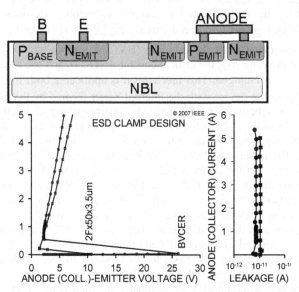

Fig. 7.19 Cross section and TLP characteristics of the internal Zener NPN and BSCR clamps [300]

The scenario of the isothermal current instability that results in the negative differential resistance in the collector-emitter $I–V$ characteristics depends on the product of the collector thickness L and the collector doping level N_D [301]. In the case of the collector-emitter breakdown with the open base (BVCEO), the base current $I_B = 0$ and the collector current is simply equal to the emitter current and is proportional to the value $(1 - \alpha M)^{-1}$.

When the emitter and base terminals are shorted to the ground (BVCES) ($U_{EB} = 0$, Fig. 7.17), the initial location of the avalanche breakdown is at the collector-base junction. In this case, the avalanche breakdown initially is not accompanied by the injection from the unbiased emitter junction. Therefore, the collector current is provided by the multiplication of the thermal generation current I_{C0}, i.e., $I_C \sim M I_{C0}$. In this case, the breakdown voltage is close to the collector-base breakdown voltage U_{CBBR}.

With the increasing avalanche current, the generated holes flow out of the base. In this negative base current regime, the hole current results in an additional voltage drop across the p-base, increasing the base potential at the emitter junction by the value $r_B I_C$, where r_B is the internal base resistance. The r_B can be extracted from the simulation or from measured data. Thus, when $r_B I_C$ is larger or equal to the potential of the emitter junction opening (~0.6 V), the injection of electrons from the emitter begins and it is followed by the positive feedback mechanism. Therefore, this voltage drop is controlling both the critical current level and the critical voltage of the instability. In the experimental data, the beginning of the instability signifies the emitter junction opening. Therefore, to estimate the critical regime the following equation may be used: $I_{CR}(R_{BE} + r_B) = 0.6\,\text{V}$, where R_{BE} is the external resistance in the base circuit (BVCER regime with $R_{BE} = 10k$, Fig. 7.15a). With increasing R_{BE}, the critical current decreases and the breakdown voltage value U_{CEBR} is changed. This mechanism is applicable both for the circuit in Fig. 7.15a and for the device with the internal Zener formed by the overlap of the PBASE and the subcollector diffusion. The distinguishing characteristic of the internal Zener bipolar device is the alternative current path at the surface.

A different triggering mechanism is realized in the enhanced Zener clamp circuit where the avalanche diode is feeding the positive base current into the snapback device. After the Zener breakdown occurs, this circuit can be considered as the common-emitter circuit with a referenced positive base current. The criterion for the current instability is the same, but in this case the conditions for the snapback characteristics are realized at a higher α and lower M., i.e., in this case the injection process is dominant. Indeed, according to Fig. 7.18 a much higher base current level is required to trigger the ESD device into the snapback mode. Thus, this type of ESD clamp not only has a larger footprint to accommodate the external reference Zener diode but is also less efficient.

Indeed, the critical conditions for snapback are related to the inception of the negative differential resistance regime, i.e., $dI/dU \leq 0$. Under the positive base current regime $I_B > 0$, the collector current, the thermal generation current being neglected, is equal to

$$I_C = \frac{\alpha M I_B}{1 - \alpha M} \tag{7.1}$$

Since the multiplication coefficient $M = M(U_{CB}, I_C)$ is a function of the collector base voltage and the collector current, (7.1) determines the device collector emitter I–V characteristic. By taking the derivative of (7.1) in U_{CE} at α = const, T = const, and $\partial M / \partial U_{CE} > 0$, the simple criterion for the instability to take place is $\dfrac{I_C}{M(1 - \alpha M)} \dfrac{\partial M}{\partial I_C} > 1$.

At the lower base currents, when $j_C < j_0 = q N_D v$, the transistor operation is simply similar to the case with a grounded base $I_B = 0$ presented above. When αM approaches unity, the collector current increases up to the critical snapback current value I_{T1} when the negative differential conductivity is formed and is followed by the snapback.

At the high base current level realized in the circuit (Fig. 7.15b) $I_C > 2 I_0$, the thickness of the space charge region W will still be lower than the thickness of n-epilayer L. In this case, $U_{CE\,CR} \approx \dfrac{E_{BR} W_{CR}}{2}$, where $W_{CR} \approx \dfrac{E_{BR} \varepsilon}{\left| j/v - q N_D \right|}$, v is the drift velocity of electrons, ε is the dielectric constant, and E_{BR} is the electric field at breakdown. Therefore, for adequate high current $I_C \gg I_0$, the isothermal stability limit must be close to the hyperbolic function $U_{CR} I_{CR} \approx \dfrac{\varepsilon E_{BR}^2 v}{2}$ (Fig. 7.17a).

Similarly, the criteria can be obtained for the BSCR devices, taking into account the additional gain from the built-in PNP α_{PNP} device. In this case, the current instability criterion in the case of the grounded base is $\alpha_{NPN} M_N + \alpha_{PNP} M > 1$, that for the case of the BSCR can be simplified to $\alpha_{NPN} M_N + \alpha_{PNP} > 1$. For a standalone BSCR device, the value $\alpha_{NPN} M_N + \alpha_{PNP} M_P$ is rather high so that a relatively low current is required to bring the BSCR device into the snapback mode (Figs. 7.17d and 7.19).

It should also be taken into account that, from the structural point of view, the NPN BJT is composed of the vertical and lateral NPN devices. At the lower current values, the vertical BJT is involved in a conductivity modulation mechanism, thus providing the initial part of the snapback characteristic (Fig. 7.18). After the subsequent current increase, the second snapback occurs when the lateral surface NPN becomes engaged.

7.5 Device-Level Physical ESD Design

Several device-level concepts for ESD protection are discussed in order to demonstrate the physical approach to ESD design.

7.5.1 Self-Protection

Device structure level

A number of changes can be made in the standard device architecture toward self-protection capabilities creation.

One of the most common examples of self-protecting solution is addition of the drain ballasting region. With simultaneous increase in the well tap spacing, the I/O NMOS devices can withstand reversibly the snapback operation similarly to the ESD clamps. In normal operation mode, the only side effects of this design are an increased space on chip, parasitic drain capacitance, and lower on-state resistance.

Another way to implement an additional self-protection capability or at least push the current instability level to the higher voltage and current range is inter-digitated well tap connection solution (Fig. 7.20a). In this case, the parasitic NPN bipolar has the minimum possible parasitic base (bulk)–emitter (source). This measure provides the lowest NMOS current gain in the internal parasitic bipolar structure. Since the avalanche-injection conductivity modulations occur at much higher critical avalanche current, both the instability boundary and the ESD protection window are changed.

This solution is especially practical in the case of fully silicided processes that provide reliable connection to the interdigitated regions with no special contact placement.

From the physical point of view, this device operation is based on reduction of the hole density in the discharge region due to reduced gain of the parasitic NPN BJT and effective extraction of the carriers by the p^+-contact regions. Respectively, the average electric field amplitude under the gate in the triggering-on state is increased due to lack of space charge compensation between electrons injected from the source junction and the holes generated in avalanche multiplication drain region. In particular, this results in increase of the holding voltage. TCAD-level validation for this solution is presented in Fig. 7.20b,c. The major parameter for comparative analysis is the distance L_{PG} between the introduced PPLUS region and the source side of the polygate region.

The mixed-mode signal waveforms (Fig. 7.20b) HBM ESD operation demonstrates that the holding voltage of the structure can be made even larger than the triggering voltage. The electron and hole currents (Fig. 7.20d,e) demonstrate the details of conductivity modulation and the holding voltage increase, described above.

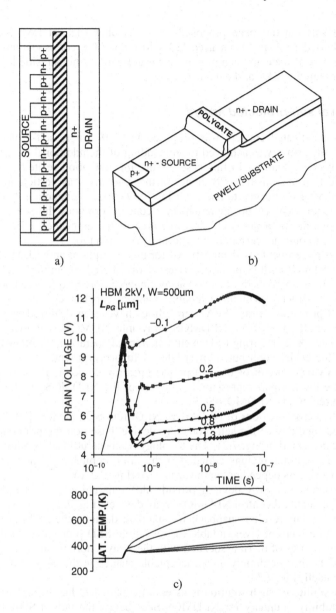

Fig. 7.20. Power array finger with interdigitated source/P-well diffusions (a), 3-D equivalent structure (b) and the results of 3-D numerical simulation analysis (c) the waveforms of mixed-mode HBM simulation upon the distance between the PPLUS region and the edge of polygate (L_{PG})

Similar solution has been proposed for high-voltage lateral DMOS devices, where P-buried layer has been used [302] to reduce the corresponding internal base-emitter resistance and common base current gain of the parasitic lateral BJT thus improving the SOA of the device.

Array-level self-protected arrays

A more sophisticated concept was proposed in [299] where self-protection capability of the arrays was achieved by embedding of the additional parasitic SCR structure inside the array. In this case, before the limitation related to the avalanche injection in parasitic NPN device of the standard array portion, an embedded ESD device is engaged to take over the ESD current conduction. Contrary to a conventional local clamp approach, this concept provides the self-protection capability within the array itself since the area of the embedded ESD device is contributing to on-state current during normal operation too.

This was experimentally demonstrated for the example of an NLDMOS power array in a 50-V BiCMOS process. Series of distributed p^+-emitter diffusion regions have been embedded into some array fingers in order to form an additional parasitic SCR structure with the reversible snapback capabilities.

The self-protecting array (SPA) capabilities have been achieved by transformation of small parts of the NLDMOS array into NDMOS-SCR that provided the current conduction during ESD events. In normal operation, the NDMOS-SCR portion of the device contributed toward the useful array current.

The equivalent circuit of the self-protecting array device is presented in Fig. 7.21, where the left part represents the fingers of the NLDMOS, while the right part of the represents NLDMOS-SCR.

Simplified cross sections of NLDMOS, NLDMOS-SCR devices [305], and the constructed topology of the self-protecting device, based on the principles listed above, are presented in Fig. 7.21. The topology has been utilized by additional interdigitated P-emitter regions added into the drain composite region of one or more device fingers depending on the array size (Fig. 7.21c).

The experimental SPA devices provide different pulsed $I-V$ characteristics compared to standard control NLDMOS array devices (Fig. 7.22). Unlike the case of irreversible failure after snapback for standard devices, the SPAs are able to sustain a larger current level suitable for reversible ESD protection. At the same time, major figures of merit associated with the array's normal on-state operation have been changed negligibly under acceptable change in the pulse ESD SOA of the device itself (Fig. 7.23).

The possibility of such solution is essentially achieved due to implementation of the instability boundary for NLDMOS-SCR below the pulsed SOA limits of the NLDMOS device. This principal implementation in general requires a precise embedded SCR design that has a much stronger function of the snapback voltage on gate bias. At the same time, one of the objectives is to preserve the SOA of the whole array at the original level.

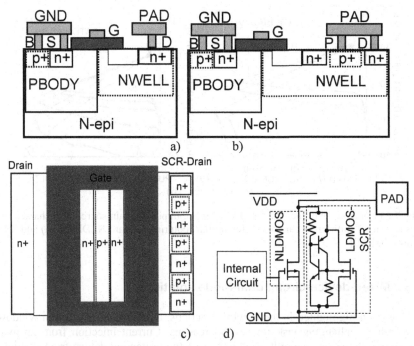

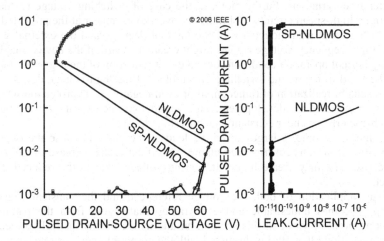

Fig. 7.21 Simplified cross section for NLDMOS (a), local clamp NLDMOS-SCR devices (b), and topology (c) for the self-protecting device with the left finger formed by a standard supported device and the right finger formed by the embedded SCR device; (d) simplified circuit diagrams for self-protecting array [300]

Fig. 7.22 Comparison of the TLP characteristics for the standard and self-protecting 50-V NLDMOS array with 400-μm width in the grounded gate configuration [300]

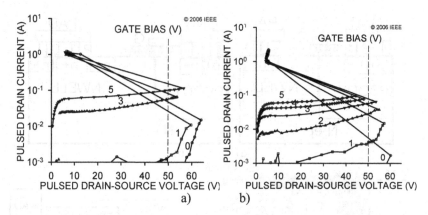

Fig. 7.23 Comparison of the pulsed SOA using output pulsed drain-source TLP characteristics measured at different gate biases for similar footprint standard NLDMOS (a) and SPA arrays (b) [300]

7.5.2 Mixed device-circuit dual-mode solutions

As has been highlighted in the previous sections, targeting the protection window is a rather challenging task for several reasons. Current injection from an avalanching junction is generally used to trigger the protection device from its high-impedance blocking state into its low-impedance conducting state. The avalanche current source may be either an internal blocking junction of the protection device, or an external avalanche diode. A precise control of the breakdown voltage is difficult due to a lack of self-aligned blocking junctions, across process and temperature variations. Furthermore, in the case of switching voltage regulators and other high-speed circuits, fast transient overshoots appear at the protected pad due to the finite load inductance of the internal package and the external wires. Such high-frequency voltage ripples might create a hazard of the device snapback during normal operation conditions due to dV/dt reduction of the trigger voltage.

The trend in improving switching capabilities of the high-voltage devices usually results in realization of rather small or even negative ESD protection window as a side effect of the development. This window can be measured using pulsed $I–V$ characteristics under constant gate bias (Fig. 7.24).

At the same time, a large capacitive load at the pad can cause the opposite problem with a transient-triggered clamp. Specifically, the increase of the HBM pulse rise time may create higher triggering voltages due to the reduced dV/dt effect.

Thus, application of both the voltage referenced and the transient triggered clamps becomes undesirable and even rather challenging from the parameter targeting point of view. In this section, an alternative means for ESD protection implementation [303, 304] is discussed with the purpose to demonstrate a physical concept of using the ESD device as an active structure with the snapback characteristics actively controlled by the circuit driver.

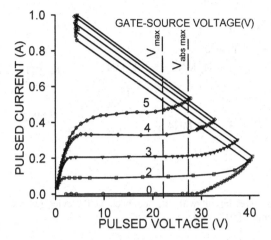

Fig. 7.24 Pulsed output characteristics and ESD protection window for the 25-V absolute maximum voltage DeMOS device. V_{max} is the maximum expected voltage on V_{in} (~ 22 V). V_{absmax} is the maximum voltage on the pin before failure (~27 V) [303]

Mixed device-circuit concept

To overcome the challenges highlighted above, the new mixed device-circuit ESD solution is based on different principles. Instead of solving the problem of targeting the ESD protection window in both the voltage and rise-time domains, the mixed device-circuit approach provides for different triggering characteristics of the device under ESD and normal operating conditions.

The snapback ESD device is first modified to provide access to a terminal that the active circuit will control. The snapback device is required to have (i) a very low trigger voltage in the case of a positive bias at, or forced current through, the control electrode, and (ii) a trigger voltage that is higher than any expected voltage in the case of a grounded control electrode. The requirements for the active control circuit are as follows: (i) to provide control electrode biasing or current forcing during the first part of an ESD event (~20 ns) in order to turn on the ESD device; (ii) to ground the control electrode after the ESD device is operating in snapback mode (after ~ 100 ns); (iii) always turn off the control electrode if VDD (~5 V) is generated.

Example of the circuit design

A simplified schematic of the active control circuit is presented in Fig. 7.25. The circuit contains a driver that controls the gate of the snapback device MESD and operates depending on the signal on the VIN and VDD pins. This circuit has been studied utilizing circuit simulation.

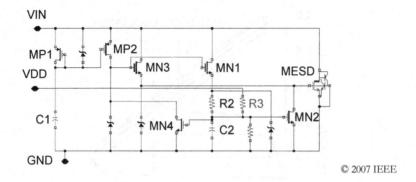

Fig. 7.25 Active driver circuit with the snapback SCR device (MESD) being controlled from the gate terminal [303, 304]

In the case of a positive ESD event on the VIN pin, MP2 will be on until the capacitor C1 is fully charged (~20 ns). During this time, MP2 and MN3 directly apply a positive bias on the MESD control electrode (gate). This creates the conditions for fast triggering of MESD into snapback at rather low VIN pad voltages. Significantly, the circuit does not depend only on capacitive coupling to pull the gate high, as in some alternative designs. MP2 also turns on MN1, which sources current for a turn-off RC circuit composed of the resistor R2 and the capacitor C2. After a corresponding RC delay of ~200 ns , the transistors MN2 and MN4 turn on. MN2 shorts the control electrode of the ESD device to ground while MN4 turns off both MN3 and MN1. MESD remains in the high-current mode until the ESD discharge is completed. The low-voltage avalanche diodes (Fig. 7.25) are used to limit the gate voltage of the MOS devices. During an ESD event, the VDD potential is near ground due to the high capacitance and leakage of the internal circuitry of the voltage regulator.

In the case of normal operation, during the relatively slow (determined by the total chip equivalent capacitance and the external parasitic) powering sequence of VIN above ~5 V, the VDD signal is generated in the internal circuitry; MN2 is then turned on through R3. MN2 holds the control electrode low. This maximizes the trigger voltage of MESD. The VDD signal will also turn on MN4. MN4 will keep MN3 and MN1 off, regardless of VIN, eliminating the chance of contention between MN3 and MN2. Holding these transistors off also provides for low leakage in the clamp circuit.

This ESD solution has been experimentally validated in [303] for a 0.35-μm CMOS process with 20-V extended drain MOS devices (DeMOS). For pulses with a 10 ns rise time, V_{ABSMAX} is 27 V. Given $V_{MAX} = 22$ V, the protection window is 22–27 V (Fig. 7.24). With a 200-ps rise time, the upper limit is lowered to ~23 V. The DeMOS-SCR lateral ESD protection device used is similar to the NLDMOS-SCR (Fig. 7.26a).

The TLP *I–V* characteristics for two different rise times and under two different operating conditions—at VDD = 0 V (the results were the same as with a floating VDD node) and at VDD = 5 V—are presented in Fig. 7.26b. Under the condition that corresponds to ESD stressing of the nonpowered circuit (VDD = 0 V), the triggering voltage is very low ~ 5 V. This guarantees reliable ESD protection not only for high-voltage devices, but also for the low-voltage devices in the internal circuit that are connected to the protected pin. However, the leakage measured at VIN = 20 V remains the same as that of a grounded-gate ESD device (Fig. 7.26b, left plot).

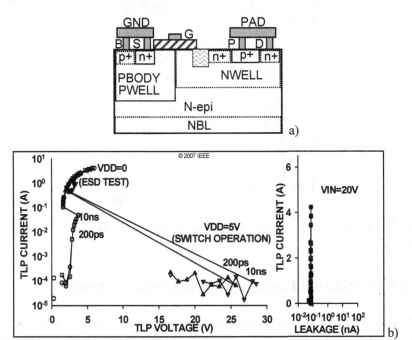

Fig. 7.26 (a) Cross section for high-voltage snapback clamp and (b) measured TLP results for the mixed device-circuit solution clamp showing both modes of operation [303]

In the case that corresponds to normal operation (VDD = 5 V), the trigger voltage of the clamp is high (~29 V) and equivalent to that of an NLDMOS-SCR device with a small gate-source resistor. Evidently, the clamp may be designed to have a triggering voltage that, under normal operation, exceeds V_{ABSMAX}. This is desirable since unreliable snapback operation and damage of devices with a high trigger voltage is observed. The ESD operation of the clamp is fully controlled by the channel current, rather than avalanche breakdown, thus always providing reversible snapback operation.

It has been shown that both overshoot and turn-off voltage can be controlled by this solution (Fig. 7.27).

This example demonstrates a cross-disciplinary aspect of the future advanced physical ESD design. The ESD device, as a component of the dual-mode clamp, should be designed with proper DC voltage tolerance, provide low sensitivity to the fast rise time and at the same time provide sufficient control of the triggering characteristics (reversible instability boundary) by the control electrode under reversible high-current operation and small footprint.

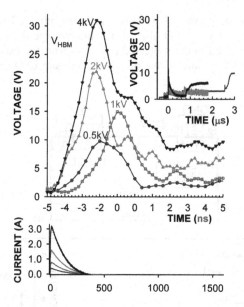

Fig. 7.27 Comparison of the measured voltage waveforms for the active gate control circuit [310]

7.6 Device-Level Positive and Negative Feedback

The final section of this chapter demonstrates the means of implementation of the device-level positive and negative feedback in the condition of the operation in the conductivity modulation.

The practical implementation is discussed on the example of ESD protection devices with dual-direction voltage tolerance. A small-footprint and dual-direction functionality is required for a number of analog applications, for example column drivers in liquid-crystal displays, RF inputs, common node voltage regulators, interface applications, digital-analog converters.

A typical requirement is that the protected dual-direction pin be tolerant to a certain positive and negative voltage range relative to the grounded epi p^+-substrate. Respectively, the ESD device is connected between the dual-direction pad and the ground bus and must provide a proper waveform under the ESD pulse discharge.

One of the most area-efficient solutions is the dual-direction silicon-controlled rectifier (SCR) device (DD-SCR) [306, 307]. The device operation is somewhat similar to a diode AC switch structure (DIAC) [308]. The corresponding control electrodes are connected to the anode and cathode regions. However, a straight-forward attempt to implement the same device structure in the case of a CMOS process reveals a number of problems.

Dual-direction device architecture

In [309], the test structures have been fabricated using a nonsilicided 0.5-µm 5-V CMOS process with a deep N-well on a P$^+$-substrate with a P-epi layer ~ 4 µm thick .

With the accuracy of the voltage overshoot during the first few nanoseconds of the switching time domain, the waveform of the device, in general, can be extrapolated using parameters of the transmission line pulse (TLP) measurements for pulsed I–V characteristics. The snapback triggering voltage V_{T1} and the holding voltage V_{OP} at the ESD current level (e.g., ~1.3 A at 2-kV HBM ESD pulse) provide the most important information for the waveform during ESD stress, while the minimum holding voltage VH projects the expected latch-up robustness of the device in the case of a system-level specification.

The initial implementation of the symmetrical version of the dual-direction SCR architecture is presented in Fig. 7.28a. This device is formed by the common contact regions to the injection and the blocking junction's diffusions formed inside two isolated P-well regions (RW). The RW regions are laterally isolated from both each other and the P-substrate using the N-well (NW) rings of the "butterfly wings"-like topology. Deep N-well region (DNW) provides the vertical isolation of the device regions from the P-epi substrate. A complete device structure also includes the outer P-well ring (PW) (Fig. 7.28a).

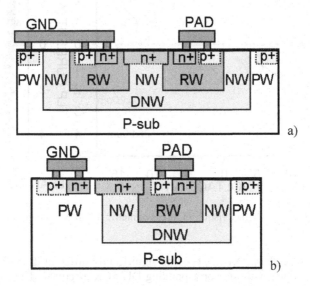

Fig. 7.28 Typical dual-direction SCR structure PN-n-NP structure with the n$^+$- Pwell (a) and the compact asymmetrical version of the PN-n-PN device (b)

Since for a typical 5-V CMOS process the P-well to N-well breakdown voltage is usually higher than 20 V, the additional floating diffusion region is used at the blocking junction to provide the waveform with the reference voltage in the desired 10–14 V voltage range. The breakdown voltage of the blocking junction is adjusted by introducing either the middle floating n^+-region (Fig. 7.28a) or two p^+-regions. One of the most space-efficient implementations of this device for a given ESD protection problem statement is the asymmetrical version (Fig. 7.28b).

There are several different combinations of the diffusion regions that are possible in the same generic device structure (Fig. 7.28). First, the PN-n-NP devices have been studied, where the region sequence represents the ground site pair of the p^+ and n^+ diffusion regions, the blocking junction n^+-diffusion and the pad side second pair of n^+ and p^+ in the RW region, respectively (Fig. 7.28a).

In the case of a floating substrate and a disconnected outer p-ring, the experimental characteristics of the device are absolutely similar to each other in both positive and negative directions. They are snapback $I–V$ characteristics typical for an SCR with a rather low holding voltage ~2 V and are similar to the negative voltage $I–V$ characteristic from Fig. 7.29.

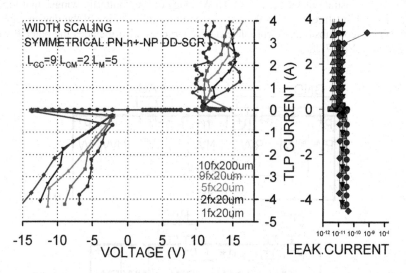

Fig. 7.29 TLP $I–V$ characteristics of the symmetric PN-n-NP structures for different array sizes

These device characteristics become significantly different if the P^+ guard ring is grounded along with the corresponding DD-SCR electrode. The asymmetry in the electrical connection of these regions immediately provides an asymmetry in the device operation. Under a negative TLP pulse, even the device design with a rather large contact region spacing of ~10 μm still provides a low holding voltage and an on-state resistance (Fig. 7.29, negative part of the $I–V$ characteristics).

However, in the case of the positive current direction, the holding voltage of the device is much higher and in general can exceed the triggering voltage, thus leading to the undesirable voltage waveforms (Fig. 7.29, positive I-VC portion). In the case of a multifinger device array with a common outer P-ring, this effect results in the incorrect width scaling of the operation holding voltage V_{OP} in the range above V_{T1} (Fig. 7.29, positive part of the $I\text{–}V$ characteristics) and in the reduction of the maximum TLP current I_{T2}.

The mechanism of device-level negative feedback

The observed asymmetry in the device operation on the phenomenological level can be understood as a result of the parasitic PNP engaging. This PNP is formed by the P-substrate acting as a collector, deep N-well acting as a base, and the upper PWELL acting as an emitter.

After the device turns on into the conductivity modulation mode, in the case of the positive current direction the holes injected from the pad junction are partially escaping through the parasitic PNP current path into the P^{+}-substrate. As a result of this effect, the level of the space charge neutralization between the electrons injected from the ground junction is reduced. This negative feedback results in the formation of the corresponding electric field distribution between the cathode and the anode of the DD-SCR structure and leads to rather high values of the holding voltage and the on-state resistance.

With increasing device size, the effective collector resistance of the PNP is increasing as well and the negative feedback level being reduced, thus providing lower holding voltages for the larger devices (Fig. 7.29a).

An opposite situation is realized in the case of the negative current direction. Here, the P-substrate may essentially contribute to the hole injection level acting as an additional P-emitter.

Device-level negative feedback compensation

In order to compensate the device-level negative feedback provided by the parasitic PNP to the substrate, the diffusion sequence has been altered toward the PN-n-PN structure (Fig. 7.30a), followed by the reduction in the contact-to-contact spacing. In this case of the positive current direction, the injection junctions are closer to each other.

This additional positive feedback mechanism compensates the negative feedback due to the parasitic PNP. At the same time, this measure does not impact the characteristics of the device in the negative current (Fig. 7.30b). Both asymmetrical and symmetrical types of the device architecture for the case of NP-n-PN DIAC perform similarly (Fig. 7.30b).

The reduction of the contact-to-contact spacing results in a further adjustment of the holding voltage within the desired range (Fig. 7.31).

Thus, in the case of ESD design, it becomes very important to utilize the knowledge about physical processes under conductivity modulation toward implementation of desired high-current characteristics of the device addressing both the desired voltage waveforms and potential dynamic latch-up issues.

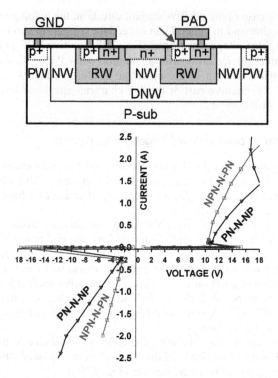

Fig. 7.30 Results of positive feedback implementation TLP *I–V* characteristics of PN-n-NP structure

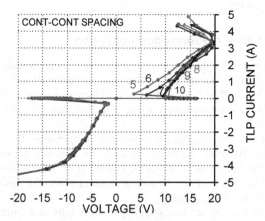

Fig. 7.31 Results of positive feedback implementation TLP *I–V* characteristics of PN-n-NP structure by scaling the device characteristics and the reduction of the contact-to-contact space

7.7 Summary

The ESD area benefits, perhaps, most from the material presented in this book. In ESD devices, multiple scenarios for current conduction at breakdown and instability are not just realized as parasitic phenomena, but directly used to achieve the desired spec for ESD device and clamp parameters. Together with different negative feedback mechanisms realized on the structure level, a combination of the conductivity modulation is used to achieve the desired mode of operation under high current density.

However, as demonstrated in this chapter, the basic phenomena realized during ESD events are the same as the corresponding phenomena discussed in Chapters 3–5. At high conductivity modulation current operation, ESD devices not only obey the same principles and regularities, but even are confined within a rather limited set of the same basic isothermal conductivity modulation mechanisms.

The 3D complexity and nonlinearity of the conductivity modulation processes in the case of ESD events gating implementation of the compact models for direct circuit analysis in ESD pulse conditions. Thus, physical level of understanding of semiconductor structure operation under ESD pulse conditions is critical for successful design of the small-footprint ESD protection solutions where high current density should be maintained uniform in the conductivity modulation mode.

Perhaps, one of the less-studied aspects is the interaction between the electrical overstress events and the degradation. For example, one of the typical requirements for integrated circuit components is some passing level confirmed by standard ESD test results. At the same time, it is required that the components after the ESD test must not further be used in the field. The intuitive reason for this is a possible impact of the test that involves operation in the conductivity modulation mode on the long-term reliability parameters.

Thus, a paradox is formed. On the one hand, the purpose of ESD protection is to provide a primary protection of the components during handling, packaging, and assembly. On the other hand, a long-term operation of the components might be impacted by the lifetime ESD events and therefore might significantly change the long-term reliability parameters of the components and the reliability of the system itself in the field.

One of the clues to future solution of this problem is possible formation of the small local areas in structure with properties similar to local inhomogeneities. For example, local melting areas might form in places of current filament formation and might act like local inhomogeneities. However, even in this case the knowledge about the impact of local inhomogeneities on long-term reliability parameters is rather limited.

The authors of this book believe that the events mentioned above might provide an explanation for some unexpected failures of the devices in the field and it is important to conduct further research in this area. Understanding and further study of these phenomena might be a matter of critical importance for different high-reliability remote applications for example in the aerospace field.

Chapter 8
Physical Approach to Reliability

From the material in this book, more or less complete regularities concerning transistor failures can be derived. Based on the knowledge of these regularities, several questions might be asked. For example: What is the application value of this knowledge? Is it possible to develop and to make each particular transistor in the production lot more reliable? What are the methods for SOA evaluation of new devices? Is it possible to predict with desired accuracy catastrophic failures of the particular devices and systems? Some opportunities are summarized in this final chapter. They are based on the physical approach to reliability.

One of the possible ways can be presented by means of fulfillment of a few complementary conditions. These conditions are expected to provide a considerable increase in the probability of reliable operation transistor. Without excessive detail, these conditions can be listed as follows:

(i)   Gain in-depth understanding of the basic physical failure mechanisms for given semiconductor device structure and process

(ii)   Maintain understanding of phenomena that are limiting safe operation regimes and long-term reliability parameters

(iii) Develop recommendations for device design using knowledge of failure mechanisms toward possible reduction, compensation, or even elimination of particular failure mechanisms while preserving the basic operation parameters within desired spec limits

(iv)   Reveal potentially "weak" places in the device architecture, process technology, and manufacturing for identified failure mechanisms

(v)   Provide competent, physically reasonable choice of the test regimes and long-term reliability tests

(vi)   Understand the limitations of diagnostic methods according to the basic failure mechanisms

(vii) Estimate real transistor reliability in physically reasonable regimes of safe operating area (SOA)

(viii) Estimate the long-term reliability parameters with the help of accelerated tests focusing on the indications of potential reliability issues

The framework of these statements for physical approach is detailed below in order to cover some modern aspects in the field of practical reliability.

8.1 Reliability Assurance at the Stage of Its Development

As soon as the subject of semiconductor device reliability had been deeply studied and the basis for a physical description had been created, a new conclusion appeared. Is it impossible to provide validated reliability parameters of transistor device at the stage of its design for example similarly to operation frequency or dissipated power? Indeed, if the failure mechanisms are well explored, why not take into account all necessary factors in particular design to guarantee, for example, the value of $\lambda = 10^{-8}\,\mathrm{h}^{-1}$. From this point of view, a major reason prohibiting this approach is not obvious.

The real difficulties are hard to overcome. At the development stage, it is practically impossible to estimate compliance of the device with the long-term reliability requirements. If the value f_T or P_{OUT} can be directly measured using experimental test samples, the direct experimental validation of the value $\lambda = 10^{-8}\,\mathrm{h}^{-1}$ will require for example a test of 10^4 devices over 10^4 hours. Thus, direct assurance of the reliability parameters at the development cycle is excluded due to the impossibility of an adequate check.

However, separating the subject from the exact numerical values, the achievements of failure physics are quite real and a rather effective path for reliability improvement.

This means the use of numerous models of failure mechanisms that were originally developed for research purposes, but then were specified and adopted to specific transistor structures for estimation of the reliability parameters. Simple analytical models are used at transistor structure design and optimization [311–316]. However, insufficient understanding of basic physical limitations, causes, scenarios, and mechanisms of failures prevent creation of highly predictable models.

The failure models are required in order to reflect the real processes, enable analysis of the expected reliability characteristics, and formulate appropriate requirements for the design components, architecture, and electrophysical parameters of the structures.

Certain minimum requirement for an equivalent of approximately 1 year of operation within 10% degradation limits is usually confirmed at the development phase using the accelerated tests.

Similar functions can carry out only numerical models and therefore the modern physical theory of reliability is migrating from analytical models toward numerical simulation and physical statistical modeling of the failures. Computer modeling for reliability is described for example in [317–320]. A number of models are presented in Chapters 3–5 of this book.

The simulation models are used for analysis of physical processes of thermoelectrical instabilities resulting in catastrophic failures. The models allow reproduction not only of the final result, but also of the kinetics of instability evolution. As shown in Chapter 3, the same models for structure inhomogeneity effect accounting are valid with various structures.

There is also the opportunity for adequate "tuning" of these models according to experimental results. Meanwhile, even complex three-dimensional models are not always able to provide satisfactory quantitative estimation of safe operation regimes. However, comparative estimation of the design versions, as shown in the example of thermal current filamentation model (see Section 3.1.6), can be quite useful.

Physical-statistical models are used for degradation failure modeling. The models take into account the probability nature of failures. These models basically allow prediction of quantitative parameters of reliability on the basis of computer experiment, for example [317, 320, 321]. Obviously, a complete reproduction of transistor structure in models is impossible. Therefore, at the creation of physical-statistical failure models, the basic critical reliability components of the structure are allocated. Such components are metallization, internal wire connections, protection dielectric, semiconductor–dielectric interface, gate oxide in MOSFETs, active region, and others.

However, the progress in degradation modeling is restrained by challenges to acquire detailed information for direct introduction into the model. The concern is related to all of the basic random factors influencing kinetics of degradation evolution. For example, there is insufficient information on the distribution function of casual parameters of structural technological defects and inhomogeneities. In accordance with accumulation of such information, there is an opportunity for perfection of degradation failure models and enhancement of the numerical simulation for quantitative parameters of transistor reliability.

Finally, further development should result in creation of a CAD system, focused on reliability.

The real opportunities of design with the reliability consideration can be demonstrated with a simple example. Let us assume that in the transistor datasheet, DC maximum ratings are stated for maximum current I_{max}, maximum dissipated power P_{max}, and maximum voltage U_{max}. The respective range of operation temperatures is given. It is required that in a given area of allowable regimes, certain quantitative parameters of reliability are provided. DC operation regime is used here for simplicity only.

Then, it is first necessary to reduce the probability of scenario for hot spot and current filament formation, since these effects inevitably result in either catastrophic failure or accelerated degradation events. To achieve this goal, the design solution must provide the conditions for thermoelectrical instability only outside the area of safe operating regimes. This can be partly achieved by additional experimentation and numerical simulation. One of the examples for thermal optimization of the bipolar transistor active region is presented in Chapter 3. For analysis of the isothermal instability mechanisms, the basic models for internal avalanche-injection processes are suitable (Chapters 4, 5). In this case, optimization of transistor structure should be done toward increasing the critical current and voltage of electrical instability.

The real boundary of thermoelectrical instability can be significantly different from the calculated version for example due to defects and inhomogeneities. Thus, it is necessary to ensure certain structural and technological margin related to casual events, doping profile scattering, and mask misalignment. Simultaneously in transistor technological process, corresponding tests for rejecting devices with a "dangerous level" of structure inhomogeneities must be introduced.

If the probability of formation of nonuniform thermoelectrical states is minimized, it is possible to proceed with estimation of quantitative parameters for transistor reliability.

The middle time before failure mainly depends on thermoelectrical state $(T, j,$ and $E)$ in the regions of structure with accelerated degradation processes, i.e., on the conditions in potentially unreliable components of the device.

With increase in temperature, the time before failure exponentially decreases practically for all degradation failures. The current density j impacts the electromigration failures through the cross section of metallization. The electric field amplitude (for example, in the region of subgate dielectric of MOSFETs or in the space charge region of blocking junction region or field domain region in GaAs MESFETs) determines the dominant mechanism and time before failure. The mechanism can be connected to diverse phenomena of charge instability. Using known models for degradation failures and corresponding experimental data, it is possible to establish conformity between the reliability parameters (t_{BF} and λ) and the thermoelectrical state of structure in such a manner that limitation of values T, j, E at the level $T_{max}, j_{max}, E_{max}$ provides desired reliability level. This way is effective only if the failure model is developed in detail and basic parameters of the model are extracted from the exact measurement.

From known failure mechanisms, the models of electromigration, mutual diffusion, and intermetal formation [317, 320, 321] are most validated. They allow predictive calculation of limiting $T_{max}, j_{max}, E_{max}$ values for the given λ and can provide a starting point for development solutions for active region topology and metallization.

Another means of prediction of reliability parameters is comparison of the designed device and the structural technological prototype. For the prototype, data availability for long-term and SOA reliability tests is assumed. Successful prediction depends on the competence and completeness of the comparison. Certainly it should consider all potentially unreliable components of the device structure.

It is necessary to emphasize that the optimization approach cannot eliminate all problems related to reliability-oriented design. Frequently, it is not possible for an optimization to provide simultaneous combination of the required high-frequency properties, power capabilities, and high-reliability requirements. There is a necessity for essentially new design and technological solutions. Examples of such solutions are: design of new metallization on the basis of Ti and Au; decrease of the semiconductor substrate thickness down to ~ 30 μm; use of ballasting emitter resistors in bipolar transistors; gate recess and ion implantation for n^+ contact regions in transistors on GaAs and many others [312–315].

At the present time, it is not possible to provide a complete automation of the process for reliable device design through technological guidelines, numerical simulation, or look-up table comparison. Much depends on intuition, expertise, and experience of the device and process engineers. One of the ways of gaining of such abilities comes from understanding of failure mechanisms and skills to apply physical and mathematical models toward practical solutions.

8.2 Reliability Assurance on a Production Phase

With respect to studies on reliability assurance for the stage of manufacturing, the following measures can be summarized:

(i)   control and certification of technological processes of device manufacture;
(ii) realization of qualification and screening tests, tests of structural and technological margins, estimation and reliability tests;
(iii) diagnostic inspection and control, special selection tests.

Excluding the general requirements, in the transistor manufacturing and process technology, minimization of the technological dispersion of the transistor structure parameters is important. For example, casual technological variation of metallization structure results in dispersion of the time before failure due to electromigration. The dispersion increase causes substantial growth of the failure rate. Direct correlation between reliability parameters and the distribution function for t_{BF} has been confirmed by numerical modeling [321]. Therefore, special attention is given to development of such a manufacturing process technology that would not only provide achievement of the required range of parameters, but provide minimal dispersion of these parameters too [311].

Various reliability tests are described in details the reliability literature, for example [322, 323]. Here, the focus is concentrated on a choice of the regimes of test up to failure. As a rule, industrial reliability tests are conducted in the DC regime. Tests conditions that are close to operational, for example, in a dynamic regime at operation frequency, are expensive and mainly conducted for research purposes only. For bipolar power microwave transistors, the test method in the quasi-dynamic regime could be used. In such test regime, input of the transistor is loaded on a sequence of half-sine wave current pulses with frequency from 100 kHz up to 1 MHz. This method allows receiving more authentic data about transistor reliability.

For pulse transistors, the conditions of failure event and parameters of permissible limiting regimes (pulsed values T_{pmax}, j_{pmax}, E_{pmax}) differ from those measured in the DC operation regime [324, 325]. Test methodology for these devices in the DC regime is not adequate. Therefore, the tests of pulsed transistors should be conducted in pulsed periodic regime with given pulse duration, duty factor, amplitude, and appropriate load characteristics.

The following approach for selection of the reliability test regime parameters is used. The DC regime parameters are determined by I, U, T_C. The problem is reduced to a selection of the specified parameters within the area within maximum rating regimes. Since the temperature increase accelerates practically all degradation failure mechanisms, the test regime is chosen at $T = T_{max}$, i.e., $P = P_{max}$ according to the determined case temperature T_C. The case temperature value is maintained constant in the range from T_{1C} up to T_{2C}, where T_{2C} is maximum rating for the case temperature. The choice of value T_C includes consideration of the most accelerating the regime with P_{max}. The conditions of failure mechanisms depend not only on the active region temperature T, but also on the parameters of electrical regime (I, U). For example, in a limiting case $T_C = T_{max}$, $P_{max} = 0$, in essence the storage test at elevated temperature is defined. In this case the problem of as-surance of reliability parameters does not cause any special difficulties.

Diagnostics of thermoelectrical states in transistor structures

The concept of technical diagnostics was generated in the field of mechanical en-gineering [326]. By definition, the technical diagnostic is a process of detection of the technical state of the object. The technical condition refers to a set of proper-ties with a number of quantitative and qualitative attributes. A problem of diag-nostics is not only detection of the object state, but also detection of the potential defects and malfunctions.

For semiconductor devices and transistors, the concept of absolute maximum limiting conditions is important. Above these limits, even short-term test opera-tion of the device may result in immediate failure. A "technical" state of the tran-sistor means first of all a thermoelectrical state. It becomes a limiting factor when the probability rate of catastrophic or degradation failure begins a sharp increase. Diagnostics of the limiting thermoelectrical states allows revealing defective struc-tures and ensuring stability of a given process of manufacturing technology, high production efficiency, and necessary quality level.

The conditions of physical limitations in transistor structure are formed at some boundary operation regime. In general, the boundary is similar to the boundary of safe operating area. The gap between these boundaries provides characterization of structural and technological margin of the transistor on the basic parameters $(I_{max}, P_{max}, U_{max})$. It is possible to allocate two types of limiting states.

The first type are quasi-uniform thermoelectrical states with the parameters, for example active region temperature, over some allowable level for given device structure. The second type are quasi-nonuniform thermoelectrical states related to realization of various instability mechanisms.

In the first case, the diagnostic problem is reduced to control of the basic pa-rameters in comparison with corresponding maximum ratings. In the second case, the problem is determination of the conditions and the operation regimes that may result in instability and nonuniform states, formation of hot spots, current fila-ment, microplasmas, and field domains.

For diagnostics of the thermoelectrical state in transistor structures, the follow-ing direct and indirect methods are used.

The direct methods are based on the measurement of for example infrared electron troluminescence from the active region surface [327], Transmission electron microscope TEM [328], different front- and back-side laser photo- and induced current scanning, liquid crystals and thermo indicators of melting [322], and others. These methods allow receiving somewhat evident information on the temperature distribution, current density, or electrical fields in the device structure. However, these methods as a rule require rather complex equipment or have rather limited volume capability to be efficiently used in high-volume production.

The indirect diagnostic methods are based on registration of external electro-physical characteristics, for example the method of thermal resistance (R_T) mea-measurements. This method provides estimation of the value T_{max} in a given operation regime. It is based on the measurement of one of the thermosensitive parameters of transistor structure in the pulsed regime.

In particular, this group of electrophysical methods for diagnostic control may include the detection of hot spots or current filament formation. Some methods and the equipment for indirect control of the thermal resistance of bipolar and field transistors are covered in [315, 316]. For example, the thermosensitive parameters of transistors are the voltage U_{EB} for bipolar, the gate-source voltage U_{GS} or the channel resistance for MOSFETs, and the voltage of forward-biased Schottky barrier U_{GS} for GaAs MESFETs and MODFETs.

For calibration of thermosensitive parameters, the dependence on structure temperature is first measured using external heating. Then the measurements of the thermal resistance itself are realized in the pulsed regime. During the first part of the regime, an appropriate dissipated power P_{DIS} is supplied to the device. Then the transistor is rapidly switched into a functional test mode regime. During the test mode, the measurement of the thermosensitive parameter is conducted. The delay time τ_D between the regime switching and the measurement must be much less than the time of thermal relaxation of the transistor structure.

The application area of the method is limited by deviation of the thermoelectrical state of the transistor from the quasi-uniform. Thus, essential underestimation of R_T may occur due to difference of the measured calibrated dependence for thermosensitive parameter from the real one. For example, the value $dU_{EB}/dT_{\max}\big|_{I_E}$ for bipolar transistors at hot spot formation is decreased from 1.7–1.9 mV/°C down to 1 mV/°C [329]. Without this factor taken into account, the relative error in R_T estimation can be up to 100%.

The second group of the diagnostic methods for limiting states can be demonstrated for bipolar transistors. For this device, the area of limiting regimes is mainly limited by hot spot formation. This process begins at the electrical regime that corresponds to some boundary $I_{CR}(U)$ and practically realized under no detectable changes in I–V characteristic at positive differential resistance. In this case (Chapter 3), the electrophysical characteristics: $U_{EB}\big(U_{CB}\big)\big|_{I_E}$ for quasistatic operation; $\Delta U_{EB}\big(t\big)\big|_{I_E,\bar{U}_{CB}}$ for the pulsed regime; $\Delta\tilde{U}_{EB}\big(U_{CB}\big)\big|_{t_P,Q,I_E}$ for the pulsed period regime give the information about transistor state (t_P is the pulse duration, Q the duty factor, I_E the amplitude, and $\Delta\tilde{U}_{EB}$ the parameter value at the end of heating pulse in the periodic regime). The application range and possibility of diagnostics on the basis of these characteristics are discussed in [324]. Each diagnostic characteristic has peculiarities regarding hot spot formation. For example, with respect to the characteristics $U_{EB}(U_{CB})$ and $\Delta\tilde{U}_{EB}\big(U_{CB}\big)$ at the critical regime (U_{CR}) for hot spot formation, step decrease of U_{EB} (in the case of abrupt excitation) or increase of dU_{EB}/dU_{CB} are observed (in the case of soft hot spot excitation). With respect to $\Delta U_{EB}(t)\big|_{I_E}$, the spot formation is visualized as a specific step on the monotonous decrease of U_{EB} during structure heating. Using these diagnostic characteristics, the boundary of the electrical regime can be defined for hot spot formation and further taken into account at the final SOA definition.

8.3 Estimation of Reliability Using Accelerated Tests

The methods of accelerated tests (AT) in forced regimes are widely used in manufacturing and at the development stage. The AT methods are applied in order to estimate the efficiency of changes in the device architecture or process technology at manufacturing thus providing data about the failure rate.

The idea of AT is based on creation of forced regimes that can accelerate the degradation process and provide corresponding reduction of the test time. Then the data are further used for extrapolation of long-term reliability parameters for the normal operation regime. Practical application of ATs has both provided the methods for realizations and revealed their advantages and disadvantages [330, 331].

The limitation of AT methods is the probability for false prediction. In some cases, it might be a result of simultaneous action of various degradation mechanisms, weak sensitivity of some degradation processes to the temperature activation, or difficulty of confirmation of the prediction by real reliability measurements. However, no alternate experimental methods have been found so far to estimate the long-term reliability parameters of semiconductor devices with better accuracy.

In general, the AT regime can be realized by increase of temperature in the active region both by elevation of the heat sink temperature and by dissipated power of the electrical regime due to elevated current I or voltage U. For the basic failure mechanisms, the temperature increase is an accelerating factor. The dependence of the time before failure on temperature is given by the Arrhenius expression $t_{BF} = A \exp\left(E_a/kT\right)$. The acceleration coefficient is given by

$k_A = \exp\dfrac{E_a}{k}\left(\dfrac{1}{T_1} - \dfrac{1}{T_2}\right)$, where E_a, T_1, T_2 are the values of activation energy

for major degradation processes in the device, the active region temperature in the application regime, and at AT, respectively.

The analytical dependence k_A from T_2 for various E_a can be presented. If the dominant failure mechanism with its activation energy is initiated, it is possible to choose T_2 for the forced regime that will provide an appropriate k_A for given T_1. Then it is possible to state the expected parameters of reliability in normal operation regime testing a certain set of samples in the forced regime at $T = T_2$ during the time $t_A = t_T/k_A$ where t_T is the operation time in normal regime or time before failure in application. This rather simple procedure provides reliable results only under the following conditions:

(i)   in both the forced and the normal operation regimes the long-term reli-
      ability is limited by the same physical failure mechanism;
(ii) acceleration factor shifts the distribution function of operation before
      failure without significant distortion. In other words, the average value of
      the operation time varies, but the dispersion σ remains unchanged;
(iii) the numerical parameters of the failure model (for example, E_a, T_1, T_2)
      are taken with appropriate accuracy.

For completion of the first condition, the AT regime should not exceed the
boundary of possible forcing (BPF). This implies a limitation to the possible accelerated test regimes (temperature, current, voltage) for the corresponding basic
failure mechanism. In general, BPF should satisfy the following requirements:

(iv) inside of BPF, no failures are observed due to formation of nonuniform
      thermoelectrical states;
(v)   there is no avalanche multiplication in the AT regime;
(vi) current density through the metallization cross section is not limited by
      electromigration, usually below 10^6 A/cm^2;
(vii) heat sink temperature T_C is below the maximum rating for the materials
      used in package design.

For BPF determination, the diagnostic method of steps for elevated temperature
and the methods of limiting state are used. At given BPF, the maximum value T_2
can be selected usually as high as possible and the acceleration factor k_A is determined automatically. In the transistors, the value $T_2 = P_{DIS}R_T + T_C$ is achieved by
the increase of both T_C and P_{DIS}. The P_{DIS} increase can be supplied by increase in
both I and U.

Taking into account the limitations, a preferable scenario for AT is achievement of T_2 by means of T_C increase. In this case, the major accelerating factor is
only the temperature and the value $\Delta T = T_2 - T_1$ that is rather simple to maintain
under the test conditions. At the same time at change only T_C in the DC electrical
regime (I, U) the probability of increasing the parameters over BPF is minimal.

However, choice of T_2 only at the expense of T_C increase is usually impractical
due to low maximum rating of T_C and dependence on current of a number of degradation mechanisms. Therefore, the desired increase is obtained by increase of
the test current I since the action of this factor is well studied and usually results
in acceleration of the electromigration and electrodiffusion failures of contact
metallization.

There are significant difficulties in adequate representation of the impact of
elevated voltage level U as an independent accelerating factor in various operation
regimes. The voltage increase can result in numerous charge instability phenomena. In particular, the electric field variation can result in change of the activation
energy and the coefficient A in the Arrhenius equation. In this case, interpretation
of the AT results becomes rather sophisticated.

Conclusion

The circle of physical phenomena connected with the physical limitations and failures of transistors is rather wide. Many of these phenomena have a fundamental nature. For this reason, the majority of reliability problems require deep research studies. On the other hand, failure physics is the field of knowledge that is first and foremost related to applications. It is fully determined by a subject of research—semiconductor devices. Thus, the conclusions are preliminarily oriented toward specific and practical problems. However, in this book priority is given to the physical aspects rather than to application problems. This is done because the former determines the latter.

The physics of semiconductor device limitations and failures, as a research field, is far from completion. Many general and particular problems remain unsolved. Several of them are listed below.

(viii) Further development of numerical simulation and modeling tools for failure analysis to avoid bulky and expensive experiments for estimation of the quantitative reliability parameters of real structures toward reliability-oriented and optimized design.

(ix) Further research of the failure mechanisms in compound semiconductor transistors including new materials for power application (GaN, SiC, diamond). The mathematical models of failures of these devices have insufficient experimental confirmation.

(x) Solution of the set of problems related to failures in nonuniform and defective structures. Experimental data are insufficient with regard to the critical operation regimes concerning parameters of structure defects in the device and their impact on long-term reliability parameters.

(xi) Creation of diagnostic methods and experimental tools for detection of potentially unreliable devices.

(xii) ESD device design for challenging analog applications under reduction of the limiting impact of the ESD devices on the product circuit performance.

There is no lack of other specific reliability and ESD problems that everyday practice presents for engineers and researchers in the area of reliability for new process and device development, circuit design, and field applications. We believe this book will be very useful in accomplishing these challenging goals.

References

[1]   Shockley W, Scarlet D (1963) Secondary breakdown and hot spots in power transistors. IEEE Int. Conv. Rec. Pt. 3 11:3–13.

[2]   Sze S (1981) Physics of semiconductor devices. Wiley, New York.

[3]   Mazel EZ (ed) (1985) Power high frequency transistors. Radio and Sviaz, Moscow.

[4]   Blicher A (1981) Field-effect and bipolar power transistor physics. Academic, New York.

[5]   Oxner ES (1985) Power field effect transistor and its application.

[6]   DiLorenzo JV, Kwandelwal DD (eds) (1982) GaAs FET principles and technology. Artech House, Norwood, MA.

[7]   Ashburn P (2003) SiGe heterojunction bipolar transistors. Wiley, New York.

[8]   Zaitcev AA, Savel'ev YN (1985) Generation microwave transistors. Radio and Sviaz, Moscow.

[9]   Chernyshev AA (1988) Basis of reliability of semiconductor devices and integrated circuits. Radio and Sviaz, Moscow.

[10] Miller YG (ed) (1976) Physical basis of integrated circuits reliability. Soviet Radio, Moscow.

[11] Konakova RV, Kordosh P, Tkhorik YA, et al (1986) Prognoses of reliability of semiconductor avalanche diodes. Naukova Dumka, Kiev.

[12] Charitat G (2001) Voltage handling capability and termination techniques of silicon power semiconductor devices. Proc. BCTM, 175–183.

[13] Deboy G, et al (1998) A new generation of high voltage MOSFET's breaks the limit line of silicon. Proc. IEDM, 683–685.

[14] Saggio M, Fagone D, Masumeci S (2000) Mdmesh: innovative technology for high voltage power MOSFET's. Proc. ISPSD, 65–68.

[15] Volkov AF, Kogan SM (1968) Physical phenomena in semiconductors with negative conductance. Sov. Phys. Usp. 96:633–672.

[16] Bonch-Bruevitch VL, Zviagin NP, Mironov AG (1975) Domain electrical instability in semiconductors. Nauka, Moscow.

[17] Nikolis R, Prigogine I (1979) Self-organisation in nonequilibrium systems. Mir, Moscow.

[18] Haken H (1980) Sinergetics. Mir, Moscow.

[19] Kerner BS, Osipov VV (1994) Autosolitons. A new approach to problems of self-organization and turbulence. Kluwer Academic, Dordrecht.

[20] Joss G, Jozef FL (1983) Elementary theory of stability and bifurcations. Mir, Moscow, 3000.

[21] Arnold VI (1983) Peculiarities, bifurcations and catastrophes. Sov. Phys. Usp. 141:569–590.

[22] Gurevitch AV, Minz RG (1984) Localised waves in nonuniform media. Sov. Phys. Usp. 142:61–98.

[23] Verner IV, Kopaev YV, Molotkov SN (1981) Nonequilibrium phase transition in semiconductors with negative differential conductivity. Sov. Solid-State Phys. 23:3021–3027.

[24] Nechaev AM, Sinkevitch VF (1984) Determination current stratification and isolated branches of current-voltage characteristics of semiconductor system. Sov. Phys. Semicond., 350–353.

[25] Nechaev AM, Sinkevitch VF (1983) Conditions of current filamentation in semiconductor structures with inhomogeneity. Elektron. Tekh.2:45–54.

[26] Nechaev AM (1982) About stability of current filaments in semiconductor systems with inhomogeneity. Radiotechnics and Electronics XXVII:1020–1026.

[27] Landau LD, Livshic EM (1963) Quantum mechanics. Nauka, Moscow.

[28] Kerner BS, Osipov VV (1978) Nonlinear theory of stationary strata in dissipate systems. JETP 74:1675–1697.

[29] Kerner BS, Osipov VV (1975) Theory of thermal breakdown in transistor. Radiotechnics and Electronics XX:1694–1703.

[30] Kerner BS, Osipov VV (1981) Stationary and traveling dissipative structures in active kinetic media. Sov. Microelectron. 10:407–432.

[31] Schaff HA (1967) Second breakdown—a comprehensive review. Proc. IEEE 55:1272–1288.

[32] Scarlett RM, Shockley W (1963) Secondary breakdown and hot spots in power transistors. IEEE Int. Conv. Rec. Pt. 3 11:3–13.

[33] Bergman F, Gerstner D (1963) Thermisch bedingte Stromeinschnurung bei Hochfrequenz-Leistungstransistoren. Arch. Elektron. Ubertragung 17:467–475.

[34] Sunchine RA, D'aiello RV (1975) Direct observation of the effect of solder voids on the current uniformity of power transistors. IEEE Trans Electron Devices 22:61–62.

[35] Kerner BS, Osipov VV (1989) Autosolitons. Sov. Phys. Usp. 137:201–266.

[36] Kerner BS, Osipov VV (1975) Theory of transistor thermal breakdown. Sov. Radiotechnics and Electronics XX:1694–1703.

[37] Kerner BS, Nechaev AM, Rubakha EA, Sinkevitch VF (1978) Computer calculation of current density and temperature in transistor structures (in Russian). Sov. Microelectron. 7:147–151.

[38] Sze S (1981) Physics of semiconductor devices. Wiley, New York.

[39] Kerner BS, Osipov VV (1977) Nonlinear theory of nonisothermal current filamentation in transistor structures. Sov. Microelectron. 6:337–353.

[40] Hower PL, Govil PK (1974) Comparison of one- and two dimensional models of transistor thermal instability. IEEE Trans. Electron Devices 21:617–623.

[41] Oettinger FF, Blackburn DL, Rubin S (1976) Thermal characterization of power transistors. IEEE Trans. Electron Devices 23:831–838.

[42] Kerner BS, Rubakha EA, Sinkevitch VF (1978) Analysis of current distribution in power transistor structure with inhomogeneity (in Russian). Elektron. Tekh. Ser. 2, 15–29.

[43] Mazel EZ (ed) (1985) Power high-frequency transistors. Radio and Sviaz, Moscow.

[44] Kozlov NA, Sinkevitch VF (1984) Kinetics of thermal current stratification in static and pulsed operation regimes of power transistors (in Russian). Elektron. Tekh. Ser. 2 2:35–45.

[45] Shafft HA,Schwuttke GH, Ruggles RL (1966) Second breakdown and crystallographic defects in transistors. IEEE Trans. Electron Devices 13:738–742.

[46] Aharoni H, Bar-Lev A (1975) The role of crystal defects in transistor operation at high power levels. Microelectronics 6:11–15.

[47] Vorob'ev NN, et al (1979) Influence of crystallographic defects in silicon defects in silicon on power transistor parameters (in Russian). Elektron. Tekh. Ser. 3, 95–99.

[48] Bosch G (1977) Anomalous current distributions in power transistors. Solid-State Electron. 20:635–640.

[49] Gaur SP, Navon DH (1976) Two-dimensional carrier flow in a transistor structure. IEEE Trans. Electron Devices 23:50–57.

[50] Gaur SP (1977) Safe-operating area for bipolar transistors. IBM J. Res. Dev., 433–442.

[51] Alwin VC, Navon DH, Turgeon LG (1977) Time-dependent carrier flow in a transistor structure under nonisothermal conditions. IEEE Trans. Electron Devices 24:1297–1303.

[52] Leturck P, Cavalier C (1974) A thermal model for high power devices design. Int. Electron Devices Meet., Washington, D.C. Tech. Dig. N.Y., 422–425.

[53] Kerner BS, Nechaev AM, Rubakha EA, Sinkevitch VF (1980) Kinetics of thermal filamentation at fluctuation instability in transistor structures. Sov. Radiotechnics and Electronics XXV:168–176.

[54] Nechaev AM, Rubakha EA, Sinkevitch VF (1981) Thermal filamentation in transistor structures with inhomogeneity. Sov. Radiotechnics and Electronics XXVI:1773–1782.

[55] Rabodzei AN (1981) Control of stability of power high voltage transistors to secondary breakdown. Foreign Electron. Tekh. 6:65–80.

[56] Kerner BS (1976) Peculiarities of thermal breakdown of transistor in pulse regime (in Russian). Sov. Microelectron. 5:237–267.

[57] Rabodzei AN (1981) Method of uniformity current distribution control in power high voltage transistors (in Russian). Electronic Industry 9:29–31.

[58] Nechaev AM, Sinkevitch VF, Kozlov NA (1988) Calculation of stationary thermal fields in power transistor structures (in Russian). Elektron. Tekh. Ser. 2 1:19–24.

[59] Kerner BS, Osipov VV, Sinkevitch VF (1976) Thermal breakdown of transistor in direct and alternative signal. Sov. Radiotechnics and Electronics XX:2172–2184.

[60] Vlasov VA, Sinkevitch VF (1975) Parameters of current filamentation in transistors. High school news of USSR Radioelectronics XVIII:52–55.

[61] Savel'ev YN, Aronov VL (1976) Analysis of operation of generator microwave transistor. Elektron. Tekh. Ser. 2.

[62] Kerner BS, Sinkevitch VF (1976) About limit dissipated power of transistor in periodic regime. Sov. Radiotechnics and Electronics XXI:2645–2646.

[63] Yoss G, Jozef L (1983) Elementary stability and bifurcation theory. Mir, Moscow.

[64] Blicher A (1981) Field-effect and bipolar power transistor physics. Academic, New York.

[65] Yoshida I, Okabe T, Katsueda M, et al (1980) Thermal stability and secondary breakdown in planar power MOSFET's. IEEE Trans. Electron Devices 27:395–398.

[66] Kerner BS, et al (1982) Peculiarities of breakdown of power MOSFET's with vertical channel (in Russian). Elektron. Tekh. Ser. 2 3:38–44.

[67] Nechaev AM, Sinkevitch VF, Kozlov NA (1984) Thermal filamentation in power MOSFET structures (in Russian). Elektron. Tekh. Ser. 2 1:29–38.

[68] Blackburn D (1977) Safe operating area limits for power transistors. NBS Special Publication 400–44, U.S. Department of Commerce, 1–15.

[69] Wemple SH, Niehous WC, Fukui H, et al (1981) Long-term and instantaneous burnout in GaAs power FET's: mechanisms and solutions. IEEE Trans. Electron Devices 28:834–840.

[70] Kohzu H, Nagasako I, Nakajima H, et al (1960) Reliability study of high power microwave GaAs MESFET's. NBC Res. Dev. 56:84–189.

[71] Kerner BS, Kozlov NA, Nechaev AM, Sinkevitch VF (1983) Lighting points and breakdown in GaAs transistor structure. Sov. Phys. Semicond. 17:1931–1934.

[72] Morizane K, Dosen M, Mori Y (1979) A mechanism of source-drain burnout in GaAs MESFET's. Inst. Phys. Conf. Ser. 45:287–294.

[73] Evdokimova NL, et al (1985) Thermal breakdown of power transistors on GaAs (in Russian). Elektron. Tekh. Ser. 2 2:42–50.

[74] Sunshine RA, Lampert MA (1972) Second-breakdown phenomena in avalanching silicon-on-sapphire diodes. IEEE Trans. Electron Devices 19:873–885.

[75] Martirosov IM (1967) About avalanche thermal breakdown of p-n junctions. Sov. Phys. Semicond. 1:1075.

[76] Tager AS, Vald-Perlov VM (1968) IMPATT diodes. Soviet Radio, Moscow.

[77] Evdokimova NL, et al (1986) Influence of the resistance in the gate circuit on output I-V characteristic of GaAs transistors. Elektron. Tekh. Ser. 2 5:118–120.

[78] Sze S (1981) Physics of semiconductor devices. Wiley, New York.

[79] Tager AS, Vald-Perlov VM (1968) IMPATT diodes. Soviet Radio, Moscow.

[80] Grekhov IV, Serezhkin YN (1980) Avalanche breakdown of p-n junction in semiconductors. Energiya, Leningrad.

[81] Purwins H-G, Radehaus C, Berkemeier J (1988) Experimental investigation of spatial pattern formation in physical systems of activator inhibitor type. Z. Naturforsch. 43a:17–29.

[82] Hower PL, Reddi VGK (1970) Avalanche injection and second breakdown in transistors. IEEE Trans. Electron Devices 17:320–335.

[83] Hane K, Suzuki T (1978) Effect of external resistance in the vicinity of current-mode second breakdown in Si PNN⁺ structures. Jpn. J. Appl. Phys. 17:857–864.

[84] Egawa H (1966) Avalanche characteristics and failure mechanism of high voltage diodes. IEEE Trans. Electron Devices 13:754–758.

[85] Grutchfield HB, Moutoux TJ (1966) Current mode second breakdown in epitaxial planar transistors. IEEE Trans. Electron Devices 13:743–747.

[86] Beatty BA, Krishna S, Adler JS (1976) Second breakdown in power transistors due to avalanche injection. IEEE Trans. Electron Devices 23:851–857.

[87] Kerner BS, Osipov VV (1975) Theory of thermal breakdown of transistor (in Russian). Sov. Radiotechnics and Electronics XX:1694–1703.

[88] Sunshine RA, Lampert MA (1972) Second-breakdown phenomena in avalanching silicon-on-sapphire diodes. IEEE Trans. Electron Devices 19:873–885.

[89] Petrov BK (1976) Effect of avalanche injection on secondary breakdown in drift transistors (in Russian). Sov. Radiotechnics and Electronics XXI: 2601–2607.

[90] Aronov VL, Rubakha EA, Savina AS, Sinkevitch VF (1980) Investigation of reliability of power microwave transistors in dynamic regimes (in Russian). Microelectronics and semiconductor devices: Collection of papers. Vasenkov AA, Fedotov YA (eds). Soviet Radio, Moscow, 5:117–131.

[91] Blicher A (1981) Field-effect and bipolar power transistor physics. Academic, New York.

[92] Asakawa T, Tsubouchi N (1966) Second breakdown in MOS transistors. IEEE Trans. Electron Devices 13:811–812.

[93] Nakakiri M, Lids K (1977) Damage introduced by second breakdown in N-channel MOS devices. Jpn. J. Appl. Phys. 7:1187–1193.

[94] Kennedy DP, Phillips A (1973) Source-drain breakdown in an insulated gate field effect transistors. Int. Electron Devices Meet., Washington, D.C.

[95] Yoshida I, Okabe T, Katsueda M, et al (1980) Thermal stability and secondary breakdown in planar power MOSFET's. IEEE Trans. Electron Devices 27:395–398.

[96] Toyabe T, Yamaguchi K, Asai S, Mock MS (1978) A numerical model of avalanche breakdown in MOSFET's. IEEE Trans. Electron Devices, 25.

[97] Ochi S,Toshiba I, Okabe T, Nagata M (1980) Computer analysis of breakdown mechanism in planar MOSFET's. IEEE Trans. Electron Devices 27: 399–400.

[98] Kozlov NA, Nechaev AM, Sinkevitch VF (1982) Investigation of isothermal current filamentation power MOSFET's (in Russian). Elektron. Tekh. Ser. 2 2:24–34.

[99] Hsu FG, Ko PK, Tam S, et al (1982) An analytical breakdown model for short-channel MOSFET's. IEEE Trans. Electron Devices 29:1735–1739.

[100] Diakonov VP (1973) Avalanche transistors and its application in pulse units (in Russian). Soviet Radio, Moscow.

[101] Kardo-Sysoev AF, Chashnikov IG (1977) Switching at impact ionization in semiconductor (in Russian). Sov. Phys. Semicond., 11.

[102] Kozlov NA, Nechaev AM, Romanko MT, Sinkevitch VF (1985) Electroluminescence in field effect transistor structure (in Russian). Elektron. Tekh. Ser. 2 3:22–33.

[103] Hu C, Chi M (1982) Second breakdown of vertical power MOSFET's. IEEE Trans. Electron Devices 29:1287–1293.

[104] Tranduc H, Rosse L, Sanchez JL (1985) Safe operating area of the MOS'T : second breakdown limitations. Physica 129B+C:286–290.

[105] Ezhov VS, Sinkevitch VF (1984) Injection breakdown in power MOSFET's with vertical channel (in Russian). Elektron. Tekh. Ser. 2 5:68–76.

[106] Russ C, Verhaege K, Bock K, Groeseneken G, Maes HE (1996) Microelectron. Reliab. 36:1739.

[107] Ker MD, Wu CN (1996) Microelectron. Reliab. 36:1726.

[108] Meneghesso G, Luchies JRM, Kuper FG, Mouthaan AJ (1996) Microelectron. Reliab. 36:1726.

[109] Hower PL, Reddy VGK (1970) IEEE Trans. Electron Devices 17:320.

[110] Kennedy DP, Phillips A (1973) Int. Electron Devices Meet., Washington, D.C.

[111] Nakageri M, Iida R (1977), Jpn. J. Appl. Phys. 16:1187.

[112] Toyabe T, Tamaguchi K, Asai S (1978) IEEE Trans. Electron Devices 25:825.

[113] Toshida I, Okabe T, Katsuda M (1980) IEEE Trans. Electron Devices 27:395.

[114] Vashchenko VA, Martynov YB, Sinkevitch VF, Tager AS (1996) Solid-State Electron. 39:1027.

[115] Vashchenko VA, Kozlov NA, Martynov YB, Sinkevitch VF, Tager AS (1996) IEEE Trans. Electron Devices 43:513.

[116] Vashchenko VA, Sinkevitch VF (1996) Solid-State Electron. 39:851.

[117] Golant EI, Martynov YB (1992) Elektron. Tekh. 466:59.

[118] Grant WN (1973) Solid-State Electron. 16:1188.

[119] Scholl E (1987) Nonequilibrium phase transitions in semiconductors: Self-organization induced by generation and recombination rocesses. Springer-Verlag, Heidelberg.

[120] Vashchenko VA, Martynov YB, Sinkevitch VF, Tager AS (1996) Microelectron. Reliab. 36:1887.

[121] Vashchenko VA, Martynov YB, Sinkevitch VF (1997) Solid-State Electron. 41:1761.

[122] Vashchenko VA, Martynov YB, Sinkevitch VF (1996) Proc. 21st Int. Conf. on Compound Semiconductors.

[123] Gaa M, Kunz RE, Scholl E (1996) Phys. Rev. B 53:15971.

[124] Kerner BS, Nechaev AM, Rubacha EA, Sinkevitch VF (1980) Sov. Radiotechnics and Electronics XXV:168.

[125] Yamaguchi K, Asai S, Kodera H (1976) Two-dimensional numerical analysis of stability criteria of FET's. IEEE Trans. Electron Devices 23:1283–1290.

[126] Fjedly TA, Johannessen JS (1983) Negative differential resistance in GaAs MESFET's. Electron. Lett. 17:649–650.

[127] Levinshtein ME, Pozhela UC, Shur MS (1975) Gunn effect (in Russian). Soviet Radio, Moscow.

[128] Grubin HL, Ferry DK, Cleason KR (1980) Spontaneous oscillations in gate field effect transistors. Solid-State Electron. 23:157–172.

[129] Frensley WR (1981) Power-limiting breakdown effects in GaAs MESFET's. IEEE Trans. Electron Devices 28:962–970.

[130] Poresh SB, Tager AS (1983) Anode static domain formation in Gunn diodes with nonlocal dependence velocity on fields. Sov. Radiotechnics and Electronics XXVIII:2448–2451.

[131] Wemple SH, Neihaus WC, Cox HM, et al (1980) Control of gate-drain avalanche in GaAs MESFET's. IEEE Trans. Electron Devices 27:1013–1018.

[132] David JPR, Sitch JE, Stern MS (1982) Gate-drain avalanche breakdown in GaAs power MESFET's. IEEE Trans. Electron Devices 29:1548–1552.

[133] Barton TM, Ladbrooke PH (1986) The role of the device surface in the high voltage behavior of the GaAs MESFET. Solid-State Electron. 20:807–813.

[134] Mizuta H, Yamaguchi K, Takahashi S (1987) Surface potential effect on gate-drain avalanche breakdown in GaAs MESFET's. IEEE Trans. Electron Devices 34:2027–2033.

[135] Cook RK, Frey J (1982) Two-dimensional numerical simulation of energy transport effects in Si and GaAs MESFET's. IEEE Trans. Electron Devices 29:970–977.

[136] Wroblewski R, Salmer G, Crosnier Y (1983) Theoretical analysis of the DC avalanche breakdown in GaAs MESFET's. IEEE Trans. Electron Devices 30:154–199.

[137] Buot FA, Sleger KJ (1984) Numerical simulation of hot-electron effects in source-drain burnout characteristics of GaAs power FET's. Solid-State Electron. 27:1067–1081.

[138] Garber GZ (1988) Two-dimensional nonlocal modeling of field effect transistors with Schottky gate on GaAs (in Russian). Sov. Rev. Electron. Tech. Ser. 2 3:36.

[139] Yamamoto R, Higashisaka A, Hasegawa F (1978) Light emission and burnout characteristics of GaAs power MESFET's. IEEE Trans. Electron Devices 25:567–573.

[140] Furutsuka T, Higashisaka A, Aono Y, et al (1979) GaAs power MESFET's with a grade recess structure. Electron. Lett. 15:417–418.

[141] Dumas JM, Paugam J, Le Mouellic C, et al (1983) Long term degradation of GaAs power MESFET's induced by surface effects. 21st Annu. Proc. Reliab. Phys. Symp., 226–228.

[142] Tsironis C (1980) Prebreakdown phenomenon in GaAs epitaxial layers and FET's . IEEE Trans. Electron Devices 27:277–282.

[143] Kocot C, Stolte CA (1982) Backgating in GaAs MESFET's. IEEE Trans. Microwave Theory Tech. 30:963–968.

[144] Goronkin H, Vaitkus RL (1981) Impact ionization of traps in ion implanted GaAs MESFET's. Int. Symp. on GaAs and Related Compounds, 287–292.

[145] Lemnios ZJ, Lau CL, Shade GF, et al (1982) Buffer layer material limitation for high power GaAs FET's. Int. Symp. on GaAs and Related Compounds, 371–378.

[146] Wemple SH, Neihaus WC (1976) Source-drain burn-out in GaAs MESFET's. Int. Symp. on GaAs and Related Compounds 33b:254–261.

[147] Kerner BS, Kozlov NA, Nechaev AM, Sinkevitch VF (1983) Study of breakdown mechanisms in the GaAs MESFET structure (in Russian). Sov. Microelectron. 12:217–224.

[148] Whalen J, Kemerley RT, Rastefano E (1982) X-band burnout characteristics of GaAs MESFET's. IEEE Trans. Microwave Theory Tech. 30:2206–2211.

[149] Wemple SH, Niehous WC, Fukui H, et al (1981) Long-term and instantaneous burnout in GaAs power FET's: mechanisms and solutions. IEEE Trans. Electron Devices 28:834–840.

[150] Tiwary S, Eastman LF, Rathbun L (1980) Physical and material limitations on burnout voltage of GaAs power MESFET's. IEEE Trans. Electron Devices 27:1045–1654.

[151] Whalen J, Calcatera N, Thorn M (1979) Microwave nanosecond pulse burnout properties of GaAs MESFET's. IEEE Trans. Microwave Theory Tech. 27:1026–1031.

[152] Wemple SM, Steinberger ML, Schlosser WO (1980) Relationship between power added efficiency and gate-drain avalanche in GaAs MESFET's. Electron. Lett. 16:459–460.

[153] Finlay HJ, Roberta BD (1985) Improvements in receiver RF burnout characteristics and reduction of post overload degradations in low noise GaAs FET's. Advances in FET Technology, 151–156.

[154] Kozlov NA, Sinkevitch VF, Vashchenko VA (1992) Isothermal current instability and local breakdown in GaAs FET. Electron. Lett. 28:1265–1267.

[155] Vashchenko VA, Sinkevitch VF (1996) Current instability and burnout of HEMT structures. Solid-State Electron. 39:851–856.

[156] Vashchenko VA, Kozlov NA, Martynov YB, Sinkevitch VF, Tager AS (1996) Negative differential conductivity and isothermal drain breakdown of the GaAs MESFET. IEEE Trans. Electron Devices 43:513–518.

[157] Vashchenko VA, Martynov YB, Sinkevitch VF (1996) Physical limitation on drain voltage of power PM HEMT. Microelectronics and Reliability.

[158] Vashchenko VA, Martynov YB, Sinkevitch VF, Tager AS (1996) Current instability in GaAs n^+-i-n^+ structures as a limitation of the maximum drain voltage of power MESFET's. Solid-State Electron. 39:1027–1031.

[159] Vashchenko VA, Martynov YB, Sinkevitch VF (1997) Simulation of multiple filaments in GaAs structures. Solid-State Electron. 41:75–80.

[160] DiLorenzo JV, Kwandelwal DD (eds) (1982) GaAs FET principles and technology. Artech House, Norwood, MA.

[161] Wada Y, Tomizawa M (1988) Drain avalanche breakdown in gallium arsenide MESFET's. IEEE Trans. Electron Devices 35:1765–1770.

[162] Yamamoto R, Higashisaka A, Hasegawa F (1978) Light emission and burnout characteristics of GaAs power MESFET's. IEEE Trans. Electron Devices 25:567–573.

[163] Jantsch W, Heinrich H (1970) Rev. Sci. Instrum. 41:228.

[164] Fainberg VI (1980) Poluprovodnikovaya tehnika i mikroelektronika (in Russian), 53.

[165] Crosnier Y (1994) Proc. Eur. Microwave Conf. Nexus, 88.

[166] Shigekawa N, Enoki T, Furuta T, Ito H (1995) IEEE Electron Devices Lett. 16:515.

[167] Zanoni E, Manfredi M, Bigliardi S, et al (1992) IEEE Trans. Electron Devices 39:1849.

[168] Tedesco C, Zanoni E, Canali C, et al (1993) IEEE Trans. Electron Devices 40:1211.

[169] Temcamani F, Crosnier Y, Lippens D, Salmer G (1990) Microwave Opt. Technol. Lett. 3:195.

[170] Chau HF, Pavlidis P, Tomizawa K (1991) IEEE Trans. Electron Devices 38:213.

[171] Bahl SR, del Alamo JA, Dickmann J, Schildberg S (1995) IEEE Trans. Electron Devices 42:15.

[172] Geiger G, Dickmann J, Wolk C, Kohn E (1995) IEEE Electron Devices Lett. 16:30.

[173] Stratton R (1962) Diffusion of hot and cold electrons in semiconductor barriers. Phys. Rev. 126:2002–2014.

[174] Widiger DJ, Kizilyalli IC, Hess K, Goleman JJ (1985) Two-dimensional transient simulation of an idealized high electron mobility transistor. IEEE Trans. Electron Devices 32:1092–1102.

[175] Blotekjaer K (1970) Transport equations for electrons in two-valley semiconductors. IEEE Trans. Electron Devices 17:38–47.

[176] Shimizu A, Koshimizu T (1981) Avalanche breakdown voltage of hyper abrupt junction. Solid-State Electron. 24:1155–1160.

[177] Shockley W, Read W (1951) Statistics of the recombination of holes and electrons. Phys. Rev. 87:835–842.

[178] Carnez B, Cappy A, Kaszynski A et al (1980) Modeling of non stationary electron dynamics. J. Appl. Phys. 51:784–790.

[179] Garmatin AV (1985)The program on the base of the Monte Carlo modeling of the non stationary process of the heating of the electrons by the electric field. Sov. Electron. Eng. Ser. 1 3:66–68.

[180] Engl WL, Dirks HK, Meinerzhagen B (1983) Device modeling. Proc. IEEE 71:10–38.

[181] Snowden CM, Loret D (1987) Two-dimensional hot-electron models for short-gate-length GaAs MESFETs. IEEE Trans. Electron Devices 34:212–223.

[182] Blatt FJ (1968) Physics of electronic conduction in solids. McGraw-Hill, New York.

[183] Golant EI, Martynov YB (1992) Fully conservative, absolutely stable finite-difference scheme for solution of the non stationary problems of the theory of semiconductor devices (in Russian). Sov. Electron. Eng. Ser. 1 2:59–63.

[184] Lampert MA, Mark P (1970) Current injection in solids. Academic, New York.

[185] Bonch-Bruevitch VL, Zviagin IP, Mironov AG (1972) Domain electrical instability in semiconductors. Nauka, Moscow.

[186] Shigekawa N, Enoki T, Furuta T, Ito H (1995) Electroluminescence of InAlAs/InGaAs HEMT's lattice-matched to InP substrates. IEEE Electron Devices Lett. 16:315–317.

[187] Sutherland JE, Hauser JR (1977) A computer analysis of heterojunction and graded composition solar cells. IEEE Trans. Electron Devices 24:363–372.

[188] Semicad Device (1994) Dawn Technologies, Inc., Sunnyvale, CA.

[189] Scholl E (1987) Nonequilibrium phase transition in semiconductors: Self-organization induced by generation and recombination processes. Springer-Verlag, Heidelberg.

[190] Mayer KM, Parisi J, Huebener RP (1988) Z. Phys. B 71:171.

[191] Niedernostheide FJ, Arps M, Dohmen R, et al (1992) Phys Status Solidi (b) 172:249.

[192] Vashchenko VA, Kerner BS, Osipov VV, Sinkevitch VF (1990) Sov. Phys. Semicond. 24:1065.

[193] Vashchenko VA, Vodakov YA, Gafiichuc VV, et al (1991) Sov. Phys. Semicond. 25:730.

[194] Kerner BS, Osipov VV (1990) Sov. Phys. Usp. 33:679.

[195] Anderson WT, Buot FA, Christou A (1986) High power pulse reliability of GaAs power FETs. IEEE/IRPS, 144–149.

[196] Franklin AJ, Dwyer VM, Campbell DS (1990) Thermal breakdown in GaAs MES diodes. Solid-State Electron. 33:1055–1064.

[197] Franklin AJ, Dwyer VM, Campbell DS (1990) Thermal failure in GaAs semiconductor devices. Solid-State Electron. 33:553–560.

[198] Vashchenko VA, Martynov YB, Sinkevitch VF, Tager AS (1996) Simulation of the gate burnout of GaAs MESFET. Microelectron Reliab. 36:1887–1890.

[199] Vashchenko VA, Martynov YB, Sinkevitch VF, Tager AS (1996) Electrical current instability at gate breakdown in GaAs MESFET. IEEE Trans. Electron Devices, 43.

[200] Vashchenko VA, Martynov YB, Sinkevitch VF (1996) Simulation of current filaments in GaAs structures. Proc. Int. Conf. on Compound Semiconductors, Sankt-Petersburg.

[201] Scholl E (1987) Nonequilibrium phase transition in semiconductors: Self-organization induced by generation and recombination processes. Springer-Verlag, Heidelberg.

[202] Vashchenko VV, Kerner BS, Osipov VV, Sinkevitch VF (1990) Excitation and evolution of microplasmas acting spike autosolitons in silicon p-i-n structures. Sov. Phys. Semicond. 24:1065–1066.

[203] Tager AS, Vald-Perlov VM (1968) IMPATT diodes. Soviet Radio, Moscow.

[204] Bowers RA (1970) Space-charge induced negative resistance in avalanche diodes. IEEE Trans. Electron Devices 15:343–350.

[205] Muller MW (1968) Current filaments in avalanching p-i-n diodes. Appl. Phys. Lett. 12:218–219.

[206] Konakova RV, et al (1986) Prognostication of Sem. Avalanche Diodes Reliability. Naukova Dumka, Kiev.

[207] Gafiichuk VV, Datsko BI, Kerner BS, Osipov VV (1990) Sov. Phys. Semicond. 24:724–730.

[208] Kerner BS, Kozlov NA, Nechaev AM, Sinkevitch VF (1983) Lighting points and breakdown in transistor structure. Sov. Phys. Semicond. 17:1931–1934; Kerner BS, Sinkevitch VF (1982) Multifilament and multidomain stationary states in hot electron-hole plasma. JETP Lett. 36:359–362.

[209] Kerner BS, Kozlov NA, Nechaev AM, Sinkevitch VF (1983) Investigation of breakdown mechanism in GaAs field effect transistor structure. Sov. Microelectron. 12:217–224.

[210] Tsironis C (1980) Microplasma effect in gallium arsenide epilayers and FET's. Solid-State Electron. 23:249–254.

[211] Kerner BS, Kozlov NA, Nechaev AM, Sinkevitch VF (1983) Study of breakdown mechanisms in the GaAs MESFET structure (in Russian). Sov. Microelectron. 12:217–224.

[212] Grekhov IV, Serezhkin YN (1980) Avalanche breakdown of p-n junction in semiconductors. Energia, Leningrad.

[213] Kerner BS, Osipov VV (1990) Sov. Phys. Usp. 33:679.

[214] Purwins HG, Radehaus C, Berkemeier J (1988) Experimental investigation of spatial pattern formation in physical systems of activator inhibitor type. Z. Naturforsch. 43:17–29.

[215] Vashchenko VA, Sinkevitch VF (1996) Current instability and burnout of HEMT structures. Solid-State Electron. 36:851–856.

[216] Mayer KM, Parisi J, Huebener RP (1988) Z. Phys. B 71:171.

[217] Niedernostheide FJ, Arps M, Dohmen R, et al (1992) Phys. Status Solidi (b) 172:249.

[218] Vashchenko VA, Kerner BS, Osipov VV, Sinkevitch VF (1990) Excitation and evolution of microplasmas acting spike autosolitons in silicon p-i-n structures. Sov. Phys. Semicond. 24:1065–1066.

[219] Vashchenko VA, Vodakov YA, Gafiichuc VV, et al (1991) Sov. Phys. Semicond. 25:730.

[220] Vashchenko VA, Martynov YB, Sinkevitch VF (1997) Simulation of multiple filaments in GaAs structures. Solid-State Electron. 41:75–80.

[221] Vashchenko VA, Martynov YB, Sinkevitch VF, Tager AS (1996) Current instability in GaAs n^+-i-n^+ structures as a limitation of the maximum drain voltage of power MESFET's. Solid-State Electron. 39:1027–1031.

[222] Vashchenko VA, Martynov YB, Sinkevitch VF, Tager AS (1995) Electrical current instability at gate breakdown in GaAs MESFET. IEEE Trans. Electron Devices 43:12.

[223] Vashchenko VA, Martynov YB, Sinkevitch VF (1996) Simulation of current filaments in GaAs structures. Proc. Int. Conf. on Compound Semiconductors, Sankt-Petersburg.

[224] David JPR, Sitch JE, Stern MS (1982) Gate-drain avalanche breakdown in GaAs power MESFET's. IEEE Trans. Electron Devices 29:1548–1552.

[225] Barton TM, Ladbrooke PH (1986) The role of the device surface in the high voltage behavior of the GaAs MESFET's. Solid-State Electron. 29:807–813.

[226] Mizuta H, Yamaguchi K, Takahashi S (1987) Surface potential effect on gate-drain avalanche breakdown in GaAs MESFET's. IEEE Trans. Electron Devices 34:2027–2033.

[227] Wroblewski R, Salmer G, Crosnier Y (1983) Theoretical analysis of the DC avalanche breakdown in GaAs MESFET's. IEEE Trans. Electron Devices 30:154–199.

[228] Gribnikov ZS (1977) About one instability mechanism of uniform avalanche breakdown of Schottky layers. Sov. Phys. Semicond. 11:2111–2117.

[229] Kozlov NA, Nechaev AM, Romanko MT, Sinkevitch VF (1985) Electroluminescence in field effect transistor structure (in Russian). Sov. Electron. Eng. Ser. 2 3:22–33.

[230] Sudzilovsky VY (1973) Impact ionization in Gunn diodes. Domain form and type of its motion. Sov. Phys. Semicond. 7:563–371.

[231] Macksey HM, Tserng HQ (1982) S-band GaAs power FET with semiinsulating gate. Int. Symp. on GaAs and Related Compounds, Albuquerque, 371–378.

[232] Fukui H, Wemple SM, Irvin JC, et al (1982) Reliability of power GaAs field-effect transistors. IEEE Trans. Electron Devices 29:395–401.

[233] Furutsuka T, Tsuli T, Hasegawa F (1978) Improvement of the drain break-down voltage of GaAs power MESFET's by a simple recess structure. IEEE Trans. Electron Devices 25:563–567.

[234] Fix VB (1969) Ion conductance in metals and semiconductors (in Russian). Nauka, Moscow.

[235] Koleshko VM, Belizkiy VF (1982) Mass-transport in thin films (in Russian). Nauka and Tekhnika, Minsk.

[236] Vavilov VS, Kiv AE, Niyazova RR (1981) Mechanisms of formation and migration of defects in semiconductors (in Russian). Nauka, Moscow.

[237] Litovchenko VG, Gorban AP (1978) Annals of physics of microelectronic system metal-dielectric-semiconductor (in Russian). Naukova Dumka, Kiev.

[238] Black JR (1969) Electromigration failure modes in aluminum metallization for semiconductor devices. Proc. IEEE 57:1587–1594.

[239] D'Orl (1972) Electrodiffusion and failures in electronics (paper collection). Mir, Moscow, 35–46.

[240] Rubakha EA, Sinkevitch VF (1976) Failures of power microwave transistors due to electrodiffusion (in Russian). Elektron. Tekh. Ser. 2 7:119–128.

[241] Attardo HJ, Rutledke R, Jack RC (1971) Statistical metallurgical model for electromigration failure in aluminum thin-film conductors. J. Appl. Phys. 42:4343–4349.

[242] Alexanian IT, Krivoshapko VM, Rubanik YT (1976) Modeling of electromigration failures of IC elements on computer (in Russian). Elektron. Tekh. Ser. 8 6:39–47.

[243] Soloviev VN, Sinkevitch VF, Liaduna GA (1985) About electrodiffusion failure micro-mechanisms of thin film metallization. Sov. J. Tech. Phys. 55:348–353.

[244] Dikovsky VI, Zhuravlev VK, Kausova AI, Sinkevitch VF (1975) Stability to electromigration of aluminum metallization in microwave transistors (in Russian). Elektron. Promst. 10:27–33.

[245] Wada T, Sukimoto M, Ajiki T (1988) Electromigration in bipolar NPN transistors. Solid-State Electron. 31:1409–1412.

[246] Nechaev AM, Rubakha EA, Sinkevitch VF (1978) Failure mechanisms and reliability of power microwave transistors (in Russian). Rev. Elektron. Tekh. Ser. 2 10.

[247] Learn AJ (1971) Effect of redundant microstructure on electromigration-induced failure. Appl. Phys. Lett. 19:292–295.

[248] Kozlov NA, Nechaev AM, Sinkevitch VF (1984) Electrodiffusion failures of field-effect transistors. Elektron. Tekh.Ser. 2 2:72–77.

[249] Irvin JC, Loya A (1978) Failure mechanisms and reliability of low-noiseGaAs FET'S. Bell Syst. Tech. J. 57:2823–2846.

[250] Abbot DA, Turner JA (1976) Some aspects of GaAs MESFET's reliability. IEEE Trans. Microwave Theory Tech. 24:317–321.

[251] Nechaev AM, Sinkevitch VF, Sokolova EI, et al (1981) Failure mechanisms and reliability GaAs transistors (in Russian). Rev. Elektron. Tekh. Ser. 2 2.

[252] Canali C, Fantini F, Scorzoni A, et al (1987) Degradation mechanisms induced by high current density in AI-gate GaAs MESFET's. IEEE Trans. Electron Devices 34:205–211.

[253] Black JR (1970) RF power transistor metallization failure. IEEE Trans. Electron Devices 17:800–803.

[254] Alexanian IT, Grigoroshvili YE, Simonov AN, et al (1976) Diffusion process in two-layer film systems Au-Al, Ti-Au, Cr-Au, V-Au, Cr-Cu, V-Cu, Ti-Cu (in Russian). Elektron. Tekh. Ser. Mater.7:21–25.

[255] Behmann F (1979). Reliability assessment of small signal GaAs FET's. Microelectron. Reliab. 19:107–115.

[256] Kozlov NA, Piskarev AB, Sinkevitch VF, et al (1983) Degradation processes in GaAs MESFET structure and ways of its reliability increase. Elektron. Tekh. Ser. 2 5.

[257] Mizuishi K, Kmrono H, Sato H, Kodera H (1979) Degradation mechanism of GaAs MESFET's. IEEE Trans. Electron Devices 26:1008–1614.

[258] Vertoprakhov VN, Kuchumov BN, Sal'man EG (1981) Structure and properties of Si-Si0₂-M structures. Nauka, Moscow.

[259] Hofstein SR (1967) Proton and sodium motion in SiO₂ films. IEEE Trans. Electron Devices 14:749–759.

[260] Perevedenzev AV, Sopov OV, Nogin VM (1978) New understanding of phenomena of boundary voltage instability in MOSFET's at negative gate voltage (in Russian). Elektron. Tekh. Ser. 2 5–6:117–135.

[261] Koatylev SA, Prokhorov EF, Ukolov AT (1986) Effect of semiinsulating substrate on parameters GaAs MESFET's (in Russian). Rev. Elektron.Tekh. Ser. 1 7.

[262] Vashchenko VA, Kozlov NA, Sinkevitch VF (1990) Identification of instability mechanisms of GaAs MESFET's (in Russian). Elektron. Tekh. Ser. 2.

[263] Itoh H, Ohata K, Hasegawa F (1991) Influence of the surface and episubstrate interface on the drain current drift of GaAs MESFET's. IEEE Trans. Electron Devices 28:871–882.

[264] Itoh T, Janai H (1980) Stability of performance and interfacial problems in GaAs MESFET's. IEEE Trans. Electron Devices 27:1037–1045.

[265] Gesch H, Leburton JP, Dorda GE (1982) Generation of interface states by hot hole injection in MOSFET's. IEEE Trans. Electron Devices 29:913–918.

[266] Bhattacharyya A, Shabde SN (1986) Degradation of short-channel MOSFET's under constant current stress across gate and drain. IEEE Trans. Electron Devices 33:1329–1333.

[267] Tsuchiya T, Kobayashi T, Nakajima S (1987) Hot-carrier-injected oxide region and hot-electron trapping as the main cause in Si nMOSFET degradation. IEEE Trans. Electron Devices 34:386–390.

[268] Qiu-Yi Y, Zrenner A, Koch F (1983) Interface degradation in Si-metal-oxide-semiconductor structures by homogeneous, microwave heating of channel carriers. Appl. Phys. Lett. 52:561–563.

[269] Mylkin AA, Prustlina S, Rubakha EA, et al (1979) Study of failure and life mechanisms of power microwave transistors at tests in dynamic regimes at frequency 1 GHz. Elektron. Tekhn. Ser. 2 9:104–115.

[270] Rubakha EA, Ivanova LI (1983) Peculiarities of failure mechanisms and reliability of power pulse transistors. Elektron. Tekh. Ser. 2 7:43–52.

[271] Mimura T, Fukuta M, Matayoshi H (1972) Al conductor failure due to thermal expansion by power pulses. IEEE Trans. Electron Devices 19:696–697.

[272] Wemple SH, Niehous WC, Fukui H, et al (1981) Long-term and instantaneous burnout in GaAs power FET's: mechanisms and solutions. IEEE Trans. Electron Devices 28:834–840.

[273] Buslaev SU, Kleinfeld YS, Sinkevitch VF (1985) Dislocation effect on limit operation regime of silicon structures (in Russian). Elektron. Tekh. Ser. 8 3:19–21.

[274] Kozlov NA, Nechaev AM, Sinkevitch VF (1988) Channel current filamentation at charge instability in MOS-structures. Elektron. Tekh. Ser. 2 1:20–23.

[275] Amerasekera A, Duvury C (1995) ESD in silicon integrated circuits. Wiley, New York.

[276] Dabral S, Maloney TJ (1998) Basic ESD and I/O design. Wiley, New York.

[277] Proceedings of EOSESD Symposiums (1997–2007).

[278] Estmark K, Gossner H, Stadler W (2003) Advanced simulation methods for ESD protection. Elsevier, Amsterdam.

[279] Merrill R, Issaq E (1993) ESD design methodology. Proc. ESD/EOS Symp., 233–237.

[280] Avery L (1983) Using SCR's as transient protection structures in integrated circuits. Proc. ESD/EOS Symp., 27–29.

[281] Vashchenko VA, Hopper P (2005) Bipolar SCR ESD devices. Microelectron. Reliab. J., 457–471.

[282] Voldman SH (2006) ESD RF technology and circuits. Wiley, New York.

[283] Voldman SH (2004) ESD physics and devices. Wiley, New York.

[284] Voldman SH (2005) ESD circuits and devices. Wiley, New York.

[285] Proceedings of 1st International ESD Workshop (2007) South Lake Tahoe.

[286] Stadler W (2007) State-of-the-art in ESD standards. 1st Int. ESD Workshop, South Lake Tahoe, 127–147.

[287] HBM MIL883-STD, method 3015.7

[288] HBM-ESD Sensitivity Testing: Human Body Model (HBM)-Component Level, ESD Association Standard, S5.1 1994.

[289] MM-ESD Sensitivity Testing: Machine Model (MM)-Component Level, ESD Association Standard, S5.2 1994.

[290] IEC 61000-4-2 Specification.

[291] Gossner H, Muller-Lynch T, Esmark K, Stecher M (1999) Wide range control of the sustaining voltage of ESD protection elements realized in a smart power technology. Proc. ESD/EOS Symp., 19–27.

[292] Vashchenko VA, Concannon A, Ter Beek M, Hopper P (2002) Emitter injection control in LVTSCR for latch-up free ESD protection. Proc. MIEL 2:741.

[293] Ker MD, Chang H (2000) Cascaded LVTSCR with tuneable holding voltage for ESD protection in bulk CMOS technology without latchup danger. Solid-State Electron. 44:425–445.

[294] Salcedo J, Liou J, Bernier J (2005) Design and integration of novel SCR-based devices for ESD protection in CMOS/BiCMOS technologies. IEEE Trans. Electron Devices 52:2682–2689.

[295] Vashchenko VA, Concannon A, ter Beek A, Hopper P (2002) Technology CAD evaluation of BiCMOS protection structures operation including spatial thermal runaway. Proc. EOS/ESD Symp, 101–110.

[296] MEDICI Two-Dimensional Device Simulation Program, Synopsys, 2002.4.0; "TSUPREM-4" "Two-Dimensional Process Simulation Program," Versions 2002.4.0. Synopsys, 2002.

[297] Vashchenko VA, Martynov YB, Sinkevitch VF (1997) Electrical instability and filamentation in ggMOS protection structures. Solid-State Electron. 41:1761–1767.

[298] Vashchenko VA, Martynov YB, Sinkevitch VF (1996) Negative differential conductivity and isothermal drain breakdown of the GaAs MESFET. IEEE Trans. Electron Devices 43:513–518.

[299] Vashchenko VA, Hopper P (2006) A new principle for a self-protecting transistor array design. ISPSD, IEEE, 1–4; (2006) Self-protecting arrays for open drain circuit. IRPS, 637–638.

[300] Vashchenko VA, Kuznetsov V, Hopper P (2007) ESD protection of fast transient pins in bipolar processes. Proc. BCTM.

[301] Grutchfield HB, Moutoux TJ (1966) IEEE Trans. Electron Devices 13:743–747.

[302] Bakeroot B, Doutreloigne J, Vanmeerbeek P, Moens P (2007) An ultrafast and latch-up free lateral IGBT with hole diverter for junction-isolated technologies. Proc. ISPSD, 21–24.

[303] Vashchenko VA, Olson N, Farrenkopf D, Kuznetsov V, Hopper P, Rosenbaum E (2007) New concept of mixed device-circuit solution for ESD protection of high-voltage fast pins. Proc. IRPS, 602–603.

[304] Vashchenko VA, Farrenkopf D, Hopper P (2007) Active control of the triggering characteristics of NPN BJT, BSCR and NLDMOS-SCR device. Proc. ISPSD, 41–44.

[305] Concannon A, Vashchenko VA, ter Beek M, Hopper P (2004) ESD protection of double-diffusion devices in submicron CMOS processes. Proc. ESSDERC, 261–264.

[306] Wang AZ, Tsay CH (1999) A compact square-cell ESD structure for BiCMOS IC. Proc. BCTM, 46.

[307] Vashchenko VA, Kindt W, ter Beek M, Hopper P (2004) Implementation of 60V tolerant dual direction ESD protection in 5V BiCMOS process for automotive application. Proc. EOS/ESD Symp., 117–124.

[308] Sze SM (1982) Physics of semiconductor devices. Wiley, New York.

[309] Vashchenko VA, Kuznetsov V, Hopper P (2007) Implementation of dual-direction SCR devices in analog CMOS process. Proc. EOS/ESD Symp., 75–79.

[310] Vashchenko VA, Jansen P, Scholz M, Hopper P, Sawada M, Nakaei T, Hasebe T, Thijs S (2007) Voltage overshoot study in 20V DeMOS-SCR devices.Proc. EOS/ESD Symp., 53–57.

[311] Zavrazhnov YV, Kaganova II, Mazel EZ, et al (1985) Power high frequency transistors. Radio and Sviaz, Moscow.

[312] Blicher A (1981) Field-effect physics and bipolar transistor physics. Academic, New York.

[313] Oxner ZS (1985) Power field effect transistors and its application. Radio and Sviaz, Moscow.

[314] DiLorenzo JV, Kwandelwal DD (eds) (1982) GaAs FET principles and technology. Artech House, Norwood, MA.

[315] Fedotov YA (ed) (1973) Silicon planar transistors (in Russian). Soviet Radio, Moscow.

[316] Zakharov AL, Asvadurova EI (1983) Calculation of thermal parameters of semiconductor devices (in Russian). Radio and Sviaz, Moscow.

[317] Alexanian IT (1981) Methodological basis of simulation in the theory of reliable devices (in Russian). Electron.a Tekh. Ser. 8 4:7–10.

[318] Grigoriashvili YE (1981) Failure model and reliability study of rigid terminal of integrated circuits (in Russian). Electron. Tekh. Ser. 2:20–24.

[319] Nechaev AM, Rubakha EA, Sinkevitch VF (1981) Imitation modeling of thermal filamentation in transistor structures. Elektron. Tekh. Ser. 2:39–45.

[320] Nechaev AM, Rubakha EA, Sinkevitch VF (1981) Cause-consequence methods at investigation of power transistor reliability. Elektron. Tekh. Ser. 2:16–20.

[321] Alexanian IT, Krivoshapko V, Rubanik YT (1976) Modelling of electromigration failures of elements IC on computer (in Russian). Elektron. Tekh. Ser. 8 6:39–47.

[322] Chernyshev AA (1988) Reliability basis of semiconductor devices and integrated circuits. Radio and Sviaz, Moscow.

[323] Miller YG (1976) Physical basis of integrated circuits reliability. Soviet Radio, Moscow.

[324] Kozlov NA, Sinkevitch VF (1984) Kinetics of thermal current stratification in static and pulse operation regimes of power transistors. Elektron. Tekh. Ser. 2 2:35–45.

[325] Rubakha EA, Ivanova LI (1983) Peculiarities of failure mechanisms and reliability of power pulse microwave transistors (in Russian). Elektron. Tekh. Ser. 2 7:43–52.

[326] Kluev VV, et al (eds) (1989) Technical diagnostic tools (Reference book in Russian). Mashinostroenie, Moscow.

[327] Koncevoi YA, Kudin VL (1973) Methods of technology control of semiconductor device manufacture. Energiya, Moscow.

[328] Gostev AE, et al (1987) Structure inhomogeneity of GaAs epitaxial films and electrophysical characteristics of Schottky diodes. Sov. Microelectron.16:302–310.

[329] Kerner BS, Nechaev AM, Rubakha EA, Sinkevitch VF (1980) Kinetics of thermal filamentation at fluctuation instability in transistor structures. Sov. Radiotechnics and Electronics XXV:168–176.

[330] Peck DS, Zierdt CH (1974) The reliability of semiconductor devices in the Bell System. Proc. IEEE 62:185–211.

[331] Reynolds FH (1974) Thermally accelerated aging of semiconductor components. Proc. IEEE 62:212–222.

About the authors

Dr. Vladislav Vashchenko received MS, Engineer-Physicist (1986) followed by "Ph.D. in Physics of Semiconductors" (1990) from Moscow Institute of Physics and Technology for the study of self-organization phenomena in semiconductor structures under breakdown. Since 1984 he was working in reliability department of State Research Institute "Pulsar" (Moscow) occupying positions from the student intern to head of laboratory. In 1997 he was awarded the "Doctor of Science in Microelectronics" degree for the cycle of studies and new solutions of the reliability problems in power GaAs MESFET's, microwave silicon devices and the developed test methods. In the period 1995-1998 he was the leader of the work on contracts for high reliability GaAs components for Russian Space Agency, commercial and military customers. In 2000 he joined Advanced Process Development Group in National Semiconductor Corp. to work on design of the ESD protection solutions for integrated analog products. Currently he is the leader and manager of R&D group responsible for ESD development for new processes and products. His current research interests are mainly focused on the power devices, device level reliability, ESD solutions, physical process and device simulation for ESD. His studies are widely presented in major device research forums. He is the author of over 90 U.S. patents and 70 research and review papers in the fields of reliability and ESD development.

Prof. Dr. Vladimir F. Sinkevitch received MS, followed by Ph.D in Physics from Moscow Institute of Physical Engineering. In 1987 he received Doctor of Science in Microelectronics degree from the same Institute for studies of the reliability and physical limitations in power semiconductor devices. Since his graduation in 1975 he joined State Research Institute "Pulsar" where his carrier began from the research collaborator in the Reliability Department up to the position of Director of Corporate Reliability Department (1983) followed by the Deputy of General Director on Quality and Reliability since 1886. He is also the leader of the group of reliability assurance and additional special selection tests of the discrete components for Russian Space Agency and other customers. He is invited Professor of Moscow Aviation Technological Institute and Member Scientific Council of the Institute. V.F. Sinkevitch is the author of above 100 research papers and numerous Soviet Union and Russian Federation patents in the field of reliability and semiconductor device technology. He is also the recipient of the State Prize of USSR in microelectronics.

Index